The Horizontal Metropolis

Rururbain
Rururbain
City-region
Sprawl
Sprawl
Sprawl
Edge City
Megalopolis
Rururbain
Périurbain
Périurbain
Agropolitan
Sprawl
Rururbain
Rururbain
Rururbain
Rururbain
Nw Met Area
Zwischenstadt
Großstadt
Conurbation
Rururbain
Città Diffusa
Città Territorio
Rururbain
Rururbain
Rururbain
Rururbain
Rururbain
Rururbain
Rururbain
Périurbain
Périurbain
Agropolitan
Rururbain
Rururbain
Sprawl
Sprawl
Rururbain
Rururbain
Konjuka
Kotadesasi
Desakota
Agropolitan
Desakota
Kotadesasi
Desakota
Rururbain
Conurbation
Rururbain
Rururbain
Sprawl
Sprawl
Edge City

Martina Barcelloni Corte · Paola Viganò
Editors

The Horizontal Metropolis

The Anthology

Springer

Editors
Martina Barcelloni Corte
ULiège
Liège, Belgium

Paola Viganò
EPFL
Lausanne, Switzerland

IUAV
Venice, Italy

ISBN 978-3-030-56400-1 ISBN 978-3-030-56398-1 (eBook)
https://doi.org/10.1007/978-3-030-56398-1

This Springer imprint is published by the registered company Springer Nature Switzerland AG
The registered company address is: Gewerbestrasse 11, 6330 Cham, Switzerland

To Bernardo Secchi

Acknowledgements

We would like to thank those whose support has been essential in the development of this book. The most heartfelt thanks goes to Bernardo Secchi, with whom we have explored the Horizontal Metropolis project and our constant inspiration.

For the production phase we would like to thank Célia Küpfer for her commitment in the construction of many of the book's materials and her involvement in legal processes; Anna Karla Almeida for her enthusiasm and precious help in finalizing the book's construction process; and to Myriam Treiber for helping in the first phases of the book's conception.

Our gratitude goes also to the book's many contributors, for their expertise and help in the building of each chapter: Farzaneh Bahrami, Chiara Cavalieri, Marine Durand, Tommaso Pietropolli, Carlo Pisano, Matthew Skjonsberg, Roberto Sega, Guillaume Vanneste, and Qinyi Zhang.

Thanks also to Steven Piccolo, Domique Ronayne, Robert Burns and Victoria Franzinetti for the quality of their translations and to Teo Vexina and Anne-Charlotte Astrup for their technical support and precious suggestions. We are particularly grateful to Chantal Blanc and her team for having guided us with precision and constancy through the copyright labyrinth. Thanks also to Valerio di Lucente for his timely and skillful graphic support.

A special thank goes also to Eugenio Barcelloni Corte, which has accepted the task of exploring, with his sharp eye and camera, some of the territories described in this book.

Contents

Part II Interpretative Concepts and Visions

Part III A Contemporary Reconsideration

Chapter 1
Genealogy, Tradition: An Anthology

Paola Viganò

The Anthology concludes the series of publications on the "Horizontal Metropolis," a research and an oxymoron whose strength resides in expanding the traditional image of the metropolis to a larger territory where urban and non-urban elements live side by side, together with functions and infrastructures that transcend the local scale and help to shape, as a whole, new territorial relationships. The idea of metropolitan horizontality encompasses, and this is the second aspect tackled by the research, a project to overcome the traditional hierarchic relationships between territories, centres and peripheries, and the hypothesis of extending the benefits of multi-functionality, independence, and the possibility of making residential and work choices typical of a metropolitan context to traditionally servant territories. In the first case, the form of the city is discussed in terms of the heterogeneity of its physical and spatial construction, its disparate densities, co-existence between natural and urban elements and between agriculture and the city. The second dimension of the research looks at the structure of the relationships between the different parts and territories, the organizational logic of strong and weak infrastructures, the territorial distribution of the functions, the density of the accessibility and ecological mesh, and the detail of the hubs that mark the space and of the margins that can be rethought in terms of a horizontal design.

1.1 The Emergence of a Tradition

Many urban scholars have, for over a century now, accumulated reflections, readings and interpretations on both dimensions and themes; it seemed to us important to select and bring them together so that their cumulative legacy could

P. Viganò (✉)
EPFL, Lausanne, Switzerland
e-mail: paola.vigano@epfl.ch

IUAV, Venice, Italy
e-mail: paola.vigano@iuav.it

M. Barcelloni Corte and P. Viganò (eds.), *The Horizontal Metropolis*,
https://doi.org/10.1007/978-3-030-56398-1_1

emerge with its strength and incisiveness in the actual debate on the urban, ecological and socio-economic transition. It is a question of identifying a tradition of research endowed with a strong interpretative and projective character that recognises the formation of a different space as a new subject of study and design, a new city form as an expression of extensive territorial transformations following on from the industrial revolution and then constructed together with the appearance of widespread forms of production and the inhabitation of territories often densely micro-infrastructured in the long-term and inhabited in a diffuse way.

The expression "Horizontal Metropolis" was coined during forward-looking reflections on the metropolitan territory of Brussels (Secchi & Viganò et al. 2010) and is underpinned by much research that went before it producing different interpretations entirely founded on in-depth knowledge of specific, site-based contexts. The Anthology explores the multitude of spatial, socio-cultural and political conditions that emerge within them, generating a rich, multifaceted, inhomogeneous and to a large extent unprecedented image of contemporary urban contexts, permitting chronological and thematic readings of the different positions and allowing us to understand the formation of a new discourse on the city. The first aim of the research on the Horizontal Metropolis was to bring this tradition to the surface.

As it was being put together, the Anthology intersected with other similar but nevertheless different projects with which it shared the attempt to give visibility to the collective accumulation of reflections on the characteristics of the contemporary urban space, starting with the expansion of its morphological and geographical horizons. A critical analysis of the traditional assumptions was proposed in 2013 by Neil Brenner in the hybrid collection of texts (between anthology, program texts and research)[1] which takes up the hypothesis of planetary urbanization formulated by Henri Lefebvre at the start of the '70s (Lefebvre 1970). The urgency for a change of focus, even if starting with a less general term than urbanisation *-suburbia*, extended to *infinite suburbia-* is found in the volume of the same title edited by Alan Berger in 2017. In this case, the emphasis is on the need to reveal and overcome the prejudice that has become rooted over time, as Robert Bruegmann recounts in "The Antisuburban Crusade," with respect to the space in which the vast majority of the global population will live in the years ahead[2] and on the multi-faceted topics and disciplines involved in its possible design.

---

[1]Brenner (2013). What drives the initiative is the expansion of the horizons… However, the point of view, here, is that of critical geography, which does not coincide with the HM research and Anthology perspective.

[2]See for example the position of Roger Keil ("The global suburb: divesting from the world's white picket fences", in the same book), who is interested in the process (suburbanization), more than in the subject (suburban) aimed at countering the prevalent idea of "urban age" which prevents us from grasping the predominantly suburban dimension of global urbanization.

In the case of this Anthology, the problems to be investigated were the time span considered (we are dealing with "prophecies" and "anticipations"), the influences and exchanges among the main authors, and the relationship between concrete situations and interpretations. The delimitation of the research is then the result of understanding a geography (physical and cultural) and a different way of conceiving of urban history designed both by case studies and authors, situations and ideas about them. The result is the Anthology where texts, diagrams and maps propose an open dialogue, possible cross-criticism and comparisons.

1.2 Genealogy. Tradition

The second aim of the Anthology, as part of the research on the Horizontal Metropolis, was to frame a genealogy capable of connecting long-term processes of territorial/political construction and contemporary urbanization, and to highlight a series of statements which, one after another, form a body of interpretative and projective rural–urban concepts. The hypothesis is that this tradition is fundamental in the contemporary understanding of city and territory, and in the development of future design images and tools.

From Rousseau, Kropotkin, Geddes, Mackaye, Wright, Gottmann and later Friedmann, McGee, Indovina, Secchi, Branzi, Smets, Sieverts, Corboz, Oswald and Baccini… the project for a horizontal metropolis has come to the fore at different times throughout the history of the city and territorial design. The selected texts build up another tradition of interpretations of the urban phenomena where the city-periphery paradigm explodes into splinters of different narratives and theory frames. They can be collected and discussed clarifying the factual or phenomenological characters and the ideological and utopian aspects that "prophecies" and "interpretations" bring to the surface, following one another over more than a century.

While putting the Anthology together, we first hesitated and then assumed two terms of reference, tradition and genealogy, aware of their distance and conflictual nature. Tradition is literally what we consider important and crucial to pass on to the future. Here we are not interested in establishing ritual continuity, observing "la loi de fidelité", (Foucault 1971:169), unique or linear paths. The idea of tradition allows us to shape, within some thematic agglutinations, a series of digressions and variations on the theme which, despite originating from distant situations and different political and cultural histories and contexts, come together in common lines of force. The anthological collection illustrates them with great clarity: these are researches often developed without a broader vision of other coeval, independent contributions, but that as a whole demonstrate, touching on research topics that have many points of convergence, the existence of just as common interpretative trajectories. It is not a question of inventing a tradition, (Hobsbawm and Ranger 1983:4), however, as Hobsbawm writes, it is at times of accelerated social change that the invention of tradition becomes a tool for overcoming old cultures incapable of adapting and inflexible with respect to changed contexts (Hobsbawm and Ranger

1983:5).[3] Nonetheless, invention, in the case of the texts here collected, concerns the formation of a new vocabulary of metaphors and metaphorical concepts that help make the shift to a new formulation of urban: "the creation, dismantling, and restructuring of the images of the past" (Hobsbawm and Ranger 1983:13).

The Anthology is the tool to enhance the existence of this coming together.

The second term, genealogy,[4] is also complex, not immune to abuse and misunderstanding and yet useful when seeking to avoid teleological paths, a gaze "qui sait d'où il regarde aussi bien que ce qu'il regarde" (Foucault 1971:163), letting the plurality of beginnings and the dispersion of origins, controversial or dead-end paths, and the heterogeneity of the research types and literary genres involved filter through. Some of the texts may seem frankly outdated, indigestible and only partially acceptable, belonging to eras and projects that are no longer our own. Yet, they add a viewpoint that somewhat subconsciously reappears, transformed, in the territory itself or in the discussions that concern it. Prophecies signal the emergence of new forces and tensions between social and spatial logics undergoing continual transformation.

Tradition and genealogy highlight the complex relationships that each interpretation contained in the Anthology constructs with the modern era it deals with: initially, in the proto-industrial epoch, Rousseau attributed an urban characteristic to the advent of new forms of production, decentralised as they were close to the source; at the end of the 19th century and start of the 20th, an author like Kropotkin imagined the possibility to design and guide modern economic and political growth along routes not followed by the orthodox line of development. In addition to overcoming the city and countryside opposition, against the social division of work, Kropotkin also overturned the ideologically-oriented social reading of Darwinian evolution as a fierce fight for survival in favour of "scientific counter-discourse" (Audier 2017) starting with the observation of relationships between animal species and emphasising the emancipatory and progressive role of "mutual aid" (Kropotkin, 1902). The stand against the virtues of competition, hierarchy, inequality is declared in favour of a principle of egalitarian solidarity, cooperation between groups and individuals even belonging to different species. A political project that radicalises the already radical positions of utopian socialism and that in any case shares with them, and likewise with Marx and Engels, a new conception of the city-countryside relationship, rethought on the basis of a new organisation of work. This was also the case of Okhitovich, who perhaps more naively designed a new acentric territory, all parts of which were at the same time productive and consumer-based. Technological and scientific progress still seems tameable, the bearer of emancipation, whose limits cannot yet be seen. In the second half of the 20th century, a small group of authors discovered, starting with the study of concrete and situated cases, paths of unexpected development, insofar as unforeseen by the great modern machine, which

[3]The new formalisations do not only occur in traditional societies, continues Hobsbawm, but in the heart of modernity itself.

[4]See the classic texts: Nietzsche (1887); Foucault.

complicate the model and from the inside fragment weaken it, but also reformulate it. These studies revealed situations in which economic and social development and the canonical forms of the modern space were not or could not be achieved. But the failure to create them left room for unforeseen and unplanned innovations, which proved fundamental (Hirschman 1971). The modern project, essential part of the modern narrative, is subjected to criticism in these readings.

Through tradition and genealogy, the Anthology does not present itself as the foundation of an alternative history of the transformation of the city or of urbanism, which is still waiting to be written, but it assumes critical responsibility, raising the veil that even today seems to prevent many eyes from seeing.

1.3 The Three Parts of the Anthology

The first part of the anthology, "Prophecies and Clues," examines some of the great texts that anticipate the recognition and interpretation of the city-territory. The letter from Jean Jacques Rousseau to Maréchal de Luxembourg (20 January 1763) announced, in the mid-eighteenth century, a reading of the Swiss territory as a single large city. Rousseau takes up these topics in the seventh walk of *Rêveries du Promeneur solitaire* (written between 1776 and 1778 and published posthumously), recalling "herborisation" in the summer of 1765 in Robaila, near Môtiers-Travers. In the letter dated 1763, Rousseau describes a territory in which, with respect to the French one, the houses are everywhere "çà et là sur le terrain avec beaucoup plus d'égalité". Well-being, freedom and equality form the foundations of an urban country, where urbanity is not produced by a high density of buildings and yet it is enough to indicate "qu'on est toujours dans la ville," thanks to its "mélange bizarre" of activities and functions. The entire Swiss territory produces "un spectacle unique en son genre, mais fait seulement pour des yeux qui savent voir". In *Les Rêveries*, he recalls the surprise of finding a factory in a precipice, or a bookseller who lives in the only house at the summit of a mountain from the top of which seven lakes can be admired in a "*mélange de la nature sauvage et de l'industrie humaine*" in which fragments of wild nature and industrial production co-exist side by side: "*La Suisse entière n'est pour ainsi dire qu'une grande ville dont les rues larges et longues plus que celle de S*[t] *Antoine, sont semées de forêts, coupées de montagnes, et dont les maisons éparses et isolées ne communiquent entre elles que par des jardins anglais*» (Rousseau 1972:169–170). Eyes that know how to see - here we can't help but recall Le Corbusier's "des yeux qui ne voient pas" - and capable of transforming Alpine meadows into a collection of English gardens, the expression of culture and civilisation. André Corboz starts with this extraordinary vision to describe the urban Switzerland contemporary to him (*Vers la Ville-Territoire*, in this book). We are in the second section of the Anthology which contains the most representative and influential authors, between the 1960s and 1990s, proposing a new look at the city. Like Corboz, Marcel Smets also uses the anticipatory thought of two great references of the past that would almost inevitably influence many of the subsequent studies on the Belgian "Banlieue Radieuse": *Land and Labour*, by Seebohm Rowntree, and *L'exode rural et le retour aux champs* by Emile Vandervelde. The former was an English scholar who saw in the land model and the organisation of Belgian rail transportation a lesson for his country and the fight against poverty. The latter, a leading political figure of the Workers' Party, carried out field research and promoted legislative actions that would be of great importance in improving workers' living conditions.

Finally, in the third part, some of the authors presented in the previous section return with reflections recorded during a series of video interviews, accompanied by maps, on different dates, of the places investigated. These are original and valuable testimonies that allow us to penetrate the authors' thinking over the course of several decades. Some appear here in the book for the first time, such as Andrea Branzi, both prophetic and post-modern, who interprets the permanent crisis of a society that no longer finds in the modern, rigid and therefore fragile apparatus the tools with which to tackle and bypass it. Design is the result of overcoming zoning and the idea of space–time stability, it contains within it "agriculture, the countryside, the landscape" and "even animals."

1.4 Conclusions

With this book the research on the Horizontal Metropolis reaches a conclusion (albeit entirely provisional[5]) after an important international conference, an exhibition held in different forms and places[6] and the two publications that preceded.[7] The Anthology was at the start of this research and it is the last to be published, after having constructed a critical debate around the hypotheses of Horizontal Metropolis and having explored it with a series of maps and design investigations in different contexts. The radical nature of its project is not rooted in the traditional avant-garde of a modernity that did not have the tools or even the interest to govern the consequences of its acts, however, the research considers the tension towards the improvement of our possibilities to inhabit the world still valid. The Anthology completes a trilogy which, pursuing the reflection on new urban forms, paves the way for a new and radical project of coexistence. This is the third great objective to which the Anthology makes a contribution, it seems to us, of a certain sophistication. This fascinating research into the roots of the Horizontal Metropolis

[5]The research also produced and inspired numerous doctoral theses, some of which are still being written.

[6]Archizoom, EPFL, Lausanne November 2015; first Architecture Biennale of Lugano, May 2016; side event of the XX Architecture Biennale of Venice, Certosa Island, 2016; Bozart, Brussels 2018; Shenzen Biennale of Urbanism/Architecture, 2019, as part of the exhibition "Porous Cities, Horizontal Metropolis", Tianyan subvenue. In 2021 the exhibition, in a new format obliged by the pandemic conditions, has been held in Hong Kong.

[7]Viganò et al. (2018); Cavalieri and Viganò (2018).

conceptualization draws from ground based explorations, where all definitions emerge from deep analysis and knowledge of territorial situations, in Europe and elsewhere. Travelling ideas, nomadic concepts, and on site re-interpretations have produced, up until now, an incomplete portrait of the Horizontal Metropolis whose project remains to be made more precise, still a formidable set of interpretations and fundamental clues for the future of living together.

References

Audier, S. (2017). *Le Société écologique et ses ennemis: pour une histoire alternative de l'émancipation*. La Découverte.

Brenner, N. (Ed.). (2013). *Implosions/explosions: Towards a study of planetary urbanization*. Jovis.

Berger, A., Kotkin, J., & Balderas Guzman, E. (Eds.). (2017). *Infinite Suburbia*. Princeton Architectural Press.

Cavalieri, C., & Viganò, P. (Eds.). (2020). *The horizontal metropolis a radical project*. Parks Books.

Foucault. (1971). Nietzsche, la généalogie, l'histoire. In *Hommage à Jean Hyppolite*. Presses Universitaires de France.

Hobsbawm, E., & Ranger, T. (Eds.). (1983). *The Invention of Tradition*, Cambridge University Press. Inventing traditions, it is assumed here, is essentially a process of formalization and ritualization, characterized by reference to the past, if only by imposing repetition.

Hirschman, A. O. (1971). *A bias for hope. Essays on development and Latin America*. Yale University Press.

Kropotkin, P. (1902). *Mutual aid: A factor of evolution*. McClure Phillips & Co.

Lefebvre, H. (1970). *La Révolution urbaine*, Paris: Gallimard (Engl. Transl., 2003, *The Urban Revolution*). University of Minnesota Press.

Nietzsche, F. (1887). *Zur Genealogie der Moral: Eine Streitschrift* (Engl. Transl.: 2013, *On the Genealogy of Morals*, edited by R. C. Holub, Penguin Classics).

Rousseau, J. J. (1972). *Les Reveries du Promeneur solitaire*. Gallimard (first publication 1782).

Secchi, B., Viganò, P., et al. (2010). *Bruxelles et ses territoires, plan régional de développement durable*. Elaboration d'une vision territoriale métropolitaine à l'horizon 2040 pour Bruxelles, first report.

Viganò, P., Cavalieri, C., & Barcelloni Corte, M. (Eds.). (2018). *The horizontal metropolis between urbanism and urbanization*. Springer.

Part I
Prophecies and Clues (Anticipations)

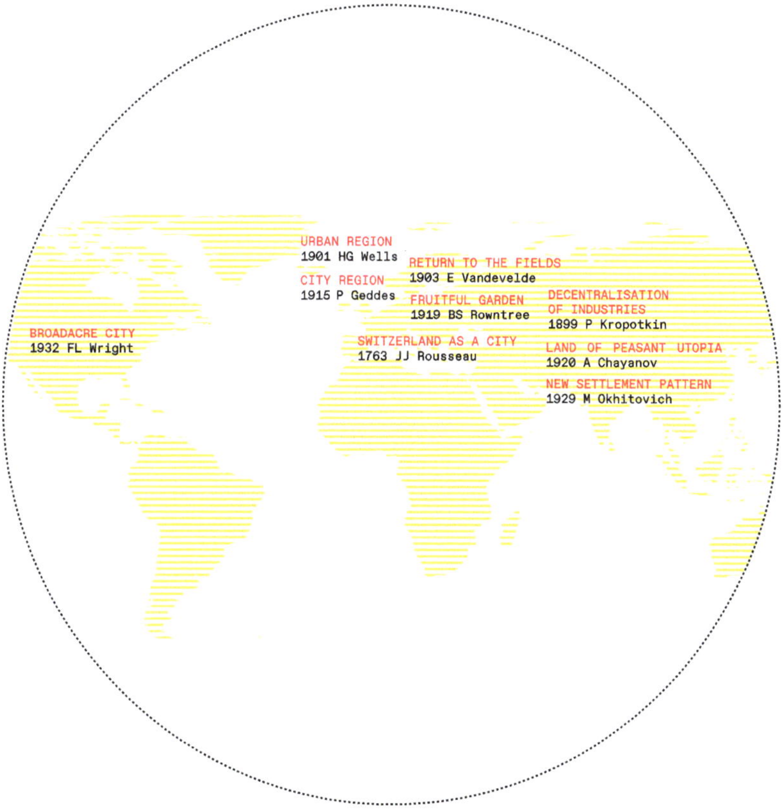

Chapter 2
Prophecies and Clues: Introduction

Martina Barcelloni Corte

The first part of the book evokes the Horizontal Metropolis' long-term tradition of thought, the emergence of which can be identified in the early "prophecies" of very different but equally visionary authors. A small selection of fragments of an alternative thought are put forward in chronological sequence to highlight the gradual but inexorable emergence of a radically new way of imagining the city and the territory, a new way of conceiving urban life. The projective imagination of this small group of minds, capable to sensing the emergence of a new territorial order, is so lucid as to be able to predict dynamics that over the decades will fully characterize entire countries around the world.

It is within an atmosphere of wonder and disquietude, developed while steam and electricity were radically modifying consolidated space–time conceptions, that many of the voices collected in this part of the book have emerged. We are at the turn of the century, when the city—synonymous with a bewildering modernity—was becoming *place-emblem* of progress but also *monstrum*[1] with which the "citizen" had to daily renegotiate his living space; a space creating a sense of emptiness and senselessness, isolation and marginalization, but which—at the same time—was able to thrust open the vast "fields of possible", to fuel hopes for economic improvement and social emancipation.

In those years, while urban layouts and interventions were being conceived by professionals to face so-called "technical problems"—as congestion or pollution—in a "gradual improvement" posture, a growing number of intellectuals, through a more lucid and forward-looking analysis, broke into the cultural *milieu* with an attitude of strong rupture. They denounced the "inadequacy of established notions"

[1]Cities were at the centre of a harsh criticism, stigmatized as places of perdition and conflict, atomization and lack of community.

M. Barcelloni Corte (✉)
ULiège, Liège, Belgium
e-mail: mbarcellonicorte@uliege.be

M. Barcelloni Corte and P. Viganò (eds.), *The Horizontal Metropolis*,
https://doi.org/10.1007/978-3-030-56398-1_2

(Geddes, 1915; see Chap. 8 of this book) and fed the imagery of many with new visions, images able to portray new ways of living and moving in the territory at large, and to propose alternative trajectories to reality.

They are the witnesses of a latent reform, of a new order, which "has been growing up naturally, yet almost unconsciously to politicians, beneath our existing, our traditional political and administrative network" (Geddes, ibid.).

In resonance with utopian novels from the past in which the idea of a city coextensive with the entire territory was already at the centre of stage,[2] the prophecies of these authors took-on the density of a dream: in these visions the "old city, (…) as obsolete as mail coach" gave the way to "wide urban regions, laced together not only by railway and telegraph but also by novel roads and a dense network of telephones, parcel-delivery tubes and nervous arterial connections"; the territory, "dotted with clumps of prosperous houses (…) school, doctor, engineers, book and provision shops" in which "the old lines of division" were replaced by "new lines of union" is widely accessible to the point that "the business-man can sit at home in his library and bargain, discuss, promise, hint, threaten and do everything that once demanded a personal encounter", from the armchair of his living room (Geddes, ibid.). The territory is described as "a curious and varied region (…) at any rate wooded, more abundantly wooded, braking continually into park and garden, and with everywhere a scattering of houses" (Wells, 1901; see Chap. 5 of this book).[3] Here "the peasant family farm, more competitive than the large-scale capitalist one", builds "a system in which labour (is) not separated from creative management, in which the freedom of individual initiative allowes each human being to develop his full spiritual potential while enabling him also to make use of the whole might of the collective large-scale economy and of public and state organisations" (Chayanov, 1920; see Chap. 9 of this book).

A society with radically new yet uncertain outlines was looming, entailing—on the one hand—strong risks of oppression and alienation but—on the other—multiple and extraordinary possibilities. This tension, inhabiting the feelings of many, produced both in art and literature, a new gaze, a "realism open to the various fields of possible" studded with images of—at the same time—"realistic precision and magical atmosphere".[4] A constellation of quasi-fantastic events emerging within the folds of everyday life, "naturalized" through ordinary scenes and perfectly realistic

[2]In "The Southern Land", utopian novel by de Foigny (1676) the protagonist, Jacques Sadeur discovers a continent divided into 15,000 squares, each comprising 16 districts with 25 houses inhabited by 16 people; in Simon Tyssot de Patot's account of a "country without towns or cities, the population is spread evenly across the whole territory, organised according to a strictly regular grid pattern of canals and roads". Both authors are quoted by André Corboz while describing the emergence of the City Territory (Corboz, 1991).

[3]See Wells (Chap. 5 of this book) "will have the choice of all England and Wales as his suburb and all the vest stretch of country from Washington to Albany will be all of it available to the active citizen of New York".

[4]We refer here to the rise of Magical Realism (see Bontempelli, 2018). High exponents of this current were painters as Felice Casorati, Edward Hopper, Antonio Donghi and writers as Dino Buzzati or Jorge Luis Borges.

techniques, began to trigger wonder and imagination: reality itself was suggesting alternative pathways.[5] While exceptional episodes were "normalised", official narratives appeared as monstrous deformations of reality: it was the naturalization of the supernatural that made everyday life fantastic and—to a certain extent—any change possible[6]: a well-oiled lever bringing out new images and desires.

The texts collected in the first part of this book contain–to a certain extent—this tension and the same urgency to create radically new—yet plausible—images, stemming directly from reality. Geddes himself, while commenting the transformation of core cities into *Conurbations*, spoke of the necessity of the collective creation of a shared imaginary, of new conceptions of place having the power to empower social capacity[7] as the "true and proper survey", for example, which simply through a careful description of reality "accustom ourselves to visualize the region", a completely new territorial figure, which lacked a clear image in people's minds.[8]

This strong imaginative thrust characterizing the first half of the twentieth century and echoed in so many, influential voices, remained -however- largely unheard by the general establishment. As clearly pointed out by Corboz (Corboz, 1991; see Chap. 24 of this book), in the following decades urbanism would react in great delay and "most often with dismay" to the ongoing social and territorial transformations, losing the opportunity of actively driving them. The general refusal for change and the diffused entrenchment on obsolete portrayals of urban life[9] represented a dangerous attitude for architects and urbanists that repeatedly refused to imagine new territorial layouts and to build -accordingly- the tools to articulate their evolution.

Here we have—thus—our small group of authors, ambassadors of "another world within their world", of another time within their time, witnessing the emergence of a new wave of reflection on the city and the territory, and—between the folds of a dominant territorial dynamic– seizing the existence of another order, working on different logics and awaiting to emerge. Similarly, but even more

[5]Never creating a true fracture with reality images of "realistic precision and magical atmosphere" began to inhabit the public arena as objective and almost *anonymous* constructions of reality, triggering-at the same time-wonder and imagination.

[6]"The use of utopia is, in its present state, a simple means of displaying a reality that has difficulty in finding realization, only because of the immaturity of the existing political and economic tools" (Tafuri et al. 1962; Chap. 14 in this book).

[7]"The truth of our condition is accessible to us only through imaginative practices, among them, ideology and utopia" (P. Ricoeur, 2006).

[8]"Does not this map (the "Well Painted Perspective of The Region" in the British gallery), shown with all its towns and connected up by railways and roads, clearly suggest that the economic and social unity of such city-regions is already becoming conscious to them"? (Geddes, ibid.).

[9]As, for example, the "walled city" or the "pristine territory" (Corboz, ibid.).

clearly, a coeval writer and painter, Carlo Levi, important Italian intellectual, described,[10] during the rise of fascism and of the centralized idol-state[11] the existence—"undertrack"—of an alternative system, deeply rooted within semi-autonomous rural communities and organized around an "organic federation"[12] of self-relying and small scale territorial entities. While warning on the risks of the construction of ever larger cities in which a "generic humanity, (..) without history and memories, uprooted from any self-determination (…) would give up the uniqueness of its world to identify with all other men",[13] Levi strongly advocated for a decentralized territorial model, made of a set of "infinite autonomies" where men would be "naturally driven towards self-determination" and self-consciousness. In the author the practical objectives of decentralization are revealed as the means to exploit not only the economic resources of each region but also the "capital of conscience that each region (coinciding with its own civilization) owned".[14]

Up beyond particular considerations on single territorial conditions, in the writing of this author, as in the ones of others collected in this first part, the "world, kept out of history suddenly makes its way in",[15] revealing to the reader different and new possibilities, places in which other reasons and logics can come into play and which—instantly—become possible.

While following the author's weave, the reader will come to doubt that the real world is the "alternative" one described in the plot and not the one he experiences every day, which—between the lines of the text—appears, for a moment, an illusion.

> "With a mind freed from the hypnosis, new unhackneyed thoughts stirred in his mind; he found it possible to think in new ways". (Okhitovich, 1930; see Chap. 10 of this book).

---

[10]We refer here to the texts: Levi (1975); and to Levi (1979).

[11]And the related concentration of population and wealth in ever larger cities, "leviathans made of unconscious bodies and minds" (…) "in which work divinizes itself into technicality and organization and technique becomes technicalism" (Levi, 2018, p. 128. First ed. 1975).

[12]The autonomy of the rural commune "cannot exist without the autonomy of the factories of the schools, of the cities, of all forms of social life". Lesson that Levi had drawn from his months of confinement in Lucania. Levi, C. (2010). *Cristo si é fermato ad Eboli*. Torino: Einaudi (pp. 222, 223).

[13]Levi (2018, pp. 126, 127. First ed. 1979).

[14]Agamben, G. in: Levi (2018). Agamben associates Levi's approach with that of Pasolini and quotes Pasolini, P. P. (ed.), *Quaderno romanzo*, n.3, Casarsa, June 1947, pp. 6, 7.

[15]Calvino on Carlo Levi in *Galleria* 3–6, 1967, p. 238.

References

Bontempelli, M. (2018). *L'avventura Novecentista*. Vallecchi.

Chayanov, A. (1920). Путешествие моего брата Алексея в страну крестьянской утопии. Moscow: State Publishing House. Source used in this publication: Chayanov, A. (1976). The Journey of My Brother Alexei to The Land of Peasant Utopia. *The Journal of Peasant Studies*, *4*(1), 63–108.

Corboz, A. (1990). Vers la Ville-Territoire. In M. Haller, H. Hauser, & R. Zaech (Eds.), *Ergänzungen, Ergebnisse derwissenschaftlichen Tagung anlässlich der Einweihung des Ergänzungsbaus der Hochschule St. Gallen* (pp. 631–635). Bern andStuttgart: Verlag Paul Haupt.

de Foigny, G. (1676). *La Terre australe connue*, Vannes.

Geddes, P. (1915). The population map and conurbations, first published. In P. Geddes (Ed.). *Cities in evolution, an introduction to the town planning movement and to the study of civics*. London: Williams & Norgate.

Levi, C. (1975). *Paura della Libertà*. Einaudi.

Levi, C. (1979). *Cristo si è fermato ad Eboli*. Einaudi.

Levi, C. (2018). *Paura della libertà*. Vicenza: Neri Pozza.

Okhitovich, M. (1930). Zametki po teorii rasseleniia. In *Sovremennaia arkhitektura*, no. 1–2, pp. 7–16.

Ricoeur, P. (2006). *The course of recognition* (D. Pellauer, Trans.). Cambridge, MA: Harvard University Press.

Tafuri, M., Piccinato, G., & Quilici, V. (1962). La città territorio: verso una nuova dimensione. *Casabella Continuità*, 270, pp. 16–25.

Wells, H. G. (1901). *Anticipations of the reaction of mechanical and scientific progress upon human life and thought*. Manhattan: Harper & Brothers and London: Chapman & Hall.

Chapter 3
Switzerland as a City (1763)

Jean-Jacques Rousseau

Introduction

Marine Durand
EPFL, Lausanne, Switzerland
marine.durand@epfl.ch

Even if not considered among his masterpieces, Rousseau's epistolary exchanges have proven to be profoundly revealing of his thinking and of his way of perceiving and criticizing the evolution of civilization.

The letter to the Marshal of Luxembourg, was written from Môtiers in 1763 after he definitively fled France, with the help of the Marshal, after being convicted for his books The Social Contract; or, Principles of Political Rights *and* Emile; or, On Education *both published in 1762 and immediately censured by the Parliament of Paris. While the letter was supposed to address the Marshal's request for a description of the small village of Môtiers, Rousseau instead provided a fascinating portrait of Switzerland, describing it as a "single large city". Through this letter, the author draws an extraordinary portrait of his country, stressing the extreme dispersion of settlements into the landscape, a figure able to make the entire territory inhabited and habitable thanks to a territorial and political vision that Switzerland has fostered for many centuries.*

This description is still surprisingly apt, which is why André Corboz, in his metaphor for Switzerland as a Hypercity[1] *refers to Rousseau, stating that his portrait is even closer to reality than the image shared by Switzerland's inhabitants today. Rousseau's portrait is based on two "territorial rationalities' that have*

[1]Corboz, A. (2000). La Suisse comme hyperville. In *L. Visiteur*, Société française des architectes (Vol. 6, pp. 112–129). Paris.

M. Barcelloni Corte and P. Viganò (eds.), *The Horizontal Metropolis*,
https://doi.org/10.1007/978-3-030-56398-1_3

strongly structured Switzerland both spatially and socially through the centuries. The first one is the extreme dissemination of human settlements, more precisely described in Rousseau's Lettre sur les spectacles *where the author praises the autonomy and the hard work of its inhabitants for their subsistence, a territorial dissemination directly related to climate and natural territorial conditions which—over time—had led the Swiss to distribute in systems of villages. Such generalized habitability of the territory is then strongly related to the second "rationality', the use of the territory and especially traveling habits: Rousseau describes them as long and frequent, saying that he has never seen as ardent travellers as the Swiss. By doing so, the inhabitants used their country as a city, a practice that in Switzerland—where the transportation infrastructure is extremely developed—has changed in its forms but is today still very present. To fully understand Rousseau's spatial and social portrait of Switzerland, we must not forget his rejection of large cosmopolitan cities. Rousseau denounces the introduction of cosmopolitan and urban values which leads to the corruption of morals across the entire country. The explicit value that Rousseau gives to the territorial dispersion that he sees as "the biggest advantage of Switzerland" is reversed by contact with the growing taste for money, brought by foreigners (the French) that contaminated the whole society. The harshness of Nature that had isolated the inhabitants couldn't resist the moral corruption.*

3.1 Letter to the Marshal of Luxembourg[2]

M. le Maréchal, you would like me to describe the land where I live. But how shall I do so? I only see what moves me; meaningless objects are nothing to my eyes. My attention is proportional to the interest they excite: and what interest can I take in these things so far from you? Trees, rocks, houses, even people, are so many isolated objects, each one on its own arousing little emotion in the observer; but the impression of all this taken together, united into a single picture, depends on our state of mind as we contemplate it. This picture, even if it is always the same, is painted in as many different ways as there are different dispositions of the hearts of the observers. And these differences, which are the fruit of our discernment, do not arise merely between one observer and another, but also in the same person at different times. This is what I feel quite strongly as I again behold this land I have loved so much. I thought I would find what had charmed me in my youth. But everything has changed; it is a different landscape, another air, another sky, another people. No longer seeing the great mountains through the eyes of a twenty-year-old,

[2]Excerpt from: Rousseau, J.-J. (1763). Deux lettres à M. le Maréchal de Luxembourg, Contenant une description du Val-de-Travers. Lettre Première. In J.-J. Rousseau (Eds.), *Collection complète des oeuvres de J. J. Rousseau, Citoyen de Genève, Tome Douzième, Contenant des Pièces sur divers sujets, & un Recueil de Lettres sur la Philosophie, la Morale, & la Politique* (pp. 283–299). Geneva.

I found them much aged. One misses the good times of old; I know this well. We attribute all the changes that have taken place in us to the things around us, and when pleasure has left us we believe it no longer exists anywhere. Others see things as we once saw them, and they will see them as we see them today. But you are asking me for descriptions, not reflections, and mine engage me as if I were a grown child pining for his old playthings. The various impressions that this country has made upon me at different ages lead me to conclude that our relations with things regard us more than they do the things themselves, and that, as we describe what we feel much more than what is objectively there, we should know the effect of a journey on the author writing about it if we are to judge how far to this or that side of the truth his depictions fall. Given this premise, do not be surprised if you see a land once so verdant, so alive, so pleasant, become cold and arid under my pen: you will sense quite easily at what time of my life and in what season of the year it has been written.

I know, *M. le Maréchal*, that to speak to you of a village, I should not begin by describing to you all of Switzerland, as if the tiny corner I inhabit needed to be framed in such a vast space. There are, however, general things which cannot be divined, and which must be known in order to judge the particulars. To know Môtiers, one must have some idea of the county of Neuchâtel, and to know the county of Neuchâtel, one must have some idea of Switzerland as a whole.

Switzerland presents the same features almost everywhere: lakes, meadows, forests, mountains. The Swiss all have more or less the same customs: a mixture of those picked up from other peoples and their own ancestral simplicity. They have ways of living that do not change at all, because they are bound to the soil, the climate, and diverse needs; the inhabitants are forced to bend to what the local nature imposes. One aspect, for example, regards the distribution of their dwellings, much less gathered into cities and towns than in France, but scattered and dispersed here and there over the land much more uniformly. Thus, although Switzerland is generally more populous than France, it has smaller cities and smaller villages. On the other hand, one finds houses everywhere; the village covers the entire parish, and the city spans the whole country. The whole of Switzerland is like a great city divided into thirteen quarters, of which some are in valleys, some on slopes, and some on mountains. Geneva, Saint-Gall, Neuchâtel are like the outskirts. Not all quarters are populated equally, but all are populated enough to indicate that one is always in the city. The houses, however, instead of being arranged in rows, are dispersed without symmetry or order, as they say were those of Ancient Rome. You can no longer believe you are penetrating the wilderness when you find bell towers among the pines, herds on the rocks, factories in gorges, workshops on torrents. This bizarre mixture has an indescribable animation and life that breathes freedom and wellbeing, and that will always make this land a spectacle unlike any other, but made only for those eyes that know how to see.

This even distribution comes from the large number of small jurisdictions into which the capitals are divided, from the ruggedness of the land, which makes transportation difficult, and from the nature of production, mainly pasturage.

This all demands that things be consumed where they are produced, and keeps men as scattered as their livestock. This is Switzerland's blessing, an advantage that her people may look upon as a hardship, but one she holds alone, that nothing can take from her. It contains or impedes progress in wealth and immorality, and will always remedy, in the long run, the astonishing loss of her men to other countries. This is the blessing. Here too is the harm brought by this blessing. When the Swiss, who once lived self-sufficient in their mountains, began to communicate with other nations, they took a liking to the way of life they found there and set out to imitate it. They gathered that money is a good thing and wanted to have some too. Without production or industries to attract it, they put themselves on the market, selling themselves at retail to the powers. They acquired just enough money to feel poor. There being almost no means for circulating the money in a country that produces nothing and has no sea ports, this money has brought them new needs without increasing their resources. Thus their first dispatch of troops abroad forced them to send even more, and never to stop. Life having become more ravenous, the country could no longer nourish the same number of inhabitants. This is the cause of the depopulation that began to be felt throughout Switzerland. She fed her many people when they stayed in their homes; now that one half has left she can barely nourish the other.

The worst thing is that of the half that left, enough came back to corrupt everything that remained by imitating the customs of other countries, and especially France, who has more Swiss troops that any other nation. I say "corrupt" without touching on the question of whether or not the French mores are good or bad within France herself, but there can be no doubt regarding Switzerland. It is not possible that the same customs are appropriate for a people who, not having the same resources, the same climate, or the same soil, will always be forced to live differently.

The combination of these two factors, one good and one bad, is felt in everything. It explains everything one particularly notices in the customs of the Swiss, and especially the strange contrast between refinement and simplicity that one senses in all their ways. It turns all their customs on their heads, not out of silliness but simply because it could not be any other way. Taking the customs of big cities back to their woods, they apply them in the most comic fashion. They don't know what country attire is anymore; they are dressed up on their rocks as if they were in Paris; they wear tassels befitting the Royal Palace under their pine trees. I have seen them returning from their hay fields wearing jackets of finely frilled muslin. Their fine manners always have something coarse about them, their wealth some roughness. They have desserts but eat rye bread; they serve imported wines but drink piquette; they accompany fine stews with rancid bacon and cabbage; they offer you cheese with coffee, ham with tea. The women wear lace with heavy linen, tasteful dresses with colored stockings. Their valets, alternatively house servants and cowherds, wear livery while serving at the table and mingle the smell of manure with that of food.

As one enjoys wealth only by showing it off, it has brought their society closer, however without depriving them of their taste for isolated abodes. No one is surprised here to see me spend the winter in the countryside; thousands of people around me do the same. We continue to live separated but do get together in long and frequent visits. To display one's furnishings and finery, one must attract his neighbors and go visit them. Since these neighbors often live at quite a distance, people are traveling continuously. I have never seen a people that travels as much as the Swiss; the French do not come close. All you see everywhere is carriages; there is no house without its own, and horses, which are abundant in Switzerland, are less than useless. However, given that when a man does mount a horse, something that is becoming increasingly rare, it is often for the purpose of visiting a woman, and so he sports pretty white stockings pulled up tight, and races along as preened as if he were going to a ball. Indeed, there is nothing so colorful as the roads of Switzerland: you always encounter young gentlemen and beautiful ladies; you see nothing but blues, greens, and pinks—you would think you were in the Jardin du Luxembourg.

One effect of this commerce is that men have almost completely lost their taste for wine; and a counter effect to this itinerant lifestyle is that cabarets have become both common and of good quality throughout Switzerland. I do not know why one brags so much about those of France; they certainly do not come close to the ones here. It is true that this makes it quite dear to live here, but the same is true even if you don't go out: it could not be otherwise in a country that produces little food and where money is not allowed to circulate.

The only three products that they have produced so far are cheeses, horses, and men. However, since being introduced to wealth, this trade is no longer enough for them, and they have added that of manufactured products, for which they are indebted to French refugees. Yet this may appear to be a greater resource than it is: since the more one buys, the more expensive goods become, and since no one wants to till the land if he can make more money at other endeavors, more money has not made them richer. We may see this in comparison with Swiss Catholics, who, not having the same resource, are poorer in money but live equally well.

It is quite odd that such a rugged country, whose inhabitants are so inclined to leaving, inspires in them a love so tender that the regret for having left it brings almost all of them back in the end, and inflicts on all those who cannot a sickness that may even be mortal, which they call, I believe, hemvé. There is a famous melody in Switzerland known as the Ranz des Vaches, heard on hills throughout the country, played on the horns of Alpine herdsmen. This melody, which is not much in itself, conjures a thousand images of the homeland in the Swiss mind and brings forth a torrent of tears when they hear it in a foreign land. It has caused such a great number to die of homesickness that it is forbidden, by order of the king, to play the Ranz des Vaches in the presence of Swiss troops. But, M. le Maréchal, you may know all of this better than I, and the reflections that this fact engenders will not have escaped you. I cannot help but note that France is certainly the best country in the world, where all the conveniences and amenities contribute to the wellbeing of the inhabitants. Nevertheless, there has never been, as far as I know, a hemvé or Ranz des Vaches that has made a Frenchman cry or die in a foreign land; and we certainly feel a great deal less homesickness in Switzerland now that we live better in their country.

Generally speaking, the Swiss are fair-minded, informal, charitable, faithful friends, brave soldiers, and good citizens, but they are also scheming, mistrustful, jealous, nosy, miserly, and their greed compromises their wellbeing more than their simplicity does. They are normally serious and phlegmatic, but are furious when ired, and seem almost inebriated when happy. I have never seen such joyful play as theirs. It astonishes me that the French dance sadly, languidly, with little grace, while the Swiss dances are bouncy and lively. The men display their natural vigor, and the girls are charmingly light-footed; you might say that the ground burns their feet.

The Swiss are shrewd and skillful in business. The French, who consider them ham-fisted, are much less astute than they are; they judge their mind by their accent. The French Court has always opted to send them shrewd people, and has always been wrong. The Swiss usually beat the French at this type of fencing match. But if you send them solid, upright men, you will get from them what you want, because naturally they will love you. The *Marquis de Bonnac*, who was quite bright but came off as being crafty, accomplished nothing in Switzerland; while earlier, Marshal Bassompierre got whatever he wanted, because he was frank, or at least so thought the Swiss. The Swiss always negotiate to their advantage—unless they are sold out by their magistrates—given that they can do without money better than the powers can do without men; while, as for your wheat, they can choose to do without it. It must be admitted that while they negotiate their agreements well, they observe them even better: good faith that one must not be too proud to acknowledge. I will say nothing, *M. le Maréchal*, of their government or their politics, as this would lead me too far afield, and I only want to tell you about what I have seen with my own eyes. As for the canton of Neuchâtel where I live, you know that it belongs to the King of Prussia. This small principality, after being separated from the Kingdom of Burgundy and passing into the houses of Châlons, Hochberg, and Longueville, finally passed in 1707 into that of Brandenburg by the decision of the Estates General, natural judges of the rights of the contenders. I will not delve into consideration of the reasons why the King of Prussia was preferred to the Prince de Conti, nor of the influence that other powers may have had in this affair; I will limit myself to noting that, in the contest between these two princes, the honour of standing one day under a great captain was not lost on the *Neuchâtelois*. Besides, they have retained under their rulers more or less the same freedoms as their fellow Swiss; although they may owe it more to their geographical position than to their skill, for I find them exceedingly boisterous for a wise people.

All that I have told you about the Swiss, in general, is even more characteristic of the people here, and the contrast between the natural and the imitation still more striking, with the difference, however, that the natural has even less substance: with the exception of some tiny corner, the gold-leaf covers the entire picture. If we exclude the cities and the lakeshores, the land is as rugged as the rest of Switzerland. Life here is equally rustic. The inhabitants, accustomed to living under princes, are even more fond of grand manners, so that here you find people everywhere using jargon and putting on airs: fine orators ploughing the fields, and courtiers in stableman's cloaks. The *Neuchâtelois* are thus the *gascons* of Switzerland. They are spirited and take pride in their vigour. They read, and reading profits them, even the villagers are educated; almost all have a small collection of selected books that they refer to as their library. They also keep up with current events and inject all this into their conversation in a way that is not at all gauche. They also keep step with the times as if they lived in Paris. Some time ago while strolling, I paused in front of a house where there were girls making lace. The mother was cradling a baby and as I watched her when a big peasant come out of the hut and approached me with an easy air, he said, "You see that we do not follow your guidance so well; but our women are as attached to old prejudices as they are in love with the new fashions." I was completely taken aback. I have heard hundreds of similar things from these people.

A lot of wit and even more pretension, but no taste: that's what struck me first about the *Neuchâtelois*. They speak quite well, quite glibly, but they write badly and without verve, especially when they want to write in a light tone, and they want to write this way all the time. They know nothing of the grace and spice of a light style, and when they string together their overly ornate sentences they think themselves so many Voltaires or Crébillons. There is a diary quality to their writing, where they seek to be kind and playful. They even incorporate some small verses in their own manner. *Mme la Maréchale* will find, if not amusement, at least a job at the Mercure, which is from one end to the other a logogriph demanding a better Oedipus than I.

The attire here is quite similar to that of the canton of Bern, but a bit more elaborate. The men dress much in French style. The women would do the same, but since they rarely travel and thus do not have first-hand experience of fashions as the men do, they abuse and disfigure them: thus, laden with frills and flounces, they appear to be adorned with rags.

As regards their character, it is difficult to judge, so cloaked it is in mannerisms. They think they are refined because they are fashionable, and merry because they are boisterous. I believe that the Chinese are the only ones who could outdo them at ceremony. You arrive tired, stressed, it doesn't matter, you have to expose your flank to an endless volley; as long as the machine is wound up, it runs, and it always winds up again with each new arrival. French courtesy involves putting people at ease, and oneself as well; courtesy of people from Neuchâtel involves annoying oneself and others. They never think about what might suit you, but only what will best show off their own would-be savoir-faire. Their overwrought offerings do not tempt at all; they always have some undefinable air of formula, something cut and dried, contrived, that invites you to decline. They are nevertheless obliging, informal, and truly hospitable, particularly to persons of quality; you can be sure to be received if you present yourself as a marquis or count. Such an easy target will not be missed by rogues, and so the rascals are often found in their city, where there are typically much fêted. A simple, honest man with misfortunes and virtues would not be treated the same: one can sport a great name without merit, but not great merit without a name. Furthermore, those whom they serve once, they serve well. They keep their promises, and do not easily abandon those in their care. They may indeed be loving and sensitive, but nothing could be further from true sentiment than the tone they take; everything they do for their fellow man they do to show off, and their vanity overshadows their good heart.

That vanity is their main vice; it comes out in everything, all the more easily because it is maladroit. They think they are all gentlemen, even though their sovereigns were nothing more than gentlemen themselves. They love hunting, less for enjoyment than because it is an aristocratic pursuit. Lastly, never do you see *bourgeois* so puffed up with their birth. They do not boast of it, but you see that they care; it is not that they are proud of it, they are merely obsessed.

Lacking dignitary or noble titles, they have military and municipal titles in such abundance that there are more people with titles than without. There is Mr. Colonel, Mr. Major, Mr. Captain, Mr. Lieutenant, Mr. Councilor; Mr. Castellan, Mr. Mayor, Mr. Justiciar, Mr. Professor, Mr. Doctor, Mr. Elder: if I had resumed my former trade here, I have no doubt I would have been Mr. Scribe. The women also bear the titles of their husbands: Madame Councilor, Madame Minister. My neighbor is Madame Major. And since we only name people by their titles, we are at a loss as to how to address people who have only their own name; it is as if they did not have one at all.

The fair sex is not fair; it is said to have degenerated. The girls have a great deal of freedom and make good use of it. They gather in social *milieux* where people come to play, dine, or chat, and where they lure the young men as best they can. But unfortunately young men are rare and they have to grab them when they can. The women live fairly dutifully; there are good households in this land, and there would be many more if life were good with the husband. Furthermore, with so many of them living in the country, reading less, and reaping fewer fruits than the men, they do not have a very florid mind. And in the idleness of their lives, they have no other resource than to make lace, spy on the affairs of others, backbite, and gamble. While there are some very kind ones, generally speaking one does not find the tone of decency and honesty in their conversation that would make it seductive, the tone that French women are so good at affecting when they want, showing sentiment and soul, and casting themselves as heroines of a novel. The conversation of the *Neuchâtelois* women is playful yet arid, it dries up as soon as the joking stops. The two sexes are not lacking in good nature, and I do not believe that they are a people without manners; but they are a people without principles, and the word virtue is as foreign and ridiculous here as it is in Italy. The religion they pride themselves on makes them more ill-tempered than good. Guided by their clergy, they hold forth on dogma; but as for morality, they know not what it is. For while they speak much of charity, they certainly have no love for their neighbour; it is only the affectation of giving alms. For them, a Christian is someone who goes to hear the sermon every Sunday; what he does the rest of the time does not matter. Their ministers, who gained great respect among the people when their princes were Catholic, would like to preserve this respect by meddling in everything, quarrelling over everything, extending the jurisdiction of the Church over everything: they do not want their time to have passed. However, they have just stirred up a state of ferment that will end up destroying them. The great issue was whether the punishment of the damned was eternal. You will have a hard time believing how hotly this dispute was argued; the Jansenism debate in France was nothing in comparison. All the troops mustered, the people ready to take up arms, ministers dismissed, magistrates silenced; everything signalled a coming civil war. And this affair has not been put to such rest that it does not leave lasting memories. When they are all ready to go to hell, they will finally stop worrying about what goes on there.

Here you have the principal observations I have made so far about the people here in this land where I live. You might find them to be a rather harsh way for a man to speak of his hosts, if I were not to tell you that I owe them no hospitality. I did not come to the gentlemen of Neuchâtel to ask for asylum, which they certainly would not have granted me, I came to Milord Marshal, and I am only here as the King of Prussia's guest. On the contrary, upon my arrival in the principality, the municipal magistrate, by way of greeting, immediately banned my book without being acquainted with it. The class of ministers then referred it to the Council of State: one has never seen people more eager to imitate the foolishness of their neighbours. Without the declared protection of Milord Marshal, I would not have been left in peace in this village. So many bandits take refuge in this land that those who govern it do not know how to distinguish the wrongdoers from the oppressed innocent, or make little effort to do so. The house where I live belongs to a niece of my old friend, Mr. Roguin. Thus, far from having any obligation to the gentlemen of Neuchâtel, I have only reason to complain about them. Besides, I have not set foot in their city, they are strangers to me in every respect. I owe them only justice in speaking of them, and I shall render it.

I render it with greater warmth to those who have filled me with affection, offerings, and courtesies of all kinds. Flattered by their esteem and touched by their goodness, I shall always make it my duty and pleasure to convey to them my affection and gratitude. But the reception they gave me has nothing to do with the Neuchâtel government, which would have treated me much differently if it had been up to them. I must repeat that while the bad will of the cabinet of ministers is not in doubt, I have much to praise the one in whose parish I live. He came to see me when I arrived, he offered me services in good faith, as he proved on a critical occasion when he exposed himself to the ill temper of more than one of his confreres for having acted as a true pastor to me. I had expected less of him, since in earlier squabbles he had played a role that indicated he was not a tolerant minister. To his credit, he is rather merry in company, not lacking in wit, sometimes giving good sermons, and often telling very good stories.

I realize that this letter is a book, and I am still only halfway through my report. I am going to let you catch your breath, *M. le Maréchal*, and put off the second volume to another time.

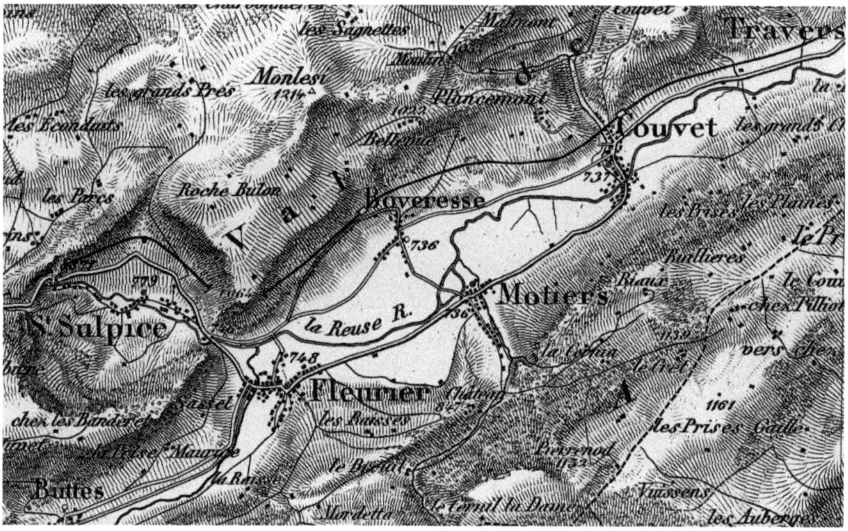

Fig. 3.1 Area sorrounding Môtiers, late 1840s, with Môtiers at the center of the frame. The map will be drawn 80 years after the drafting of JJ Rousseau's text, the "diffuse" territorial condition—highlighted in his description—is already clearly recognizable.
Source: Excerpt from *Topographische Karte der Schweiz* (Doufour Map), 1849, Sheet 11 (Pontalier, Yverdon). © Office fédéral de topographie Swisstopo

Fig. 3.2 "View of Môtiers and its surroundings" (1780–1788) by Samuel Hieronymus Grimm (drawer) and Pierre Philippe Choffard (engraver).
Original title: *«Vue de Motier-Travers et de ses environs, dans le Comté de Neuchatel, avec le Tableau de la fermeté du Philosophe de Genêve»* Scene showing Jean-Jacques Rousseau giving alms to a poor man while Pastor Montmollin excites rascals. In the background, Môtiers, Mont Sassel and the Areuse river flowing in the valley.
Source: de Laborde, J.-B., von Zurlauben, B. F. (1780–1788). *Tableaux topographiques, pittoresques, physiques, historiques, moraux, politiques, littéraires, de la Suisse*, 350 illustrations. Paris: De Clousier. © Bibliothèque cantonale et universitaire de Lausanne

Chapter 4
The Decentralization of Industries (1899)

Pëtr Alekseevič Kropotkin

Introduction

Roberto Sega
EPFL, Lausanne, Switzerland
roberto.sega@epfl.ch

Considered one of the fathers of anarchism, Pëtr Alekseevič Kropotkin was born in Moscow in 1842 from a noble family. After a period spent in Siberia as explorer geographer, he developed a critical position towards totalitarian power, first approaching liberal ideas, then socialist ones. Later, while staying in the Swiss Jura region, he approached the anarchist ideal, undertaking a revolutionary activity that forced him to escape from Russia and take refuge between England and Switzerland.

His theories, between the nineteenth and twentieth centuries, influenced a series of key figures of his time such as Patrick Geddes, Elisée Reclus and Ebenezer Howard. His writings challenge with ironic simplicity totalitarian thought and its pyramidal construction of space and social imaginary. Kropotkin's words have been a reference to generations of attentive readers who, as happens today, were concerned about the need for a new economy of resources, in order to meet the growing needs of human life.[1]

Fields, Factories and Workshops *can be considered one of the great prophetic works of the nineteenth century, through this text Kropotkin clearly criticizes the division of labour and society's multiple levels of separation, considering them among the main causes of unhappiness and social injustice of his time. Kropotkin*

[1]Kropotkin, P. (1974). *Fields, Factories and Workshops Tomorrow*, note from editor Colin Ward for the reissue of the second edition. New York: Harper & Row. First ed. 1899.

M. Barcelloni Corte and P. Viganò (eds.), *The Horizontal Metropolis*,
https://doi.org/10.1007/978-3-030-56398-1_4

argues the need to reverse the dangerous trend of specialisation of labour by radically intervening on the internal structure of the production system, being attentive to meeting local needs, thus overcoming the economic and social irrationalities that lead States to dependence upon foreign trade. The integration between producer and consumer, city and countryside, agriculture and industry, manual and intellectual work, are in his view the directions to be taken for a desirable social progress and development of the territory.

The entire work is comprised of a myriad of detailed examples of alternative practices and trends already underway which, together with precise economic and statistical data, scientifically enrich the author's thesis. As an attentive traveller, Kropotkin used to visit agricultural companies, workshops and industries with a notebook in which he—incessantly—took note of data and good practices. From this and other statistical studies, the first chapter of the book (reproduced below) shows how—all around the world—industry tends towards spatial decentralisation. According to Kropotkin:

> *Variety is the distinctive feature, both of the territory and its inhabitants; and that variety implies a variety of occupations. [...] A reorganised society will have to abandon the fallacy of nations specialized for the production of either agricultural or manufactured products. It will have to rely on itself for the production of food and many, if not most, of the raw materials; it must find the best means of combining agriculture with manufacture - the work in the field with a decentralised industry; and it will have to provide for "integrated education", which education alone, by teaching both science and handicraft from earliest childhood, can give to society the men and women it really needs.*

In Kropotkin's opinion, progress should be sought within a greater integration between city and countryside, manual and intellectual work, through a self-sufficiency project in which the production system is more focused on the small scale and privileges an ecological perspective. Such a vision has certainly enormous spatial consequences on the organization of the territory, that becomes increasingly oriented towards a horizontal distribution of settlements.

Today, it seems interesting to contextualize Kropotkin's thought in the light of the third industrial revolution we are experiencing,[2] *and within the context of the ecological transition. Could today a cleaner, smaller, sustainable, resource-friendly industry have more opportunities than in the past to integrate with other economies present in the territory? Would it be possible to imagine an integrated socio-environmental economic system able to guide a more horizontal organization of human settlements, based on local balances and on the valorisation of environmental and human resources? Could the concepts of circular economy and*

[2]We are referring here to industry 4.0 and to the economic activities linked to the third industrial revolution. See: Rifkin, J. (2011). *The third industrial revolution: How lateral power is transforming energy, the economy, and the world*. St. Martin's Press.

industrial ecology—whose objectives are respectively to extend the life of products[3] *and to have the production system evolve in line with the valorisation of the environment and its resources—be some of the tools that Kropotkin would use in order to translate his vision into practice in our time?*

4.1 The Decentralisation of Industries (from: *Fields, Factories and Workshops Tomorrow*)[4]

Who does not remember the remarkable chapter by which Adam Smith opens his inquiry into the nature and causes of the wealth of nations? Even those of our contemporary economists who seldom revert to the works of the father of political economy, and often forget the ideas which inspired them, know that chapter almost by heart, so often has it been copied and recopied since. It has become an article of faith; and the economical history of the century, which has elapsed since Adam Smith wrote has been, so to speak, an actual commentary upon it.

'Division of labour' was its watchword. And the division and subdivision—the permanent subdivision—of functions has been pushed so far as to divide humanity into castes which are almost as firmly established as those of old India. We have, first, the broad division into producers and consumers: little-consuming producers on the one hand, little-producing consumers on the other hand. Then, amidst the former, a series of further subdivisions: the manual worker and the intellectual worker, sharply separated from one another to the detriment of both; the agricultural labourers and the workers in the manufacture; and amidst the mass of the latter, numberless subdivisions again—so minute, indeed, that the modern ideal of a workman seems to be a man or a woman, or even a girl or a boy, without the knowledge of any handicraft, without any conception whatever of the industry he or she is employed in, who is only capable of making all day long and for a whole life the same infinitesimal part of something: who from the age of 13 to that of 60 pushes the coal cart at a given spot of the mine or makes the spring of a penknife, or 'the eighteenth part of a pin'. Mere servants to some machine of a given description: mere flesh-and-bone parts of some immense machinery: having no idea how and why the machinery performs its rhythmical movements.

Skilled artisanship is being swept away as a survival of a past condemned to disappear. The artist who formerly found aesthetic enjoyment in the work of his

[3]'The circular economy' requires that all stages of production (starting from the use of the raw material) are organised in such a way that the waste produced in one stage becomes a resource for the next one.

This new paradigm allows the activation of new synergies between different types of production, opening up new possibilities of integration between different professions and expertise.

[4]Excerpt from: Kropotkin, P. (1974). *Fields, Factories and Workshops Tomorrow*. London: George Allen & Unwin Ltd. pp. 23–40. First published in: Kropotkin, P. (1899). *Fields, Factories and Workshops*. Boston: Houghton-Mifflin and London: Hutchinson.

hands is substituted by the human slave of an iron slave. Nay, even the agricultural labourer, who formerly used to find a relief from the hardships of his life in the home of his ancestors—the future home of his children—in his love of the field and in a keen intercourse with nature, even he has been doomed to disappear for the sake of division of labour. He is an anachronism, we are told: he must be substituted, in a Bonanza farm, by an occasional servant hired for the summer, and discharged as the autumn comes: a tramp who will never again see the field he has harvested once in his life. 'An affair of a few years', the economists say, 'to reform agriculture in accordance with the true principles of division of labour and modern industrial organisation.'

Dazzled with the results obtained by a century of marvellous inventions, especially in England, our economists and political men went still further in their dreams of division of labour. They proclaimed the necessity of dividing the whole of humanity into national workshops having each of them its own speciality. We were taught, for instance, that Hungary and Russia are predestined by nature to grow corn in order to feed the manufacturing countries: that Britain had to provide the world-market with cottons, iron goods, and coal: Belgium with woollen cloth; and so on. Nay, within each nation, each region had to have its own speciality. So it has been for some time since: so it ought to remain. Fortunes have been made in this way, and will continue to be made in the same way. It being proclaimed that the wealth of nations is measured by the amount of profits made by the few, and that the largest profits are made by means of a specialisation of labour, the question was not conceived to exist as to whether human beings *would* always submit to such a specialisation; whether nations could be specialised like isolated workmen. The theory was good for today—why should we care for tomorrow? Tomorrow might bring its own theory!

And so it did. The narrow conception of life which consisted in thinking that *profits* are the only leading motive of human society, and the stubborn view which supposes that what has existed yesterday would last for ever, proved in disaccordance with the tendencies of human life; and life took another direction. Nobody will deny the high pitch of production which may be attained by specialisation. But, precisely in proportion as the work required from the individual in modern production becomes simpler and easier to be learned, and, therefore, also more monotonous and wearisome—the requirements of the individual for varying his work, for exercising all his capacities, become more and more prominent. Humanity perceives that there is no advantage for the community in riveting a human being for all his life to a given spot, in a workshop or a mine; no gain in depriving him of such work as would bring him into free intercourse with nature, make of him a conscious part of the grand whole, a partner in the highest enjoyments of science and art, of free work and creation.

Nations, too, refuse to be specialised. Each nation is a compound aggregate of tastes and inclinations, of wants and resources, of capacities and inventive powers. The territory occupied by each nation is in its turn a most varied texture of soils and climates, of hills and valleys, of slopes leading to a still greater variety of territories and races. Variety is the distinctive feature, both of the territory and its inhabitants;

and that variety implies a variety of occupations. Agriculture calls manufactures into existence, and manufactures support agriculture. Both are inseparable; and the combination, the integration of both brings about the grandest results. In proportion as technical knowledge becomes everybody's virtual domain, in proportion as it becomes international, and can be concealed no longer, each nation acquires the possibility of applying the whole variety of her energies to the whole variety of industrial and agricultural pursuits. Knowledge ignores artificial political boundaries. So also do the industries; and the present tendency of humanity is to have the greatest possible variety of industries gathered in each country, in each separate region, side by side with agriculture. The needs of human agglomerations correspond thus to the needs of the individual and while a *temporary division* of functions remains the surest guarantee of success in each separate undertaking, the *permanent* division is doomed to disappear, and to be substituted by a variety of pursuits—intellectual, industrial, and agricultural—corresponding to the different capacities of the individual as well as to the variety of capacities within every human aggregate.

When we thus revert from the scholastics of our text-books, and examine human life as a whole, we soon discover that, while all the benefits of a temporary division of labour must be maintained, it is high time to claim those of the *integration of labour*. Political economy has hitherto insisted chiefly upon division. We proclaim *integration*; and we maintain that the ideal of society—that is, the state towards which society is already marching—is a society of integrated, combined labour. A society where each individual is a producer of both manual and intellectual work: where each able-bodied human being is a worker, and where each worker works both in the field and the industrial workshop; where every aggregation of individuals, large enough to dispose of a certain variety of natural resources—it may be a nation, or rather a region—produces and itself consumes most of its own agricultural and manufactured produce.

Of course, as long as society remains organised so as to permit the owners of the land and capital to appropriate for themselves, under the protection of the state and historical rights, the yearly surplus of human production, no such change can be thoroughly accomplished. But the present industrial system, based upon a permanent specialisation of functions, already bears in itself the germs of its proper ruin. The industrial crises, which grow more acute and protracted, and are rendered still worse and still more acute by the armaments and wars implied by the present system, are rendering its maintenance more and more difficult. Moreover, the workers plainly manifest their intention to support no longer patiently the misery occasioned by each crisis. And each crisis accelerates the day when the present institutions of individual property and production will be shaken to their foundations with such internal struggles as will depend upon the more or less good sense of the now privileged classes.

But we maintain also that any socialist attempt at remodelling the present relations between Capital and Labour will be a failure, if it does not take into account the above tendencies towards integration. These tendencies have not yet received, in our opinion, due attention from the different socialist schools—but they must. A reorganised society will have to abandon the fallacy of nations specialised for the production of either agricultural or manufactured produce. It will have to rely on itself for the production of food and many, if not most, of the raw materials; it must find the best means of combining agriculture with manufacture—the work in the field with a decentralised industry; and it will have to provide for 'integrated education', which education alone, by teaching both science and handicraft from earliest childhood, can give to society the men and women it really needs.

Each nation—her own agriculturist and manufacturer; each individual working in the field and in some industrial art: each individual combining scientific knowledge with the knowledge of a handicraft—such is, we affirm, the present tendency of civilised nations.

The prodigious growth of industries in Great Britain, and the simultaneous development of the international traffic which now permits the transport of raw materials and articles of food on a gigantic scale, have created the impression that a few nations of West Europe were destined to become *the* manufacturers of the world. They need only—it was argued—to supply the market with manufactured goods, and they will draw from all over the surface of the earth the food they cannot grow themselves, as well as the raw materials they need for their manufactures. The steadily increasing speed of trans-oceanic communications and the steadily increasing facilities of shipping have contributed to enforce the above impression. If we take the enthusiastic pictures of international traffic, drawn in such a masterly way by Neumann Spallart—the statistician and almost the poet of the world-trade—we are inclined indeed to fall into ecstasy before the results achieved. 'Why shall we grow corn, rear oxen and sheep, and cultivate orchards, go through the painful work of the labourer and the farmer, and anxiously watch the sky in fear of a bad crop, when we can get, with much less pain, mountains of corn from India, America, Hungary, or Russia, meat from New Zealand, vegetables from the Azores, apples from Canada, grapes from Malaga, and so on?' exclaim the West Europeans. 'Already now,' they say, 'our food consists, even in modest households, of produce gathered from all over the globe. Our cloth is made out of fibres grown and wool sheared in all parts of the world. The prairies of America and Australia; the mountains and steppes of Asia; the wildernesses of the Arctic regions; the desert of Africa and the depths of the oceans; the tropics and the lands of the midnight sun are our tributaries. All races of men contribute their share in supplying us with our staple food and luxuries, with plain clothing and fancy dress, while we are sending them in exchange the produce of our higher intelligence, our technical knowledge, our powerful industrial and commercial organising capacities! Is it not a grand sight, this busy and intricate exchange of produce all over the earth which has suddenly groan up within a few years?'.

Grand it may be, but it is not a mere nightmare? Is it necessary? At what cost has it been obtained, and how long will it last?

Let us turn a hundred years back. France lay bleeding at the end of the Napoleonic wars. Her young industry, which had begun to grow by the end of the eighteenth century, was crushed down. Germany, Italy were powerless in the industrial field. The armies of the great Republic had struck a mortal blow to serfdom on the Continent; but with the return of reaction efforts were made to revive the decaying institution, and serfdom meant no industry worth speaking of. The terrible wars between France and England, which wars are often explained by merely political causes, had a much deeper meaning—an economical meaning. They were wars for the supremacy on the world market, wars against French industry and commerce, supported by a strong navy which France had begun to build—and Britain won the battle. She became supreme on the seas. Bordeaux was no more a rival to London; as to the French industries, they seemed to be killed in the bud. And, aided by the powerful impulse given to natural sciences and technology by a great era of inventions, finding no serious competitors in Europe, Britain began to develop her manufactures. To produce on a large scale in immense quantities became the watchword. The necessary human forces were at hand in the peasantry, partly driven by force from the land, partly attracted to the cities by high wages. The necessary machinery was created, and the British production of manufactured goods went on a gigantic pace. In the course of less than seventy years—from 1810 to 1878—the output of coal grew from 10 to 133,000,000 tons; the imports of raw material rose from 30 to 38,000,000 tons; the exports of manufactured goods from 46 to 200,000,000 lb. The tonnage of the commercial fleet was nearly trebled. Fifteen thousand miles of railways were built.

It is useless to repeat now at what a cost the above results were achieved. The terrible revelations of the parliamentary commissions of 1840–2 as to the atrocious condition of the manufacturing classes, the tales of 'cleared estates', and kidnapped children are still fresh in the memory. They will remain standing monuments for showing by what means the great industry was implanted in this country. But the accumulation of wealth in the hands of the privileged classes was going on at a speed never dreamed of before. The incredible riches which now astonish the foreigner in the private houses of England were accumulated during that period; the exceedingly expensive standard of life which makes a person considered rich on the Continent appear as only of modest means in Britain was introduced during that time. The taxed property alone doubled during the last thirty years of the above period, while during the same years (1810–78) no less than £1,112,000,000—nearly £2,000,000,000 by this time—was invested by English capitalists either in foreign industries or in foreign loans.

But the monopoly of industrial production could not remain with England for ever. Neither industrial knowledge nor enterprise could be kept for ever as a privilege of these islands. Necessarily, fatally, they began to cross the Channel and spread over the Continent. The Great Revolution had created in France a numerous class of peasant proprietors, who enjoyed nearly half a century of a comparative well-being, or at least, of a guaranteed labour. The ranks of homeless town workers increased slowly. But the middle-class revolution of 1789–93 had already made a distinction between the peasant householders and the village *prolétaires*, and, by favouring the former to the detriment of the latter, it compelled the labourers who had no household nor land to abandon their villages, and thus to form the first nucleus of working classes given up to the mercy of manufacturers. Moreover, the peasant-proprietors themselves, after having enjoyed a period of undeniable prosperity, began in their turn to feel the pressure of bad times, and their children were compelled to look for employment in manufactures. Wars and revolution had checked the growth of industry; but it began to grow again during the second half of our century; it developed, it improved; and now, notwithstanding the loss of Alsace, France is no longer the tributary to England for manufactured produce which she was sixty years ago. Today her exports of manufactured goods are valued at nearly one-half of those of Great Britain, and two-thirds of them are textiles; while her imports of the same consist chiefly of the finer sorts of cotton and woollen yarn—partly reexported as stuffs—and a small quantity of woollen goods. For her own consumption France shows a decided tendency towards becoming entirely a self-supporting country, and for the sale of her manufactured goods she is tending to rely, not on her colonies, but especially on her own wealthy home market.

Germany follows the same lines. During the last fifty years, and especially since the last war, her industry has undergone a thorough reorganisation. Her population having rapidly increased from 40 to 60,000,000, this increment went entirely to increase the urban population—without taking hands from agriculture—and in the cities it went to increase the population engaged in industry. Her industrial machinery has been thoroughly improved, and her new-born manufacturers are supplied now with a machinery which mostly represents the last word of technical progress. She has plenty of workmen and technologists endowed with a superior technical and scientific education; and in an army of learned chemists, physicists and engineers her industry has a most powerful and intelligent aid, both for directly improving it and for spreading in the country serious scientific and technical knowledge. As a whole, Germany offers now the spectacle of a nation in a period of *Aufschwung*, of a sudden development, with all the forces of a new start in every domain of life. Fifty years ago she was a customer to England. Now she is already a competitor in the European and Asiatic markets, and at the present speedy rate of growth of her industries, her competition will soon be felt even more acutely than it is already felt.

At the same time the wave of industrial production, after having had its origin in the north-west of Europe, spreads towards the east and south-east, always covering a wider circle. And, in proportion as it advances east, and penetrates into younger countries, it implants there all the improvements due to a century of mechanical and chemical inventions: it borrows from science all the help that science can give to industry; and it finds populations eager to grasp the last results of modern knowledge. The new manufactures of Germany begin where Manchester arrived after a century of experiments and gropings; and Russia begins where Manchester and Saxony have now reached. Russia, in her turn, tries to emancipate herself from her dependency upon Western Europe, and rapidly begins to manufacture all those goods she formerly used to import, either from Britain or from Germany.

Protective duties may, perhaps, sometimes help the birth of new industries: always at the expense of some other growing industries, and always checking the improvement of those which already exist; but the decentralisation of manufactures goes on with or without protective duties—I should even say, notwithstanding the protective duties. Austria, Hungary and Italy follow the same lines—they develop their home industries—and even Spain and Servia are going to join the family of manufacturing nations. Nay, even India, even Brazil and Mexico, supported by English, French, and German capital and knowledge, begin to start home industries on their respective soils. Finally, a terrible competitor to all European manufacturing countries has grown up of late in the United States. In proportion as technical education spreads more and more widely, manufactures grow in the States; and they do grow at such a speed—an American speed—that in a very few years the now neutral markets will be invaded by American goods.

The monopoly of the first corners on the industrial field has ceased to exist. And it will exist no more, whatever may be the spasmodic efforts made to return to state of things already belonging to the domain of history. New ways, new issues must be looked for: the past *has* lived, and it will live no more.

The chief reason for the success of Germany in the industrial field is the same as it is for the United States. Both countries have only lately entered the industrial phase of their development, and they have entered it with all the energy of youth. Both countries enjoy a widely-spread scientifically-technical—or, at least, concrete scientific—education. In both countries manufactories are built according to the newest and best models which have been worked out elsewhere; and both countries are in a period of awakening in all branches of activity—literature and science industry and commerce. They enter now on the same phase in which Great Britain was in the first half of the nineteenth century, when British workers took such a large part in the invention of the wonderful modern machinery.

We have simply before us a fact of *the consecutive development of nations*. And instead of decrying or opposing it, it would be much better to see whether the two pioneers of the great industry—Britain and France—cannot take a new initiative and do something new again; whether an issue for the creative genius of these two nations must not be sought for in a new direction—namely, the utilisation of both the land and the industrial powers of man for securing well-being to the whole nation instead of to the few.

Volumes have been written about the crisis of 1886–7, a crisis which, to use the words of the Parliamentary Commission, lasted since 1875, with but ‘a short period of prosperity enjoyed by certain branches of trade in the years 1880 to 1883’, and a crisis, I shall add, which extended over all the chief manufacturing countries of the world. All possible causes of the crisis have been examined; but whatever the cacophony of conclusions arrived at, all unanimously agreed upon one, namely, that of the Parliamentary Commission, which could be summed up as follows: ‘The manufacturing countries do not find such customers as would enable them to realise high profits.’ Profits being the basis of capitalist industry, low profits explain all ulterior consequences.

Low profits induce the employers to reduce the wages, or the number of workers, or the number of days of employment during the week, or eventually compel them to resort to the manufacture of lower kinds of goods, which, as a rule, are paid worse than the higher sorts. As Adam Smith said, low profits ultimately mean a reduction of wages, and low wages mean a reduced consumption by the worker. Low profits mean also a somewhat reduced consumption by the employer; and both together mean lower profits and reduced consumption with that immense class of middlemen which has grown up in manufacturing countries, and that, again, means a further reduction of profits for the employers.

A country which manufactures to a great extent for export, and therefore lives to a considerable amount on the profits derived from her foreign trade, stands very much in the same position as Switzerland, which lives to a great extent on the profits derived from foreigners who visit her lakes and glaciers. A good 'season' means an influx of from £1,000,000 to £2,000,000 of money imported by the tourists, and a bad 'season' has the effects of a bad crop in an agricultural country: a general impoverishment follows. So it is also with a country which manufactures for export. If the 'season' is bad, and the exported goods cannot be sold abroad for twice their value at home, the country which lives chiefly on these bargains suffers. Low profits for the inn-keepers of the Alps mean narrowed circumstances in large parts of Switzerland; and low profits for the Lancashire and Scotch manufacturers, and the wholesale exporters, mean narrowed circumstances in Great Britain. The cause is the same in both cases.

For many decades past we had not seen such a cheapness of wheat and manufactured goods as we saw in 1883–4, and yet in 1886 the country was suffering from a terrible crisis. People said, of course, that the cause of the crisis was over-production. But over-production is a word utterly devoid of sense if it does not mean that those who are in need of all kinds of produce have not the means for buying them with their low wages. Nobody would dare to affirm that there is too mush furniture in the crippled cottages, too many bedsteads and bedclothes in the workmen's dwellings, too many lamps burning in the huts, and too much cloth on the shoulders, not only of those who used to sleep (in 1886) in Trafalgar Square between two newspapers, but even in those households where a silk hat makes a part of the Sunday dress. And nobody will dare to affirm that there is too much food in the homes of those agricultural labourers who earn twelve shillings a week, or of those women who earn from fivepence to sixpence a day in the clothing trade and other small industries which swarm in the outskirts of all great cities. Over-production means merely and simply a want of purchasing powers amidst the workers. And the same want of purchasing powers of the workers was felt everywhere on the Continent during the years 1885–7.

After the bad years were over, a sudden revival of international trade took place; and, as the British exports rose in fours years (1886–90) by nearly 24%, it began to be said that there was no reason for being alarmed by foreign competition; that the decline of exports in 1885–7 was only temporary, and general in Europe; and that England, now as of old, fully maintained her dominant position in the international trade. It is certainly true that if we consider exclusively the money value of the export for the years 1876–95, we see no permanent decline, we notice only fluctuations. British exports, like commerce altogether, seem to show a certain periodicity.

This periodicity being a fact, Mr Giffen could make light in 1886 of 'German competition' by showing that exports from the United Kingdom had not decreased. It can even be said that, per head of population, they had remained unchanged until

1904, undergoing only the usual ups and downs. However, when we come to consider the *quantities* exported, and compare them with the *money values* of the experts, even Mr Giffen had to acknowledge that the price of 1883 were so low in comparison with those of 1873 that in order to reach the same money value the United Kingdom would have had to export 4 pieces of cotton instead of 3, and 8 or 10 tons of metallic goods instead of 6.

We thus see that while the total value of the exports from the United Kingdom, in proportion to its growing population remains, broadly speaking, unaltered for the last thirty years, the high prices which could be got for the exports thirty years ago, and with them the high profits, are gone. And no amount of arithmetical calculations will persuade the British manufacturers that such is not the case. They know perfectly well that the home markets grow continually overstocked; that the best foreign markets are escaping; and that in the neutral markets Britain is being undersold. This is the unavoidable consequence of the development of manufactures all over the world.

Great hopes were laid, some time ago, in Australia as a market for British goods; but Australia will soon do what Canada already does. She will manufacture. And the colonial exhibitions, by showing to the 'colonists' what they are able to do, and how they must do, are only accelerating the day when each colony *farà da sé* in her turn. Canada and India already impose protective duties on British goods. As to the mush-spoken-of markets on the Congo, and Mr Stanley's calculations and promises of a trade amounting to £26,000,000 a year if the Lancashire people supply the Africans with loincloths, such promises belong to the same category of fancies as the famous nightcaps of the Chinese which were to enrich England after the first Chinese war. The Chinese prefer their own home-made nightcaps; and as to the Congo people, four countries at least are already competing for supplying them with their poor dress: Britain, Germany, the United States and, last but not least, India.

There was a time when this country had almost the monopoly in the cotton industries; but already in 1880 she possessed only 55% of all the spindles as work in Europe, the United States and India and a little more than one-half of the looms. In 1893 the proportion was further reduced to 49% of the spindles, and now the United Kingdom has only 41% of all the spindles. It was thus losing ground while the others were winning. And the fact is quite natural: it might have been foreseen. There is no reason why Britain should always be the great cotton manufactory of the world, when raw cotton has to be imported into this country as elsewhere. It was quite natural that France, Germany, Italy, Russia, India, Japan, the United States, and even Mexico and Brazil, should begin to spin their own yarns and to weave their own cotton stuffs. But the appearance of the cotton industry in a country, or, in fact, of any textile industry, unavoidably becomes the starting point for the growth of a series of other industries; chemical and mechanical works, metallurgy and mining feel at once the impetus given by a new want. The whole of the home industries, as also technical education altogether, *must* improve in order to satisfy that want as soon as it has been felt.

What has happened with regard to cottons is going on also with regard to other industries. Great Britain, which stood in 1880 at the head of the list of countries producing pig-iron, came in 1904 the third in the same list, which was headed by the United States and Germany; while Russia, which occupied the seventh place in 1880, comes now fourth, after Great Britain. Britain and Belgium have no longer the monopoly of the woollen trade. Immense factories at Verviers are silent; the Belgian weavers are misery-stricken, while Germany yearly increases her production of woollens, and exports nine times more woollens than Belgium. Austria has her own woollens and exports them; Riga, Lodz and Moscow supply Russia with fine woollen cloths; and the growth of the woollen industry in each of the last-named countries calls into existence hundreds of connected trades.

For many years France has had the monopoly of the silk trade. Silkworms being reared in Southern France, it was quite natural that Lyons should grow into a centre for the manufacture of silks. Spinning, domestic weaving, and dyeing works developed to a great extent. But eventually the industry took such an extension that home supplies of raw silk became insufficient, and raw silk was imported from Italy. Spain and Southern Austria, Asia Minor, the Caucasus and Japan, to the amount of from £9,000,000 to £11,000,000 in 1875 and 1876, while France had only £800,000 worth of her own silk. Thousands of peasant boys and girls were attracted by high wages to Lyons and the neighbouring district: the industry was prosperous.

However, by-and-by new centres of silk trade grew up at Basel and the peasant houses round Zürich. French emigrants imported the trade into Switzerland, and it developed there, especially after the civil war of 1871. Then the Caucasus administration invited French workmen and women from Lyons and Marseilles to teach the Georgians and the Russians the best means of rearing the silkworm, as well as the whole of the silk trade; and Stavropol became a new centre for silk-weaving. Austria and the United States did the same; and what are now the results?

During the years 1872–81 Switzerland more than doubled the produce of her silk industry; Italy and Germany increased it by one-third; and the Lyons region, which formerly manufactured to the value of 454 million francs a year, showed in 1887 a return of only 378 millions. And it is reckoned by French specialists that at present no less than one-third of the silk stuffs used in France are imported from Zürich, Crefeld and Barmen. Nay, even Italy, which has now 191,000 persons engaged in the industry, sends her silk to France and competes with Lyons.

Like examples could be produced by the score. Greenock no longer supplies Russia with sugar, because Russia has plenty of her own at the same prices as it sells at in England. The watch trade is no more a speciality of Switzerland: watches are now made everywhere. India extracts from her 90 collieries two-thirds of her annual consumption of coal. The chemical trade which grew up on the banks of the Clyde and Tyne, owing to the special advantages offered for the import of Spanish pyrites and the agglomeration of such a variety of industries along the two estuaries, is now in decay. Spain, with help of English capital, is beginning to utilise her own pyrites for herself; and Germany has become a great centre for the manufacture of sulphuric acid and soda—nay, she already complains about over-production.

But enough! I have before me so many figures, all telling the same tale, that examples could be multiplied at will. It is time to conclude, and, for every unprejudiced mind, the conclusion is self-evident. Industries of all kinds decentralise and are scattered all over the globe; and everywhere a variety, an integrated variety, of trades grows, instead of specialisation. Such are the prominent features of the times we live in. Each nation becomes in its turn a manufacturing nation; and the time is not far off when each nation of Europe, as well as the United States, and even the most backward nations of Asia and America, will themselves manufacture nearly everything they are in need of. Wars and several accidental causes may check for some time the scattering of industries: they will not stop it: it is unavoidable. For each new-comer the first steps only are difficult. But, as soon as any industry has taken firm root, it calls into existence hundreds of other trades; and as soon as the first steps have been made, and the first obstacles have been overcome, the industrial growth goes on at an accelerated rate.

The fact is so well felt, if not understood, that the race for colonies has become the distinctive feature of the last twenty years. Each nation will have her own colonies. But colonies will not help. There is not a second India in the world, and the old conditions will be repeated no more. Nay, some of the British colonies already threaten to become serious competitors with their mother country; others, like Australia, will not fail to follow the same lines. As to the yet neutral markets, China will never be a serious customer to Europe: she can produce much cheaper at home; and when she begins to feel a need for goods of European patterns, she will produce them herself. Woe to Europe, if on the day that the steam engine invades China she is still relying on foreign customers! As to the African half-savages, their misery is no foundation for the well-being of a civilised nation.

Progress must be looked for in another direction. *It is in producing for home use.* The customers for the Lancashire cottons and the Sheffield cutlery, the Lyons silks and the Hungarian flour-mills, are not in India, nor in Africa. The true consumers of the produce of our factories must be our own populations. And they *can* be that, once we organise our economical life so that they might issue from their present destitution. No use to send floating shops to New Guinea with British or German millinery, when there are plenty of would-be customers for British millinery in these very islands, and for German goods in Germany. Instead of worrying our brains by schemes for getting customers abroad, it would be better to try to answer the following questions: Why the British worker, whose industrial capacities are so highly praised in political speeches; why the Scotch crofter and the Irish peasant, whose obstinate labours in creating new productive soil out of peat bogs are occasionally so much spoken of, are no customers to the Lancashire weavers, the Sheffield cutlers and the Northumbrian and Welsh pitmen? Why the Lyons weavers not only do not wear silks, but sometimes have no food in their attics? Why the Russian peasants sell their corn, and for four, six, and sometimes eight months every year are compelled to mix bark and auroch grass to a handful of flour for baking their bread? Why famines are so common amidst the growers of wheat and rice in India?

Under the present conditions of division into capitalists and labourers, into property-holders and masses living on uncertain wages, the spreading of industries over new fields is accompanied by the very same horrible facts of pitiless oppression, massacre of children, pauperism, and insecurity of life. The Russian Fabrics Inspectors' Reports, the Reports of the Plauen Handelskammer, the Italian inquests, and the reports about the growing industries of India and Japan are full of the same revelations as the Reports of the Parliamentary Commissions of 1840 to 1842, or the modern revelations with regard to the 'sweating system' at Whitechapel and Glasgow, London pauperism, and York unemployment. The Capital and Labour problem is thus universalised; but, at the same time, it is also simplified. To return to a state of affairs where corn is grown, and manufactured goods are fabricated, *for the use of those very people who grow and produce them*—such will be, no doubt, the problem to be solved during the next coming years of European history. Each region will become its own producer and its own consumer of manufactured goods. But that unavoidably implies that, at the same time, it will be its own producer and consumer of agricultural produce; and that is precisely what I am going to discuss next.

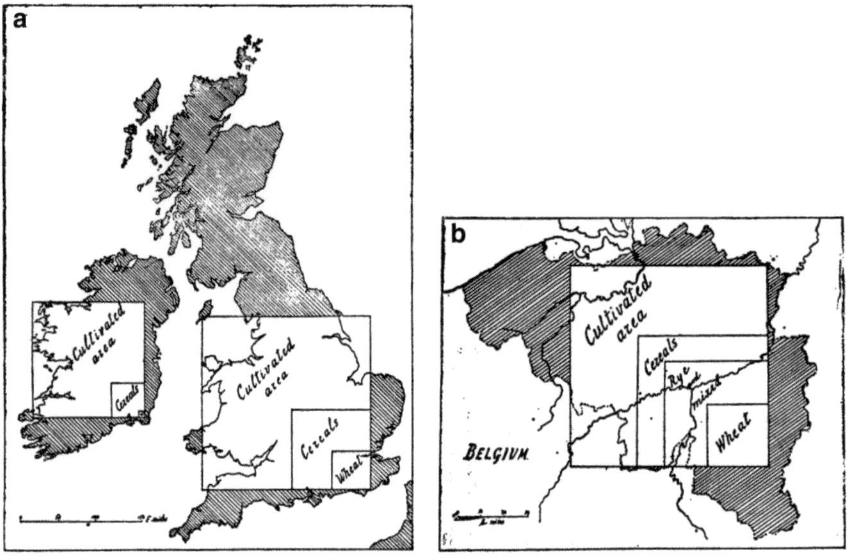

Fig. 4.1 "Proportion of the cultivated area which is given to cereal altogether, and to wheat in Great Britain and Ireland and Belgium (below). For Belgium, the square which encloses the wheat square represents the area given to both wheat and a mixture of wheat and rye".
Source: Kropotkin, P. (1919). *Fields, factories and workshops: Or, Industry combined with agriculture and brain work with manual work.* T. Nelson and Sons, Ltd., pp. 102 and 111.
"If we intend to have a correct appreciation of British agriculture, we must not base it upon what is obtained on a few selected and well-manured plots; we must inquire what is done with the territory taken as a whole." (ibid., p. 101)

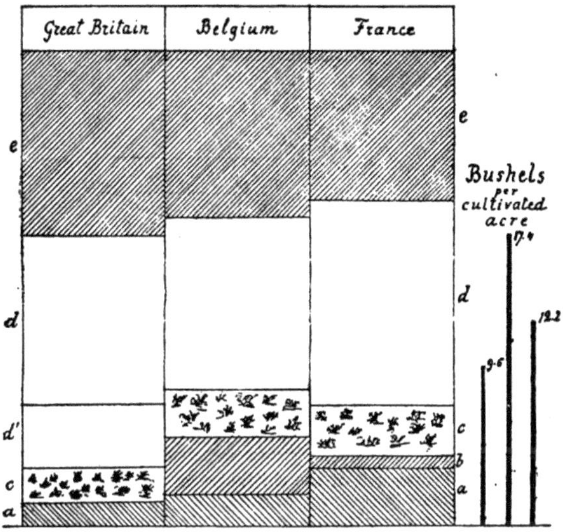

Fig. 4.2 "Proportion of cultivated and uncultivated areas in Great Britain, Belgium and France". Source: ibid., p. 112

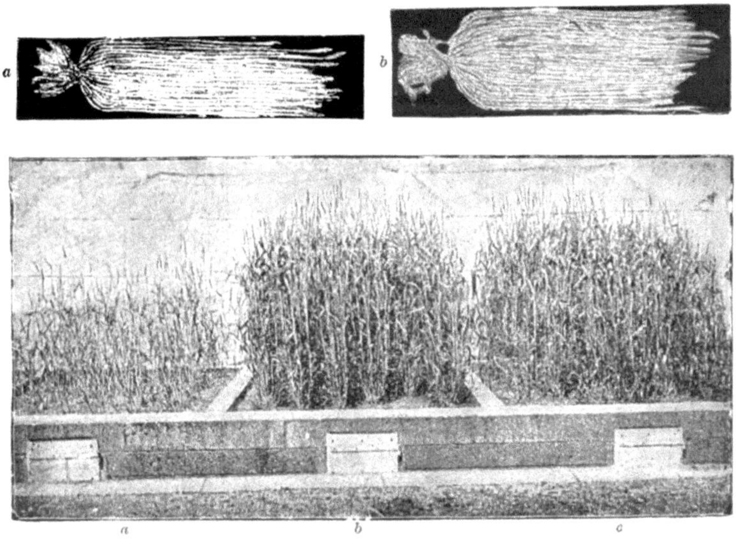

Fig. 4.3 "Wheat Plants. a, Has given 17 ears from each planted grain. Soil manured with chemical manure only. b, Has given 26 ears from each planted grain. Soil manured with both stable and chemical manure." (above) "Squares at Professor Grandeau's experimental station, planted with grains of wheat, in three soils: a, pure sand; b and c, manured arable soil; each grain 12 inches apart." (below) Source: ibid., p. 184.
«Prof. Grandeau, Director of the French Station Agronomique de l'Est, has made, since 1886, experiments on Major Hallett's method, and he obtained similar results. "In a proper soil," he wrote, "one single grain of wheat can give as much as fifty stems (and ears), and even more, and thus cover a circle thirteen inches in diameter." (ibid., p. 184)

Chapter 5
The Urban Region (1901)

Herbert George Wells

Introduction

Martina Barcelloni Corte
Université de Liège, Liège, Belgium
mbarcellonicorte@uliege.be

At the turn of the twentieth century when the power and attraction of the "great cities" was at its peak, the well-known science fiction writer H.G. Wells prophesized—in countertrend with the narratives of his time—that the technological forces that had created the "industrial Metropolis", inside and outside Europe, were now moving to dismantle it.

In "the probable diffusion of great cities", second chapter to his well-known book Anticipations, *written at the age of 34, Wells launched what he himself named "the general prophecy of urban regions", a quasi-fictional essay in which the seemingly inexorable concentration of people and resources in large cities was reversed to the benefit of vast, diffusely inhabited and equipped territories, in which the "old antithesis between city and countryside" had become obsolete.*

Taking the revolution in transport and the related acceleration of land locomotion as his point of departure, Wells announced a radical reorganization of human society and the related upheaval of its distribution in space: the increase in the distance a worker could travel in an hour, rather than producing even larger cities, would create—in his "prophecy"—an extensively inhabited territory. "New centrifugal forces" as, for example, the "passion for nature" or the "charm for cultivation and gardening" would be liberated and a new space of freedom produced, in which "citizens of all ranks would be able to choose their own site" and "build their own home by and for themselves" giving vent to a growing desire for self-determination and emancipation.

M. Barcelloni Corte and P. Viganò (eds.), *The Horizontal Metropolis*,
https://doi.org/10.1007/978-3-030-56398-1_5

Wells based his prediction on a deep and situated analysis of past and emerging trends in transportation and communication networks: throughout the nineteenth century, rail transportation had been a relatively simple system, favouring direct access to large centers, however, with the spread of branch-lines and electric tramways, a complex rail network had been created that could easily serve as the basis for decentralized urban regions. Communication and transportation systems "were to become increasingly easier, swifter and altogether better organized".

Wells pictures the "great city of the year 2000" as a series of villages sprinkled with small homes and factories set in the open fields yet connected by high-speed rail transportation and communication systems to any other point in the region: "before the end of the next century", he prophesied, "every open space of mountain and heather will be dotted (...) with clumps of prosperous houses about school, doctor engineers, book and provision shops" and the "London citizen will have a choice of nearly all England and Wales South of Nottingham and East of Exeter as his suburb". Old cities would not completely disappear but gradually lose their financial and industrial functions; "centripetal forces" as the "human love for crowds" would keep them alive, still transforming them into great bazars, "galleries of shops and places of concourse and rendez-vous, *a pedestrian place, its pathways reinforced by lifts and moving platforms, and shielded from the weather, altogether a very spacious, brilliant, and entertaining agglomeration". In the essay's premonitions, the old Metropolis would be eroded to an entertainment center while the productive life would take place in the urban region.*

Inviting his readers to "think and act in new ways" and highlighting how the current social order was dangerously disintegrating "without preparing another one to take its place", Wells proves a growing interest of the intellectual milieu for economic geography and for the study of the forces making for concentration and de-concentration of population, interests that will soon be brought forward and formalized by theorists such as Alfred Weber, August Lösch or Walter Christaller.

5.1 The Urban Region (from: *Anticipations of the reaction of mechanical and scientific progress upon human life and thought*)[1]

NOW, the velocity at which a man and his belongings may pass about the earth is in itself a very trivial matter indeed, but it involves certain other matters not at all trivial, standing, indeed, in an almost fundamental relation to human society. It will be the business of this chapter to discuss the relation between the social order and the available means of transit, and to attempt to deduce from the principles

[1]Excerpt from: Wells, H. G. (1901). The probable diffusion of great cities. In H. G. Wells, *Anticipations of the reaction of mechanical and scientific progress upon human life and thought.* Harper & Brothers, Chapman & Hall.

elucidated the coming phases in that extraordinary expansion, shifting and internal redistribution of population that has been so conspicuous during the last hundred years.

Let us consider the broad features of the redistribution of the population that has characterised the nineteenth century. It may be summarised as an unusual growth of great cities and a slight tendency to depopulation in the country. The growth of the great cities is the essential phenomenon. These aggregates having populations of from eight hundred thousand upward to four and five millions, are certainly, so far as the world outside the limits of the Chinese empire goes, entirely an unprecedented thing. Never before, outside the valleys of the three great Chinese rivers, has any city—with the exception of Rome and perhaps (but very doubtfully) of Babylon—certainly had more than a million inhabitants, and it is at least permissible to doubt whether the population of Rome, in spite of its exacting a tribute of sea-borne food from the whole of the Mediterranean basin, exceeded a million for any great length of time.[2] But there are now ten town aggregates having a population of over a million, nearly twenty that bid fair to reach that limit in the next decade, and a great number at or approaching a quarter of a million. We call these towns and cities, but, indeed, they are of a different order of things to the towns and cities of the eighteenth-century world.

Concurrently with the aggregation of people about this new sort of centre, there has been, it is alleged, a depletion of the country villages and small townships. But, so far as the counting of heads goes, this depletion is not nearly so marked as the growth of the great towns. Relatively, however, it is striking enough.

Now, is this growth of large towns really, as one may allege, a result of the development of railways in the world, or is it simply a change in human circumstances that happens to have arisen at the same time? It needs only a very general review of the conditions of the distribution of population to realise that the former is probably the true answer.

---

[2]It is true that many scholars estimate a high-water mark for the Roman population in excess of two millions; and one daring authority, by throwing out suburbs ad libitum into the Campagna, suburbs of which no trace remains, has raised the two to ten. The Colosseum could, no doubt, seat over 80,000 spectators; the circuit of the bench frontage of the Circus Maximus was very nearly a mile in length, and the Romans of Imperial times certainly used ten times as much water as the modern Romans. But, on the other hand, habits change, and Rome as it is defined by lines drawn at the times of its greatest ascendancy—the city, that is, enclosed by the walls of Aurelian and including all the *regiones* of Augustus, an enclosure from which there could have been no reason for excluding half or more of its population—could have scarcely contained a million. It would have packed very comfortably within the circle of the Grands Boulevards of Paris—the Paris, that is, of Louis XIV., with a population of 560,000; and the Rome of to-day, were the houses that spread so densely over the once vacant Campus Martius distributed in the now deserted spaces in the south and east, and the Vatican suburb replaced within the ancient walls, would quite fill the ancient limits, in spite of the fact that the population is under 500,000. But these are incidental doubts on a very authoritative opinion, and, whatever their value, they do not greatly affect the significance of these new great cities, which have arisen all over the world, as if by the operation of a natural law, as the railways have developed.

It will be convenient to make the issue part of a more general proposition, namely, that *the general distribution of population in a country must always be directly dependent on transport facilities*. To illustrate this point roughly we may build up an imaginary simple community by considering its needs. Over an arable country-side, for example, inhabited by a people who had attained to a level of agricultural civilisation in which war was no longer constantly imminent, the population would be diffused primarily by families and groups in farmsteads. It might, if it were a very simple population, be almost all so distributed. But even the simplest agriculturists find a certain convenience in trade. Certain definite points would be convenient for such local trade and intercourse as the people found desirable, and here it is that there would arise the germ of a town. At first it might be no more than an appointed meeting-place, a market square, but an inn and a blacksmith would inevitably follow, an altar, perhaps, and, if these people had writing, even some sort of school. It would have to be where water was found, and it would have to be generally convenient of access to its attendant farmers.

Now, if this meeting-place was more than a certain distance from any particular farm, it would be inconvenient for that farmer to get himself and his produce there and back, and to do his business in a comfortable daylight. He would not be able to come and, instead, he would either have to go to some other nearer centre to trade and gossip with his neighbours or, failing this, not go at all. Evidently, then, there would be a maximum distance between such places. This distance in England, where traffic has been mainly horse-traffic for many centuries, seems to have worked out, according to the gradients and so forth, at from eight to fifteen miles, and at such distances do we find the country towns, while the horseless man, the serf, and the labourer and labouring wench have marked their narrow limits in the distribution of the intervening villages. If by chance these gathering places have arisen at points much closer than this maximum, they have come into competition, and one has finally got the better of the other, so that in England the distribution is often singularly uniform. Agricultural districts have their towns at about eight miles, and where grazing takes the place of the plough, the town distances increase to fifteen.[3] And so it is, entirely as a multiple of horse and foot strides, that all the villages and towns of the world's country-side have been plotted out.[4]

A third, and almost final, factor determining town distribution in a world without railways, would be the sea-port and the navigable river. Ports would grow into dimensions dependent on the population of the conveniently accessible coasts (or river-banks), and on the quality and quantity of their products, and near these

---

[3]It will be plain that such towns must have clearly defined limits of population, *dependant finally on the minimum yearly produce of the district they control*. If ever they rise above that limit the natural checks of famine, and of pestilence following enfeeblement, will come into operation, and they will always be kept near this limit by the natural tendency of humanity to increase. The limit would rise with increasing public intelligence, and the organisation of the towns would become more definite.

[4]I owe the fertilising suggestion of this general principle to a paper by Grant Allen that I read long ago in *Longman's Magazine*.

ports, as the conveniences of civilisation increased, would appear handicraft towns —the largest possible towns of a foot-and-horse civilisation—with industries of such a nature as the produce of their coasts required.

It was always in connection with a port or navigable river that the greater towns of the pre-railway periods arose, a day's journey away from the coast when sea attack was probable, and shifting to the coast itself when that ceased to threaten. Such sea-trading handicraft towns as Bruges, Venice, Corinth, or London were the largest towns of the vanishing order of things. Very rarely, except in China, did they clamber above a quarter of a million inhabitants, even though to some of them there was presently added court and camp. In China, however, a gigantic river and canal system, laced across plains of extraordinary fertility, has permitted the growth of several city aggregates with populations exceeding a million, and in the case of the Hankow trinity of cities exceeding five million people.

In all these cases the position and the population limit was entirely determined by the accessibility of the town and the area it could dominate for the purposes of trade. And not only were the commercial or natural towns so determined, but the political centres were also finally chosen for strategic considerations, in a word—communications. And now, perhaps, the real significance of the previous paper, in which sea velocities of fifty miles an hour, and land travel at the rate of a hundred, and even cab and omnibus journeys of thirty or forty miles, were shown to be possible, becomes more apparent.

At the first sight it might appear as though the result of the new developments was simply to increase the number of giant cities in the world by rendering them possible in regions where they had hitherto been impossible—concentrating the trade of vast areas in a manner that had hitherto been entirely characteristic of navigable waters. It might seem as though the state of affairs in China, in which population has been concentrated about densely-congested "million-cities," with pauper masses, public charities, and a crowded struggle for existence, for many hundreds of years, was merely to be extended over the whole world. We have heard so much of the "problem of our great cities"; we have the impressive statistics of their growth; the belief in the inevitableness of yet denser and more multitudinous agglomerations in the future is so widely diffused, that at first sight it will be thought that no other motive than a wish to startle can dictate the proposition that not only will many of these railway-begotten "giant cities" reach their maximum in the commencing century, but that in all probability they, and not only they, but their water-born prototypes in the East also, are destined to such a process of dissection and diffusion as to amount almost to obliteration, so far, at least, as the blot on the map goes, within a measurable further space of years.

In advancing this proposition, the present writer is disagreeably aware that in this matter he has expressed views entirely opposed to those he now propounds; and in setting forth the following body of considerations he tells the story of his own disillusionment. At the outset he took for granted—and, very naturally, he wishes to imagine that a great number of other people do also take for granted—that the future of London, for example, is largely to be got as the answer to a sort of rule-of-three sum. If in one hundred years the population of London has been multiplied by seven, then in two hundred years—! And one proceeds to pack the answer in gigantic tenement houses, looming upon colossal roofed streets, provide it with moving ways (the only available transit appliances suited to such dense multitudes), and develop its manners and morals in accordance with the laws that will always prevail amidst over-crowded humanity so long as humanity endures. The picture of this swarming concentrated humanity has some effective possibilities, but, unhappily, if, instead of that obvious rule-of-three sum, one resorts to an analysis of operating causes, its plausibility crumbles away, and it gives place to an altogether different forecast—a forecast, indeed, that is in almost violent contrast to the first anticipation. It is much more probable that these coming cities will not be, in the old sense, cities at all; they will present a new and entirely different phase of human distribution.

The determining factor in the appearance of great cities in the past, and, indeed, up to the present day, has been the meeting of two or more transit lines, the confluence of two or more streams of trade, and easy communication. The final limit to the size and importance of the great city has been the commercial "sphere of influence" commanded by that city, the capacity of the alluvial basin of its commerce, so to speak, the volume of its river of trade. About the meeting-point so determined the population so determined has grouped itself—and this is the point I overlooked in those previous vaticinations—in accordance with *laws that are also considerations of transit.*

The economic centre of the city is formed, of course, by the wharves and landing places—and in the case of railway-fed cities by the termini—where passengers land and where goods are landed, stored, and distributed. Both the administrative and business community, traders, employers, clerks, and so forth, must be within a convenient access of this centre; and the families, servants, tradesmen, amusement purveyors dependent on these again must also come within a maximum distance. At a certain stage in town growth the pressure on the more central area would become too great for habitual family life there, and an office region would differentiate from an outer region of homes. Beyond these two zones, again, those whose connection with the great city was merely intermittent would constitute a system of suburban houses and areas. But the grouping of these, also, would be determined finally by the convenience of access to the dominant centre. That secondary centres, literary, social, political, or military, may arise about the initial trade centre, complicates the application but does not alter the principle here stated. They must all be within striking distance. The day of twenty-four hours is an inexorable human condition, and up to the present time all intercourse and business has been broken into spells of definite duration by intervening nights. Moreover, almost all effective intercourse has involved personal presence at the point where intercourse occurs. The possibility, therefore, of going and coming and doing that day's work has hitherto fixed the extreme limits to which a city could grow, and has exacted a compactness which has always been very undesirable and which is now for the first time in the world's history no longer imperative.

So far as we can judge without a close and uncongenial scrutiny of statistics, that daily journey, that has governed and still to a very considerable extent governs the growth of cities, has had, and probably always will have, a maximum limit of two hours, one hour each way from sleeping-place to council chamber, counter, work-room, or office stool. And taking this assumption as sound, we can state precisely the maximum area of various types of town. A pedestrian agglomeration such as we find in China, and such as most of the European towns probably were before the nineteenth century, would be swept entirely by a radius of four miles about the business quarter and industrial centre; and, under these circumstances, where the area of the feeding regions has been very large the massing of human beings has probably reached its extreme limit.[5] Of course, in the case of a navigable river, for example, the commercial centre might be elongated into a line and the circle of the city modified into an ellipse with a long diameter considerably exceeding eight miles, as, for example, in the case of Hankow.

If, now, horseflesh is brought into the problem, an outer radius of six or eight miles from the centre will define a larger area in which the carriage folk, the hackney users, the omnibus customers, and their domestics and domestic camp followers may live and still be members of the city. Towards that limit London was already probably moving at the accession of Queen Victoria, and it was clearly the absolute limit of urban growth—until locomotive mechanisms capable of more than eight miles an hour could be constructed.

And then there came suddenly the railway and the steamship, the former opening with extraordinary abruptness a series of vast through-routes for trade, the latter enormously increasing the security and economy of the traffic on the old water routes. For a time neither of these inventions was applied to the needs of intra-urban transit at all. For a time they were purely centripetal forces. They worked simply to increase the general volume of trade, to increase, that is, the pressure of population upon the urban centres. As a consequence the social history of the middle and later thirds of the nineteenth century, not simply in England but all over the civilised world, is the history of a gigantic rush of population into the magic radius of—for most people—four miles, to suffer there physical and moral disaster less acute but, finally, far more appalling to the imagination than any famine or pestilence that ever swept the world. Well has Mr. George Gissing named nineteenth-century London in one of his great novels the "Whirlpool," the very figure for the nineteenth-century Great City, attractive, tumultuous, and spinning down to death.

But, indeed, these great cities are no permanent maëlstroms. These new forces, at present still so potently centripetal in their influence, bring with them, nevertheless, the distinct promise of a centrifugal application that may be finally equal to the complete reduction of all our present congestions. The limit of the pre-railway city was the limit of man and horse. But already that limit has been exceeded, and each day brings us nearer to the time when it will be thrust outward in every direction with an effect of enormous relief.

---

[5]It is worth remarking that in 1801 the density of population in the City of London was half as dense again as that of any district, even of the densest "slum" districts, to-day.

So far the only additions to the foot and horse of the old dispensation that have actually come into operation, are the suburban railways, which render possible an average door to office hour's journey of ten or a dozen miles—further only in the case of some specially favoured localities. The star-shaped contour of the modern great city, thrusting out arms along every available railway line, knotted arms of which every knot marks a station, testify sufficiently to the relief of pressure thus afforded. Great Towns before this century presented rounded contours and grew as a puff-ball swells; the modern Great City looks like something that has burst an intolerable envelope and splashed. But, as our previous paper has sought to make clear, these suburban railways are the mere first rough expedient of far more convenient and rapid developments.

We are—as the Census Returns for 1901 quite clearly show—in the early phase of a great development of centrifugal possibilities. And since it has been shown that a city of pedestrians is inexorably limited by a radius of about four miles, and that a horse-using city may grow out to seven or eight, it follows that the available area of a city which can offer a cheap suburban journey of thirty miles an hour is a circle with a radius of thirty miles. And is it too much, therefore, in view of all that has been adduced in this and the previous paper, to expect that the available area for even the common daily toilers of the great city of the year 2000, or earlier, will have a radius very much larger even than that? Now, a circle with a radius of thirty miles gives an area of over 2800 square miles, which is almost a quarter that of Belgium. But thirty miles is only a very moderate estimate of speed, and the reader of the former paper will agree, I think, that the available area for the social equivalent of the favoured season-ticket holders of to-day will have a radius of over one hundred miles, and be almost equal to the area of Ireland.[6] The radius that will sweep the area available for such as now live in the outer suburbs will include a still vaster area. Indeed, it is not too much to say that the London citizen of the year 2000 A.D. may have a choice of nearly all England and Wales south of Nottingham and east of Exeter as his suburb, and that the vast stretch of country from Washington to Albany will be all of it "available" to the active citizen of New York and Philadelphia before that date.

This does not for a moment imply that cities of the density of our existing great cities will spread to these limits. Even if we were to suppose the increase of the populations of the great cities to go on at its present rate, this enormous extension of available area would still mean a great possibility of diffusion. But though most great cities are probably still very far from their maxima, though the network of feeding railways has still to spread over Africa and China, and though huge areas are still imperfectly productive for want of a cultivating population, yet it is well to remember that for each great city, quite irrespective of its available spaces, a maximum of population is fixed. Each great city is sustained finally by the trade and production of a certain proportion of the world's surface—by the area it commands

[6]Be it noted that the phrase "available area" is used, and various other modifying considerations altogether waived for the present.

commercially. The great city cannot grow, except as a result of some quite morbid and transitory process—to be cured at last by famine and disorder—beyond the limit the commercial capacity of that commanded area prescribes. Long before the population of this city, with its inner circle a third of the area of Belgium, rose towards the old-fashioned city density, this restriction would come in. Even if we allowed for considerable increase in the production of food-stuffs in the future, it still remains inevitable that the increase of each city in the world must come at last upon arrest.

Yet, though one may find reasons for anticipating that this city will in the end overtake and surpass that one and such-like relative prophesying, it is difficult to find any data from which to infer the absolute numerical limits of these various diffused cities. Or perhaps it is more seemly to admit that no such data have occurred to the writer. So far as London, St. Petersburg, and Berlin go, it seems fairly safe to assume that they will go well over twenty millions; and that New York, Philadelphia, and Chicago will probably, and Hankow almost certainly, reach forty millions. Yet even forty millions over thirty-one thousand square miles of territory is, in comparison with four millions over fifty square miles, a highly diffused population.

How far will that possible diffusion accomplish itself? Let us first of all consider the case of those classes that will be free to exercise a choice in the matter, and we shall then be in a better position to consider those more numerous classes whose general circumstances are practically dictated to them. What will be the forces acting upon the prosperous household, the household with a working head and four hundred a year and upwards to live upon, in the days to come? Will the resultant of these forces be, as a rule, centripetal or centrifugal? Will such householders in the greater London of 2000 A.D. still cluster for the most part, as they do to-day, in a group of suburbs as close to London as is compatible with a certain fashionable maximum of garden space and air; or will they leave the ripened gardens and the no longer brilliant villas of Surbiton and Norwood, Tooting and Beckenham, to other and less independent people? First, let us weigh the centrifugal attractions.

The first of these is what is known as the passion for nature, that passion for hillside, wind, and sea that is evident in so many people nowadays, either frankly expressed or disguising itself as a passion for golfing, fishing, hunting, yachting, or cycling; and, secondly, there is the allied charm of cultivation, and especially of gardening, a charm that is partly also the love of dominion, perhaps, and partly a personal love for the beauty of trees and flowers and natural things. Through that we come to a third factor, that craving—strongest, perhaps, in those Low German peoples, who are now ascendant throughout the world—for a little private *imperium* such as a house or cottage "in its own grounds" affords; and from that we pass on to the intense desire so many women feel—and just the women, too, who will mother the future—their almost instinctive demand, indeed, for a household, a separate sacred and distinctive household, built and ordered after their own hearts, such as in its fulness only the country-side permits. Add to these things the healthfulness of the country for young children, and the wholesome isolation that is possible from much that irritates, stimulates prematurely, and corrupts in crowded centres, and the chief positive centrifugal inducements are stated, inducements that no progress of inventions, at any-rate, can ever seriously weaken. What now are the centripetal forces against which these inducements contend?

In the first place, there are a group of forces that will diminish in strength. There is at present the greater convenience of "shopping" within a short radius of the centre of the great city, a very important consideration indeed to many wives and mothers. All the inner and many of the outer suburbs of London obtain an enormous proportion of the ordinary household goods from half a dozen huge furniture, grocery, and drapery firms, each of which has been enabled by the dearness and inefficiency of the parcels distribution of the post-office and railways to elaborate a now very efficient private system of taking orders and delivering goods. Collectively these great businesses have been able to establish a sort of monopoly of suburban trade, to overwhelm the small suburban general tradesman (a fate that was inevitable for him in some way or other), and—which is a positive world-wide misfortune—to overwhelm also many highly specialised shops and dealers of the central district. Suburban people nowadays get their wine and their novels, their clothes and their amusements, their furniture and their food, from some one vast indiscriminate shop or "store" full of respectable mediocre goods, as excellent a thing for housekeeping as it is disastrous to taste and individuality.[7] But it is doubtful if the delivery organisation of these great stores is any more permanent than the token coinage of the tradespeople of the last century. Just as it was with that interesting development, so now it is with parcels distribution: private enterprise supplies in a partial manner a public need, and with the organization of a public parcels and goods delivery on cheap and sane lines in the place of our present complex, stupid, confusing, untrustworthy, and fantastically costly chaos of post-office, railways, and carriers, it is quite conceivable that Messrs. Omnium will give place again to specialised shops.

It must always be remembered how timid, tentative, and dear the postal and telephone services of even the most civilised countries still are, and how inexorably the needs of revenue, public profit, and convenience fight in these departments against the tradition of official leisure and dignity. There is no reason now, except that the thing is not yet properly organised, why a telephone call from any point in such a small country as England to any other should cost much more than a postcard. There is no reason now, save railway rivalries and retail ideas—obstacles some able and active man is certain to sweep away sooner or later—why the post-office should not deliver parcels anywhere within a radius of a hundred miles in a few hours at a penny or less for a pound and a little over,[8] put our newspapers in our letter-boxes direct from the printing-office, and, in fact, hand in nearly every

[7]Their temporary suppression of the specialist is indeed carried to such an extent that one may see even such things as bronze ornaments and personal jewellery listed in Messrs. Omnium's list, and stored in list designs and pattern; and their assistants will inform you that their brooch, No. 175, is now "very much worn," without either blush or smile.

[8]The present system of charging parcels by the pound, when goods are sold by the pound, and so getting a miserly profit in the packing, is surely one of the absurdest disregards of the obvious it is possible to imagine.

constant need of the civilised household, except possibly butcher's meat, coals, greengrocery, and drink. And since there is no reason, but quite removable obstacles, to prevent this development of the post-office, I imagine it will be doing all these things within the next half-century. When it is, this particular centripetal pull, at anyrate, will have altogether ceased to operate.

A second important centripetal consideration at present is the desirability of access to good schools and to the doctor. To leave the great centres is either to abandon one's children, or to buy air for them at the cost of educational disadvantages. But access, be it noted, is another word for transit. It is doubtful if these two needs will so much keep people close to the great city centres as draw them together about secondary centres. New centres they may be—compare Hindhead, for example—in many cases; but also, it may be, in many cases the more healthy and picturesque of the existing small towns will develop a new life. Already, in the case of the London area, such once practically autonomous places as Guildford, Tunbridge Wells, and Godalming have become economically the centres of lax suburbs, and the same fate may very probably overtake, for example, Shrewsbury, Stratford, and Exeter, and remoter and yet remoter townships. Indeed, for all that this particular centripetal force can do, the confluent "residential suburbs" of London, of the great Lancashire-Yorkshire city, and of the Scotch city, may quite conceivably replace the summer lodging-house watering-places of to-day, and extend themselves right round the coast of Great Britain, before the end of the next century, and every open space of mountain and heather be dotted—not too thickly—with clumps of prosperous houses about school, doctor, engineers, book and provision-shops.

A third centripetal force will not be set aside so easily. The direct antagonist it is to that love of nature that drives people out to moor and mountain. One may call it the love of the crowd; and closely allied to it is that love of the theatre which holds so many people in bondage to the Strand. Charles Lamb was the Richard Jefferies of this group of tendencies, and the current disposition to exaggerate the opposition force, especially among English-speaking peoples, should not bind us to the reality of their strength. Moreover, interweaving with these influences that draw people together are other more egotistical and intenser motives, ardent in youth and by no means—to judge by the Folkestone Leas—extinct in age, the love of dress, the love of the crush, the hot passion for the promenade. Here, no doubt, what one may speak of loosely as "racial" characteristics count for much. The common actor and actress of all nationalities, the Neapolitan, the modern Roman, the Parisian, the Hindoo, I am told, and that new and interesting type, the rich and liberated Jew emerging from his Ghetto and free now absolutely to show what stuff he is made of, flame out most gloriously in this direction. To a certain extent this group of tendencies may lead to the formation of new secondary centres within the "available" area, theatrical and musical centres— centres of extreme Fashion and Selectness, centres of smartness and opulent display—but it is probable that for the large number of people throughout the world who cannot afford to maintain households in duplicate these will be for many years yet strictly centripetal forces, and will keep them within the radius marked by whatever will be the future equivalent in length of, say, the present two-shilling cab ride in London.

And, after all, for all such "shopping" as one cannot do by telephone or postcard, it will still be natural for the shops to be gathered together in some central place. And "shopping" needs refreshment, and may culminate in relaxation. So that Bond Street and Regent Street, the Boulevard des Capucines, the Corso, and Broadway will still be brilliant and crowded for many years for all the diffusion that is here forecast—all the more brilliant and crowded, perhaps, for the lack of a thronging horse-traffic down their central ways. But the very fact that the old nucleus is still to be the best place for all who trade in a concourse of people, for novelty shops and art shops, and theatres and business buildings, by keeping up the central ground-values will operate against residence there and shift the "masses" outwardly.

And once people have been driven into cab, train, or omnibus, the only reason why they should get out to a residence here rather than there is the necessity of saving time, and such a violent upward gradient of fares as will quite outbalance the downward gradient of ground-values. We have, however, already forecast a swift, varied, and inevitably competitive suburban traffic. And so, though the centre will probably still remain the centre and "Town," it will be essentially a bazaar, a great gallery of shops and places of concourse and rendezvous, a pedestrian place, its pathways reinforced by lifts and moving platforms, and shielded from the weather, and altogether a very spacious, brilliant, and entertaining agglomeration.

Enough now has been said to determine the general nature of the expansion of the great cities in the future, so far as the more prosperous classes are concerned. It will not be a regular diffusion like the diffusion of a gas, but a process of throwing out the "homes," and of segregating various types of people. The omens seem to point pretty unmistakably to a wide and quite unprecedented diversity in the various suburban townships and suburban districts. Of that aspect of the matter a later paper must treat. It is evident that from the outset racial and national characteristics will tell in this diffusion. We are getting near the end of the great Democratic, Wholesale, or Homogeneous phase in the world's history. The sport-loving Englishman, the sociable Frenchman, the vehement American will each diffuse his own great city in his own way.

And now, how will the increase in the facilities of communication we have assumed affect the condition of those whose circumstances are more largely dictated by economic forces? The mere diffusion of a large proportion of the prosperous and relatively free, and the multiplication of various types of road and mechanical traction, means, of course, that in this way alone a perceptible diffusion of the less independent classes will occur. To the subsidiary centres will be drawn doctor and schoolmaster, and various dealers in fresh provisions, baker, grocer, butcher; of if they are already established there they will flourish more and more, and about them the convenient home of the future, with its numerous electrical and mechanical appliances, and the various bicycles, motor-cars, photographic and phonographic apparatus that will be included in its equipment will gather a population of repairers, "accessory" dealers and working engineers, a growing class which from its necessary intelligence and numbers will play a very conspicuous part in the social development of the twentieth century. The much more elaborate post-office and telephone services will also bring intelligent ingredients to these suburban nuclei, these restorations of the old villages and country towns. And the sons of the cottager within the affected area will develop into the skilled vegetable or flower gardeners, the skilled ostler—with some veterinary science—and so forth, for whom also there will evidently be work and a living. And dotted at every convenient position along the new roads, availing themselves no doubt whenever possible of the picturesque inns that the old coaching days have left us, will be wayside restaurants and tea houses, and motor and cycle stores and repair places. So much diffusion is practically inevitable.

In addition, as we have already intimated, many Londoners in the future may abandon the city office altogether, preferring to do their business in more agreeable surroundings. Such a business as book publishing, for example, has no unbreakable bonds to keep it in the region of high rent and congested streets. The days when the financial fortunes of books depended upon the colloquial support of influential people in a small Society are past; neither publishers nor authors as a class have any relation to Society at all, and actual access to newspaper offices is necessary only to the ranker forms of literary imposture. That personal intercourse between publishers and the miscellaneous race of authors which once justified the central position has, I am told, long since ceased. And the withdrawing publishers may very well take with them the printers and binders, and attract about them their illustrators and designers. So, as a typical instance, one—now urban—trade may detach itself.

Publishing is, however, only one of the many similar trades equally profitable and equally likely to move outward to secondary centres, with the development and cheapening of transit. It is all a question of transit. Limitation of transit contracts the city, facilitation expands and disperses it. All this case for diffusion so far is built up entirely on the hypothesis we attempted to establish in the first paper, that transit of persons and goods alike is to become easier, swifter, and altogether better organised than it is at present.

The telephone will almost certainly prove a very potent auxiliary indeed to the forces making for diffusion. At present that convenience is still needlessly expensive in Great Britain, and a scandalously stupid business conflict between telephone company and post-office delays, complicates, and makes costly and exasperating all trunk communications; but even under these disadvantages the thing is becoming a factor in the life of ordinary villadom. Consider all that lies within its possibilities. Take first the domestic and social side; almost all the labour of ordinary shopping can be avoided—goods nowadays can be ordered and sent either as sold outright, or on approval, to any place within a hundred miles of London, and in one day they can be examined, discussed, and returned—at any rate, in theory. The mistress of the house has all her local tradesmen, all the great London shops, the circulating library, the theatre box-office, the post-office and cab-rank, the nurses' institute and the doctor, within reach of her hand. The instrument we may confidently expect to improve, but even now speech is perfectly clear and distinct over several hundred miles of wire. Appointments and invitations can be made; and at a cost varying from a penny to two shillings anyone within two hundred miles of home may speak day or night into the ear of his or her household. Were it not for that unmitigated public nuisance, the practical control of our post-office by non-dismissable Civil servants, appointed so young as to be entirely ignorant of the unofficial world, it would be possible now to send urgent messages at any hour of the day or night to any part of the world; and even our sacred institution of the Civil Service can scarcely prevent this desirable consummation for many years more. The business man may then sit at home in his library and bargain, discuss, promise, hint, threaten, tell such lies as he dare not write, and, in fact, do everything that once demanded a personal encounter. Already for a great number of businesses it is no longer necessary that the office should be in London, and only habit, tradition, and minor considerations keep it there. With the steady cheapening and the steady increase in efficiency of postal and telephonic facilities, and of goods transit, it seems only reasonable to anticipate the need for that expensive office and the irksome daily journey will steadily decline. In other words, what will still be economically the "city," as distinguished from the "agricultural" population, will probably be free to extend, in the case of all the prosperous classes not tied to large establishments in need of personal supervision, far beyond the extreme limits of the daily hour journey.

But the diffusion of the prosperous, independent, and managing classes involves in itself a very considerable diffusion of the purely "working" classes also. Their centres of occupation will be distributed, and their freedom to live at some little distance from their work will be increased. Whether this will mean dotting the country with dull, ugly little streets, slum villages like Buckfastleigh in Devon, for example, or whether it may result in entirely different and novel aspects, is a point for which at present we are not ready. But it bears upon the question that ugliness and squalor upon the main road will appeal to the more prosperous for remedy with far more vigour than when they are stowed compactly in a slum.

Enough has been said to demonstrate that old "town" and "city" will be, in truth, terms as obsolete as "mail coach." For these new areas that will grow out of them we want a term, and the administrative "urban district" presents itself with a convenient air of suggestion. We may for our present purposes call these coming town provinces "urban regions." Practically, by a process of confluence, the whole of Great Britain south of the Highlands seems destined to become such an urban region, laced all together not only by railway and telegraph, but by novel roads such as we forecast in the former chapter, and by a dense network of telephones, parcels-delivery tubes, and the like nervous and arterial connections.

It will certainly be a curious and varied region, far less monotonous than our present English world, still in its thinner regions, at any rate, wooded, perhaps rather more abundantly wooded, breaking continually into park and garden, and with everywhere a scattering of houses. These will not, as a rule, I should fancy, follow the fashion of the vulgar ready-built villas of the existing suburb, because the freedom people will be able to exercise in the choice of a site will rob the "building estate" promoter of his local advantage; in many cases the houses may very probably be personal homes, built for themselves as much as the Tudor manor-houses were, and even, in some cases, as æsthetically right. Each district,

I am inclined to think, will develop its own differences of type and style. As one travels through the urban region, one will traverse open, breezy, "horsey" suburbs, smart white gates and palings everywhere, good turf, a Grand Stand shining pleasantly; gardening districts all set with gables and roses, holly hedges, and emerald lawns; pleasant homes among heathery moorlands and golf links, and river districts with gaily painted boat-houses peeping from the osiers. Then presently a gathering of houses closer together, and a promenade and a whiff of band and dresses, and then, perhaps, a little island of agriculture, hops, or strawberry gardens, fields of grey-plumed artichokes, white-painted orchard, or brightly neat poultry farm. Through the varied country the new wide roads will run, here cutting through a crest and there running like some colossal aqueduct across a valley, swarming always with a multitudinous traffic of bright, swift (and not necessarily ugly) mechanisms; and everywhere amidst the fields and trees linking wires will stretch from pole to pole. Ever and again there will appear a cluster of cottages—cottages into which we shall presently look more closely—about some works or workings, works, it may be, with the smoky chimney of to-day replaced by a gaily painted wind-wheel or water-wheel to gather and store the force for the machinery; and ever and again will come a little town, with its cherished ancient church or cathedral, its school buildings and museums, its railway-station, perhaps its fire-station, its inns and restaurants, and with all the wires of the countryside converging to its offices. All that is pleasant and fair of our present countryside may conceivably still be there among the other things. There is no reason why the essential charm of the country should disappear; the new roads will not supersede the present highroads, which will still be necessary for horses and subsidiary traffic; and the lanes and hedges, the field-paths and wild-flowers, will still have their ample justification. A certain lack of solitude there may be perhaps, and—

Will conspicuous advertisements play any part in the landscape?

But I find my pen is running ahead, an imagination prone to realistic constructions is struggling to paint a picture altogether prematurely. There is very much to be weighed and decided before we can get from our present generalisation to the style of architecture these houses will show, and to the power and nature of the public taste. We have laid down now the broad lines of road, railway, and sea transit in the coming century, and we have got this general prophecy of "urban regions" established, and for the present that much must suffice.

And as for the world beyond our urban regions? The same line of reasoning that leads to the expectation that the city will diffuse itself until it has taken up considerable areas and many of the characteristics, the greenness, the fresh air, of what is now country, leads us to suppose also that the country will take to itself many of the qualities of the city. The old antithesis will indeed cease, the boundary-lines will altogether disappear; it will become, indeed, merely a question of more or less populous. There will be horticulture and agriculture going on within the "urban regions," and "urbanity" without them. Everywhere, indeed, over the land of the globe between the frozen circles, the railway and the new roads will spread, the network of communication-wires and safe and convenient ways. To receive the daily paper a few hours late, to wait a day or so for goods one has ordered, will be the extreme measure of rusticity save in a few remote islands and inaccessible places. The character of the meshes in that wider network of roads that will be the country, as distinguished from the urban district, will vary with the soil, the climate and the tenure of the land—will vary, too, with the racial and national differences. But throughout all that follows, this mere relativity of the new sort of town to the new sort of country over which the new sorts of people we are immediately to consider will be scattered, must be borne in mind.

[At the risk of insistence, I must repeat that, so far, I have been studiously taking no account of the fact that there is such a thing as a boundary-line or a foreigner in the world. It will be far the best thing to continue to do this until we can get out all that will probably happen universally or generally, and in particular the probable changes in social forces, social apparatus and internal political methods. We shall then come to the discussion of language, nationality and international conflicts, equipped with such an array of probabilities and possibilities as will enable us to guess at these special issues with an appearance of far more precision than would be the case if we considered them now].

Fig. 5.1 "Map of London and Environs" (chromolithographic map, engraving).
London's environs are diffusely inhabited and already served by a dense rail network (in black).
Source: K. Baedeker's travel guide (1900 circa)

"'This here Progress,' said Tom Smallways, 'it keeps on'"

Fig. 5.2 "This here progress (…) it keeps on".
Throughout his career H. G. Wells celebrated the technological developments that were revolutionizing life, but also feared they might lead to degeneration.
Source: Illustration by Eric Pape from: Wells, H. G. (1908). *The war in the air, and particularly how Mr. Bert Smallways fared while it lasted*. The Macmillan company

Chapter 6
The Return to the Fields (1903)

Emile Vandervelde

Introduction

Guillaume Vanneste
UCLouvain-LOCI, Louvain-la-Neuve, Belgium
guillaume.vanneste@uclouvain.be

At the end of nineteenth and at the dawn of the twentieth century, Belgium underwent an important period of socio-political change, in an already profoundly urbanized environment engendered by one of the first industrial economies of Europe. At that time workers could easily access factories and work facilities facilitated by tramways while continuing to live in "rural areas". Facing a deep agricultural crisis, inhabitants of the countryside could use the densely developed tramway and railway network to commute to factories and work facilities,[1] *rooting the geography of the proletariat in a background that was much more rural than, say, in England or Germany. Concurrently, defence of social rights, concerns for living conditions and questions of social organisation of these working forces were raised by socialist groups and intellectuals, founding the basis of the Belgian socialist party.*

[1]Vandervelde attributes one of the main causes of work transformation in Belgium to transport policy of 1870 on cheap railway season tickets, see Vandervelde, E. (1901) *"l'exode rural"* p. 143.

M. Barcelloni Corte and P. Viganò (eds.), *The Horizontal Metropolis*,
https://doi.org/10.1007/978-3-030-56398-1_6

Born in 1866, Emile Vandervelde[2] *grew up in Brussels in a bourgeois and liberal family. After his studies at the Université Libre de Bruxelles, he discovered socialist thinkers while involved in the initial development of socialist associations. Recognized as a mediator, he quickly emerged within those structures until he became the founding father of the Worker's Party* (Parti Ouvrier Belge), *and was one of the first socialist members and indeed the youngest member to be elected to the Belgian parliament in 1894. From there, Vandervelde assumed the character he would embody over a long political and militant career. Vandervelde was an intellectual, acting both on the parliamentary stage, and among people, his expertise being recognized both at national and international level*[3] *via a network of socialist personalities no lesser than Marx, Lasalle and Engels. Oscillating between reformism and revolution, hailed as a charismatic figure, he would constantly be praised and accused for his sense and ability to compromise.*

Aside from a brilliant political career, Vandervelde dedicated himself to a life of research underscored by detailed and documented case studies. Highly prolific in his writing, in his early career, from 1897 to 1908, he published six monographs on the rural world[4] *in which L'exode rural et le retour aux champs occupies a prominent position. Together, these works draw the attention of the general public towards the socio-economic forces of the processes of industrialisation and raised concerns on working class living conditions in Belgium at the beginning of the twentieth century, at a crucial time in the urbanization of the country. In L'exode rural et le retour aux champs, he describes the reterritorialization of labour—the forms of migration—as a result of the agricultural transformation by industrialisation and capitalism. As announced in its title, L'exode rural et le retour aux champs is a two-part dissertation. In an alarming tone, the first part explains the causes and warns as to the consequences of the rural exodus. Under the combined action of the attractivity of cities, facilitated transport networks and overpopulation of countryside, a large part of the Belgian rural population migrated from rural to urban, from periphery to center.*

---

[2]For further biographical input, see Polasky, J. (1995). Emile Vandervelde was born in 1866 in Ixelles and studied law, social sciences and biology, psychology and psychiatry (Polasky, 1995, pp. 19–20), leading him to some positivist approach in his early career that he would later abandon. In 1891 he defended a thesis on *Enquête sur les associations professionnelles d'artisans et d'ouvriers en Belgique.*

[3]Emile Vandervelde held the position of president of the Second Socialist International from 1900 to 1918.

[4]In his researches about rural proletarian condition, he published the following books: Vandervelde, E. (1897). *La question agraire en Belgique*. V. Giard et E. Brière; Vandervelde, E. (1899). *L'influence des villes sur les campagnes* (Vol. 1. "Brabant"). Annales de l'Institut des Sciences sociales; Vandervelde, E. (1899). *Les villes tentaculaires*. éd. Bellais; Vandervelde, E. (1900). *La propriété foncière en Belgique*. Schleicher frères éditeurs, Bibliothèque internationale des sciences sociologiques; Vandervelde, E. (1903). *L'exode rural et le retour aux champs*. Felix Alcan; Vandervelde, E. (1908). *Le socialisme agraire ou le Collectivisme et l'évolution agricole*. V. Giard et E. Brière.

*In the second part, Vandervelde endorses visionary hypotheses, borrowed from the Garden City theoretical models and elaborated from a lucid reading of the contemporary condition, yielding a movement heading back to the land, benefiting workers' living conditions, disclosing the premises of an emancipated metropolitan project. With reference to industrial novelists and economists, he envisions in his conclusion a new agency of society within a metropolitan-urban realm, nevertheless underscoring the difficulties of a "*retour au champ*" today.*

After the nineteenth century's paternalist concerns for hygienism and social living conditions of the poor classes in the city centers, Vandervelde shifts the locus of the debate to the countryside as industrialization and demographic growth increase urbanization. Denouncing capitalism's effects on land ownership structure and urbanization, going through a relevant set of European examples, he sketches the premises of a mechanism of globalization whose descriptions still seem relevant today. Beyond a dual vision of city and countryside, probably embracing too much of a deurbanization path,[5] *a peripheral banlieue is highlighted as an intermediate place (*les campagnes urbanisées[6]*). Here Vandervelde enlarges his conceptual reflexions onto a metropolitan scale, superseding the binary statement of the title. "[...] the countryside of the future, no longer resemble the countryside of the "good old days". The people who live there have nothing in common with the peasants of Labruyère: they have passed through the towns; they remain in constant contact with them; they have returned to the fields, but with the advantages of a social transformation, of which urban centralization was the prerequisite and the decisive factor." (p. 293). As De Meulder pointed out (2009), in the mind of Vandervelde these exchanges and movements from city to countryside were to also benefit socialism and the socialist party. Nevertheless, by recognizing the specificities of the urban and rural living conditions, and the merging of those territories, he unfolds what one would describe today as spatial justice.*

Vandervelde's legacy has been transmitted to a broad number of Belgian socialist thinkers.Today, Belgian urbanists and urbanism are also accountable to Vandervelde's contribution as he remains a strong witness of the urban transformation of his time.[7] *Often quoted together with Benjamin Seebohm Rowntree*

[5]De Meulder, B., Schreurs, J., Cock, A., & Notteboom, B. (2009). Sleutelen aan het Belgische stadslandschap. Patching up the Belgian Urban Landscape. *Oase: Tijdschrift voor Architectuur, 52*, 78–113.

[6]Vandervelde uses the term *campagnes urbanisées* to coin the urbanization of his time, but adding a potential meaning to the words, opening a vision for future.

[7]Belgian researchers and academics often quote Vandervelde as witness of the industrialization revolution. Here are some of the works that refer to the works of Vandervelde. Smets, M. (1986). *La Belgique ou la banlieue radieuse.* In *Paysages d'architectures (catalogue d'exposition)* (pp. 33–35). Fondation de l'architecture; Uyttenhove, P. (2011). *Stadland België, hoofdstukken uit de geschiedenis van de stedenbouw in België.* A&S Books; Grosjean, B. (2010). *Urbanisation sans urbanisme, une histoire de la ville diffuse.* Mardaga.; De Block, G. (2011). *Engineering the Territory. Technology, Space and Society in 19th and 20th Century Belgium (Infrastructuur als inzet voor de organisatie van het territorium. Technologie, ruimte en maatschappij in België sinds het begin van de 19de eeuw)* [PhD, KUL].

(Chap. 7) Vandervelde provides a lucid, eye-opening view onto a very particular stage of the construction of the contemporary urban condition.

6.1 The Rural Exodus and the Return to the Countryside[8]

Nothing strikes travellers journeying from London to Brussels more than the contrast between the lonely meadows of Kent and the bustling plains surrounding our main cities.

Wherever you go, be it Hesbaye or the Flanders, the countryside is dotted by a host of white, red-roofed houses, isolated or clustered in populous villages.

However, if you were to stop by day in any of these villages—I mean those where there is no local industry—you will find hardly any adult workers, and you will be led to believe that the local residents are mainly old people, women and children.

However, as evening falls, the picture changes.

Let us say, for instance, that we are 20 kms from Brussels, in a small station in Walloon Brabant, say Rixensart, Genval or La Hulpe.

An engine with an extremely large number of 3rd class carriages pulls in: as the doors open very suddenly, workers wearing dust-covered or mud-stained overalls burst out, like a river gushing towards the exit, knocking down everything and anything on their way to get home earlier, where the evening meal awaits them.

As dusk turns to night more trains follow, unloading their human cargo, teams of workers—bricklayers, plasterers, tilers and carpenters carrying their woven bags- in all the villages along the railway line.

Elsewhere, it is coalmen, metal and steel mill or glass workers, from the Mons, Charleroi or Liège districts who are going back home, sometimes 50 or 60 kms away, to some remote corner of Flanders or Limburg.

In other cases, in the Kempen, Flanders or Ardennes railways, it is the dock workers from the Antwerp Port, weavers from the textile mills of Roubaix or Tourcoing, metal workers, who commute to Longwy in France every day, returning to their rural villages—mere dormitories—after a day's work.

Most villages in Belgium have a core of industrial workers, who are employed outside and often at great distance from their homes.

Before the establishment of the railways and reduced fares for workers, these 'migrants' existed in some regions, albeit in very small numbers, but instead of being daily commutes, they were on a weekly basis.

Specifically, in Walloon Brabant a certain number of masons, tilers, plasterers, miners and metalworkers, left every Sunday evening, sometimes in a *patache* (a horse drawn carriage), more often on foot, and went to work, either in Brussels or in

[8]Excerpt from: Vandervelde, E. (1903). *L'exode rural et le retour aux champs* (pp. 138–150). Felix Alcan.

the outskirts of Charleroi. They would only return to the village on Saturday evening and, during the week, they stayed in a lodging house, for a few cents a day.

A report of the *Comité de Patronage* (Board) of the Workers' Homes in Saint-Gilles-les-Brussels states these houses, some of which still exist today, were:

> Almost all the lodging houses have the same fault: they are overcrowded. Cramped rooms and very low attics with up to six beds, each occupied by two men. The rooms are stifling. When the Board examined the premises, the boarders had only been in bed for a couple of hours, and already it was suffocating... In a respectable, middle class building, on the second floor, there were six beds in a 3.50 by 6 -metre room.; four beds in a smaller room and, on the third floor, in the attic, six beds. Bear in mind, the attic has only one window for ventilation, which must necessarily remain closed at night, because it is 50 centimetres above one of the beds.... We noticed that each of the beds was occupied by two people. Under such conditions, it is easy to see how the stench of the attic prevented visitors from examining the back of the room.

This deplorable state of affairs can only be found as an exception in the Brussels area and is only used by workers who live too far from a railway station and are unable to return home every night and are thus forced to resort to this type of accommodation. In parts of the country where railway communications are not so good, such working and housing conditions are still the rule.

A case in point are the Bas-Luxembourg workers who work on the other side of the French border, beyond Mont-Saint-Martin and Rodange, and who cannot return home due to the lack of trains if they live beyond Arlon. As a result they stay away from home for a fortnight or even a month.

> They bring one or two large loaves of rye bread, each weighing 10 to 42 lb, as big as cartwheels, which have been baked at home; one or two "*bichets*" of potatoes (1 bichet = 20 L). They also try to smuggle in a little coffee, fatback and butter. They rarely eat much other than what they have brought, except for occasionally a little low grade meat or *charcuterie* (prepared cooked meat products). The disadvantages of such a diet with mouldy potatoes and bread as the staples are self evident, as it hits the men hard from the first week. Accommodation is not much better: they pay five francs a month for a room, sleep two in a bed and often six, eight or even ten in the same room.
>
> However, housing conditions have slightly improved as a result of competition between landlords.[9]

A life that the above described one means a worker sees his family at rare intervals replacing family intimacy with the often ghastly promiscuity of lodgings, disastrous for both moral and physical health. It is understandable that country people, who have to earn a living far from their villages which they do not want to leave, consider the special train fares and running times as a real blessing as it makes it possible for them to commute, to go to their work every morning and to return home at night.

Now, although the bourgeois state does not act for philanthropic reasons, it was done in a more comprehensive manner than elsewhere, to facilitate the flow of an undemanding labour force to the industrial centres.

[9]Notes from a visit to the Luxemburg border area in 1901.

When opening a timetable at random (*Indicateur des chemins de fer*), which the Belgian government, for unknown reasons has regaled with the strange name of *Guide des chemins de fer* (Railway Guide) one is struck by the astonishing number of trains on almost all lines, running before the opening and after the closing of workshops and factory plants. What is even more striking is the extremely special low fares for a worker's week ticket (see Annex III). A weekly pass, with six journeys, costs workers only 2.25 Francs for 50 kms, while ordinary tickets cost 3.05 Francs for the same journey, round trip, in 3rd class which means they pay more for a single ticket than a worker for a whole week!

Special fees dated back from 1870, in the wake of the Franco-German war. The Secretary for Transport who introduced them seemed to have had no other concern than to cope with the temporary job shortage. The ministerial decree passed almost unnoticed and was only seen as an expedient but in fact it led to a the most profound revolution in Belgium's labour market for a quarter of a century.

This is apparent by looking at the tremendous increase in the number of weekly tickets issued by the railways as from 1870.

The trend was uninterrupted for thirty years.

It became more marked from 1895, during the industrial growth at the end of the Nineteenth century, and for the first time a slight drop in figures in 1901, at the beginning of the crisis at the beginning of the new century. Most of the tickets (3,072,980 + 406,450 in 1901) were issued to workers who made six trips a week and therefore went back home every day. Relatively few resorted to the one trip a week season ticket returning on a Saturday, to spend Sunday with their families (659,707 tickets).

The large number of season tickets issued for the 1901 tax year (Annex IV) shows that the number issued does not vary greatly from month to month. The same can be said for the 1899 and 1900 data: hardly any drop in figures, and even then negligible, except in the two months of January and February, as a result of unemployment in seasonal industries—partially offset by the activity of the sugar refineries and the increase in the number of workers in the mines.

Unlike the given wisdom, and to what we have written, before seeing the monthly[10] summaries, the vast majority of rural workers working in the city appeared to have held regular jobs, and that the number of weekly tickets they bought a year was between 45 and 50.

These figures suggest there were approximately 90,000–100,000 workers who travelled every day on the state railways.

One should also add season tickets for out of town trolleybuses, city trams and inter-village transport for short distances that increased year on year to the limited number of train lines, such as *Nord Belge* used by industry: that will give about 100,000 out of about 900,000 who lived in the. Advantages exceed the

---

[10]In our study of Land Ownership in Belgium (*La propriété foncière en Belgique*) on the basis of personal observations made in a region dominated by the building industry, we stated that employed workers only went to the city during two-thirds of the year and therefore purchased between 30 and 35 weeklies.

disadvantage of daily travel, if it was not too long and, basically the industrial workers living the country were better off than both agricultural and urban workers.

Nevertheless, in weighing the advantages and disadvantages of remaining in the city and in suburban areas, the personal factor played a very important part so no satisfactory generalizations could be made.

This is why we preferred to select a few practical examples from the personal observations we made over the past years.

First of all, a carpenter, aged 18, whose workshop was in Brussels and lived in the municipality of La Hulpe, three leagues from the capital. To get to work, every day he travelled two kilometres each way to the station, two half hours by train, not counting the journey into town. His weekly ticket was 1.35 F, or just under 23 cents a day. This young man lived on a farm tenancy with his parents and his two younger sisters, under the MS Castle estate, now a small bar described his family's living conditions:

> At 46 my father had been working for the Baron de R., for twenty-three years when M.S purchased the property. He now works for the new owner and earns 2 Fr in winter and 3 Fr in the summer. As a carpenter, I currently earn 3.50-4 Fr a day but am currently out of work as happens at this time of year (December). I much prefer my condition to my father's work as he suffers from rheumatisms at 46, inevitable because of his work in the woods, in all weather. Unlike him, I work indoors in a warm workshop in bad weather. It's just that I don't want to live in town in spite of having to take to the train which is tiring. The air is much better in the countryside, we have better and cheaper accommodation and can have fun, work in the garden, on a Sunday morning or on holidays
>
> Our rent is 150 Fr a year for the house; four rooms, a stable, a pigsty, two cellars, a barn and a green garden that supplies us with vegetables throughout the year. We also rent a few parcels of land from MS for our wheat and potatoes. We have chickens, a pig, a cow and a heifer which give us about three pounds of butter a day. In summer, when the cow has a high milk yield, we sell part of the butter, but now we just have enough for our needs. We bake our own bread which is much nicer. We hire horses from a farmer to work our land: that costs us 14 Fr plus 1 Fr plus 1 Fr tip to the farm help, better spent than that what they do in hamlet of Gallemarde, where the women till land, with a spade. I harvest with a scythe on a Sunday. My mother and two sisters take care of the garden, the animals and the sales.

Most of the carpenters, masons, plasterers and other construction workers, employed in Brussels and living in La Hulpe, or in the surrounding hamlets, more or less lead the same kind of life[11]; and, for people who do are not shy of work, nor picky about their intellectual pursuits and leisure, the benefits of low rents and additional farm work, are a powerful draw for staying in the countryside. As a result with very few exceptions, workers do not move, and there is little evidence that they settle in Brussels other than the foremen who set up small enterprises and want to be close their businesses.

If we move on now to another section of the Brussels outskirts, a 49-year-old metalworker, Van I..., works in one of the neighbourhoods in Brussels and lives in the countryside near the small town of Hal, 500 m from the railway station:

---

[11]Vandervelde, E. (1900). *La proprieté foncière en Belgique*. Schleicher. Ist part, Monographies La Hulpe, Rixensart, Geval.

> From a financial point of view, there is no reason for me to live in Hal: the cost of living is not lower than in Brussels; bread is more expensive. True, my rent is only I2.50 Fr as month; others pay 14 Fr. because landlords increase rents as demand increases and that means I pay 150 Fr. a year. In Brussels I would pay 20 Fr. a month for quarters, that is to say 240 Fr. a year; However, my weekly costs 1.35 Fr × 52 weeks = 70.20 Fr. There is That is less than a 20 Fr difference and that is not counting the rough number of hours I lose at work every year due to my trains arriving quite late a common event on the line from Hal to in Brussels. Furthermore, I lose more than two hours a day in travel; at least in winter, journeys are tiring and difficult; that, for my school-age children, the education is not what they would receive in Brussels.
>
> The only real advantage of my living in Hal was that my accommodation is much roomier and more convenient for my wife: we have a whole house, with two rooms on the ground floor and two rooms upstairs, a barn, a cellar, a vegetable garden, whereas in Brussels, for our twenty francs, we would only have a room on the 3rd or 4th floor: no cellar, for the coal; narrow rooms, with the six unmarried children still living with us, my wife was having to walk up and down several floors constantly, to fetch buckets of water or attending to other household chores. Here, on the other hand, she has everything at hand and it is mainly because of her that we came back to Hal, where I had been living since I was 7 years old, after having spent some time in Brussels.

One of Van I…'s sons, married and a father, lives 200 m away and also works in Brussels, but he says he much prefers to go back to the countryside every evening:

> The air is better, it is quieter and, as there is little traffic, small children can play in the street without the need for our supervision.

Especially for a young worker these advantages more than made up for the weariness of a twenty-minute railway journey, especially since metalworkers had a relatively short day and thus came home early not having to leave very early in the morning. However, this was not the case for the workers who lived in villages around Hal, such as Pepinghen, Bellinghen, Esschene. Since their working day was generally longer than that of the metalworkers, they had to take the 5.55 am train in winter, the 4.30 a.m. in summer, and return by the 8.06 p.m. train. Since most of them lived half or three-quarters of a mile from the station, they were left with 7 or 8 out of the 24 h left for anything other than work and the time needed to get to their job. Most of them are builders, factory workers, and handymen. Only a few put up with the wear and tear and after a time they tended to settle permanently in the Brussels area as soon as they manage to find a regular job such as tramway collectors or shop assistants.

On the contrary, in the building industry usually resigned to commuting indefinitely between Brussels and their villages, because during the off-season the countryside offers advantages: some farmers threshed by hand; the two sugar mills in the region employ a number of people.

Moreover, let us not forget, that these journeys were not exceptional not limited to a small number of villages.

According to data by the Local Authority (Annex V), in 1897 more than five thousand workers travelled at least fifty kilometres every day (return rail journeys) and, nearly four hundred of them made travelled more than one hundred kilometres!

Most of these workers, whose number has increased considerably since then, were Flemish and worked in the Hainaut collieries. Where the pits closed at the end of the day, the 3rd class train compartments were full of these poor people, who did not take the time to wash before leaving piled up one on top of the other, black with coal dust trying to make up for the lack of sleep at night by dozing on the train. They were lucky if they do not have a long night's walk through the countryside after four hours on the train and twelve hours at work before getting back to their lodgings.

To gauge the overtiredness of some rural industrial workers, living several kilometres from the railway, let us also consider the following example, relating to a worker from Luxembourg, working at the Athus blast furnaces.

> X…domiciled in Habay-la-Neuve, has to travel 3 kms to catch the 3.33 am train and therefore has to get up before three o'clock which tales him to Athus (30 kms) at 4.42 am. This is followed by twelve hours of work, from 6 am to 6 pm. He then has to wait for the 7.34 am train at Athus station, or hang around in the *assomoirs* (cheap taverns) close by;
>
> he arrives in Habay at 8.30 and is home at 9 am. That amounts to 19 h of work and travel, leaving 5 h for rest and family life!

If the working day were shorter and train times tallied with the opening and closing hours of the factories, this deplorable state of affairs would improve considerably. However, under current conditions there is no doubt that the exhaustion of the journey, in addition to the hard work, inevitably led to overexertion and, in any case, was an insurmountable obstacle to their intellectual development.

If people accept they are going to live like this, it is only they have absolutely no other means of earning a living in their own village.

On the other hand, if they do not settle in the industrial centres near their factories, it is because, all things considered, they still found it better to keep a piece of common land, a parcel of field, a vegetable garden which their wives tend to and which supplemented their wages in kind.

Instead of emigrating permanently, or working in urban agglomerations, keeping their domicile in the countryside, rural workers may still, in order to make up for the insufficiency of their resources, be absent periodically, to look for work in other regions, or in other countries.

Year	Number of Tickets	Year	Number of Tickets
1870	14,223	1886	714,408
1871	32,972	1887	798,135
1872	71,245	1888	935,039
1873	97,789	1889	1,018,383
1874	133,442	1890	1,188,415
1875	193,675	1891	1,337,730
1876	212,759	1892	1,404,370
1877	216,712	1893	1,518,777
1878	241,552	1894	1,633,780
1879	249,752	1895	1,759,025
1880	355,556	1896	2,204,613
1881	440,465	1897	2,699,594
1882	523,832	1898	3,267,588
1883	585,283	1899	3,933,525
1884	616,866	1900	4,515,214
1885	667,522	1901	4,412,723

Table 6.1 Summary of workers' season tickets issued since their creation, 10 February 1810–1901. *Source* Vandervelde (1903), p. 143

Fig. 6.1 Paul Delvaux, 1960, *La Gare forestière*, Oil on canvas, 160 × 220 cm. St Idesbald. Source: Collection Fondation Paul Delvaux

Chapter 7
The Fruitful Garden (1910)

Benjamin Seebohm Rowntree

Introduction

Paola Viganò
EPFL, Lausanne, Switzerland
IUAV, Venice, Italy
paola.vigano@epfl.ch

At the beginning of the 20th century, Seebohm Rowntree visited Belgium and carefully described its territory, to gain an in-depth understanding of the strengths of this peculiar country which, despite its small size, could offer a few lessons to Great Britain. The purpose of Seebohm Rowntree, an English sociologist, also author of Poverty. A study of Town Life *(1901), was to contribute to the solution of poverty in Britain by studying the systems of land tenure in Europe. In* Land & Labour. Lessons from Belgium *(1910), the author relates the possibility of working in the city and living in the countryside to the existence of a dense network of railways, light railways and tramways. Here we recognize the specific shape assumed by the industrial revolution in Belgium: following the idea, not only to disperse the working class outside the city, but to connect the peasants (supposedly more docile and less costly) to the new industrial economy without inurbating them, engendering a rural exodus. In Seebohm Rowntree's words: "There is no country in the world with so great a length of railway in proportion to its size as Belgium…".*[1] *The construction of an infrastructure on a regional scale will be, almost eighty years later in Marcel Smets words, "logic responds to a specific problem" (Smets 1986, see Chap. 19).*

[1]Rowntree, B. S. (1910). *Land and labour: Lessons from Belgium* (pp. 5–7, 283–288). MacMillan and Co., p. 283.

B. S. Rowntree (Deceased) (✉)
Buckinghamshire, England

M. Barcelloni Corte and P. Viganò (eds.), *The Horizontal Metropolis*,
https://doi.org/10.1007/978-3-030-56398-1_7

Today, the dense network of public transport remains a fundamental layer of territorial organization in Belgium. Nevertheless, it is important to clarify that agricultural organization—"The average size of the holdings in Belgium is smaller than in any other country of Europe"[2]*—constituted the underlying dispersed structure of small villages and cottages on top of which, by a process of stratification, the new and original industrial structure was laid. The description of the sandy landscape of East Flanders and of the enormous amount of labour needed to transform it into a "fruitful garden", together with the full recycling of organic waste to enrich a poor soil, is neatly sketched. This includes the attitude towards chemical manure, a foreshadowing the following high use of fertilisers and pesticides. The soil of Flanders, although not highly fertile, but frequently manured, witnessed the most intensive agriculture of Europe, with the exception of the market-gardens of Paris—"Manure of all kinds is carefully preserved and used on the land; nothing is wasted"*[3]*; a poor soil able to maintain a dense population and proving the capacity of small holdings to organize crop rotation. The long-term dispersed character has been there for centuries, but the "fruitful gardens" here described have today disappeared. The decline in soil organic matter, the soil contamination by pesticides are widespread in agricultural land, as well as the loss of soil biodiversity;*[4] *still their regeneration is a possible choice.*

7.1 Land and Labour. Lessons from Belgium[5]

Sandy Region[6]

To the south of the Polders lies a sandy region, stretching almost across Belgium from east to west. It is divided into two parts, the Campine lying to the east, and Flanders to the west. The latter is often described, particularly by foreigners, in glowing language, as "the fertile plains and fat meadows of Flanders." They are, indeed, productive, but they owe it to the labour of man. There is scarcely any soil in Europe so infertile by nature, and if abandoned for only two or three years it returns to barren waste. Constant care and enormous quantities of manure are

[2]*Ivi*, p. 107.

[3]*Ivi*, p. 6.

[4]Gentile, A. R., Barceló-Cordón, S., Van Liedekerke, M. (2009). Soil country analyses. EUR 23959EN/2.

[5]Excerpts from: Rowntree, B. S. (1910). *Land and labour: Lessons from Belgium* (pp. 5–7, pp. 283–288). MacMillan and Co.

[6]Excerpt from "Chapter 1—Geographical and Physical Description of Belgium" In Rowntree, B. S. (1910). *Land and labour: Lessons from Belgium* (pp. 5–7). MacMillan and Co.

required if good crops are to be obtained. A proprietor whose farm has been neglected by a bad tenant is glad to let it to a capable farmer rent-free for some years, that it may be brought into good heart again.

This region has four characteristics: a great variety of crops, intensive cultivation, very small farms, and second harvests (*récoltes dérobées*).

The aspect of the country is very different from that of the average agricultural region in Britain. There are many small villages, and far and near among the fields are dotted little cottages, often whitewashed and with red roofs, in which the cultivators live close to their work. Here are no great pasture farms employing scarcely any labour, but everywhere the peasants are busily engaged in cultivating the soil. The main crops are wheat, rye, barley, potatoes, flax, hemp, colza, fodder, and beetroot, but many others are grown in smaller quantities. The culture here is the most intensive in Belgium, and, with the exception of the market-gardens around Paris, the most intensive in Europe. Manure of all kinds is carefully preserved and used on the land; nothing is wasted. Fields have to be remanured two and three times a year, and the second crops, and even the seeds in the spring-time, are watered with liquid manure. Thanks to the diffusion of agricultural science, the Flemish peasant, although very apt to work in a rut, is becoming increasingly alive to the value of chemical manures, and every year he uses more and more of them.

According to official statistics, nearly one-third of the soil in East Flanders yields two crops annually, and in the vicinity of Ghent the proportion rises to one-half. It is solely doe to the excellent methods employed that such poor soil is able to maintain so dense a population. Almost the whole of the work is done with the spade, as it cannot be done properly by ploughs. The cultivation in Flanders contradicts the statement sometimes made that only by farming on a large scale can a suitable rotation of crops be maintained, and that small cultivators cannot devote enough capital to their land to obtain the best results. While, however, the sandy soil of Flanders has been made extraordinarily fertile, this cannot be said of the eastern portion of the sandy zone. This is a large region covering much of the two provinces of Limbourg and Antwerp, and small part of Brabant, and is known as La Campine. Its soil is fine sand and sandy clay, with very little cohesion. It is one great plain, nowhere attaining an elevation of more than 240 feet, the depressions in it being occupied by marshes and peat bogs. It is the poorest part of Belgium; the population is sparse, and there are many signs of destitution. Around the villages the patient toil of the people has brought some of the land under cultivation, but there still remain great waste tracts. Occasionally, as one travels through these, one comes upon a little homestead, very primitive in construction, occupied by a struggling peasant who is waging a hard battle against "the niggardliness of nature."

Railways[7]

The industrial and agricultural prosperity of a country—indeed, its whole social economy—is vitally affected by its facilities for transport. A country well served with railways, canals, and roads is likely to develop its resources much more swiftly than one comparatively poor in them; and the provision of cheap and rapid means of passenger transit not only makes labour more mobile, but has an important bearing on housing and rural exodus. It is therefore important to study the question of transport facilities in Belgium, and see how for the social condition of her people, and the development of her agriculture and industry, are affected by them.

There is no country in the world with so great a length of railway in proportion to its size as Belgium, with her 2859 miles, or 30·29 for every 100 square miles of country.[8] Next comes Great Britain, with 22·38 per 100 square miles; then Germany, with 15·72; and France, with 11·72. But in addition to her main lines, Belgium has to truly wonderful system of light railways, which play so important a part that they will be described in detail later it this chapter.

With the exception of 215 miles still in private hands, all the main railways of Belgium are national property, managed directly by the State, and it is probable that in time the remainder will be nationalised. Of course, there are a certain number of people who complain of the State management of the railways, but the general opinion is in favour of it.

The fares for passenger traffic are low—just under 1½d per mile first class, about one penny second class, and a halfpenny third class. The second-class carriages are on a level, from the point of view of comfort, with the best third-class carriages in Britain, or possibly a little better. The third-class carriages have wooden seats and small windows. Much use is made of season tickets, which are exceedingly cheap. A ticket for a fortnight, available on all lines in Belgium, including those in private hands, costs £2:9:2 first class, £1:12 s second class, and 18 s 10d third class, while one available for five days costs £1:4:7, 16 s 5d, or 9 s 5d in the respective classes.

Although it is notoriously difficult to state clearly and briefly the charges made in any country for the transport of goods, we must endeavour to do this for Belgium, since the question so vitally affects the interests of both agriculture and industry that it cannot be neglected. An attempt will also be mode to compare Belgian with English rates.

It is said that there are about three million different rates charged on the English railways. Each railway has its own, and these vary considerably over different sections of it, and sometimes even in one and the same section, according to the direction in which the goods are sent. For instance, there are cases in which it costs mom to send goods from station "A" to station "B" than from "B" to "A". Thus, to

[7]Excerpt From "Chap. 21—Means of Transport" In Rowntree, B. S. (1910). *Land and labour: Lessons from Belgium* (pp. 283–288). MacMillan and Co.

[8]This refers to the year 1907.

make an absolutely full and complete comparison between the goods rates in Belgium and in England is impossible, and we can aim only at a rough general statement.

The English rates quoted in this chapter are those of one of the largest companies, and are not higher, indeed on the whole they are probably lower, than the average charged by other English lines. The charges in Belgium are uniform throughout the whole system, except that under certain circumstances goods consigned in large quantities for export are conveyed at a lower rate. In comparing the two countries upon this basis, the following facts emerge. For parcels up to 14 lbs. there is little difference in the charges, and this holds good of heavier parcels, say up to 28 lbs., when these are sent short distances only; but after the 14 lb. limit is passed, the Belgian can send small parcels much more cheaply for a long distance than is possible in England. Thus, a Belgian farmer or market gardener can send 44 lbs. (20 kg) of vegetables 94 miles (150 km) at company's risk for seven pence. This would cost the Englishman 2 s 1d if sent at company's risk and 1 s 5d if sent at his own risk.[9]

The charges for express goods train in Belgium may be compared with those for passenger train in England, though the delivery will probably be somewhat more rapid in the latter case. It will be noted that for short distances, say up to 12½ miles, the rates in the two countries are practically the same, unless the Belgian prepays the cost of carriage, in which case he gets a reduction of about 16%. But for long distances, say 94 miles, the charges iv England are from two to three times as high as in Belgium. Even if the English charges per goods train be compared with the Belgian charges per express goods train, they are generally considerably higher.[10]

If we take the charges for 1 ton lots, and compare the Belgian and English rates per goods train, the latter, gain, are much higher, whether for short or long distances. In many cases the English charge is from two to three times as high as the Belgian, and it is the same with charges for full truck loads (5 tons). In the case of cattle, the difference is not nearly so great, the charges in most cases being pretty equal in the two countries. On the other hand, the Belgian who sends milk to a dairy has a decided advantage over the Englishman. In order to help the small holder, the Government has arranged a series of extraordinarily low rates for the conveyance of milk. A can of milk weighing 22 lbs. can be sent 31 miles for a penny, and 9 3¾ miles for twopence; and other rates are proportionately low. The minimum charge

---

[9]In the Appendix, pp. 585–590, will be found various tables giving further details of the rates charged in Belgium and England for agricultural produce. For the sake of simplicity only rates upon agricultural produce are considered. Investigations made by the writer, however, indicate that the relation found to exist between the rates on agricultural produce in the two countries applies more or less to industrial goods.

[10]The writer is aware that when traffic is passing regularly and in considerable quantities over the English lines, the railway companies arrange special rates somewhat lower than those which they publish. But these apply principally to traders, and others who consign large quantities of goods, and not to those who only send goods irregularly or in small quantities. Therefore, even if it were possible to obtain particulars of these special and privately arranged rates, they could hardly be compared with the rates in Belgium, which are charged to every one alike.

on the English railway is sixpence, for which amount 104 lbs. can be sent 18¾ miles. To send 22 lbs. for 31 miles would, however, cost ninepence, and to send it 9 3½ miles a shilling. On the other hand, these rates include the free returns of empties, which is not the case in Belgium. If the milk is not sent to a dairy, the Belgian rates for ordinary goods apply.

Thus we see that although it is true that the English passenger service is somewhat more expeditious than the Belgian express goods service, the rates charged are so much higher as to counterbalance this advantage. This is an important fact which must not be forgotten when discussing the relative prosperity and development of agriculture in the two countries. Cheap transport is essential to the farmer.[11]

Certain writers complain of the bed management of railways in Belgium. Mr. E. A. Pratt, for instance, in his book, *Slate Railways*, condemns it very strongly. To some extent he is justified, but what railway system is there with whose management fault could not be found? The present writer's opinion, after travelling widely over the Belgian railways at different times of the year, and making careful enquiry into the facilities offered for the transport of goods, is that while, of course, details might be much improved, the system generally is not open to the severe condemnation of such critics as Mr. Pratt. Without entering into details, it may be said that the chief criticisms of it which can be justified are that the management is too bureaucratic; that the railways are not run on strictly business lines; and that too little authority is given to those who are directly responsible for the detailed management.

There is no need to enter in detail into the finances of the Belgian railway system, but a few of the most important figures may be given. According to official statistics, the capital invested in national railways, at the end of 1905, amounted to £91,012,771; and since the establishment of the railways in 1836 a total profit of £1,440,000 has been earned.

It is, however, important to note how this figure is arrived at. Each year the railway accounts are virtually closed; and if there has been a loss on the year's working, the State pays it off, if a profit, the State takes it. No interest is charged against the railways on account of payments made by the State to wipe off losses, nor allowed on account of profits paid into the National Exchequer by the railways.

The total profit referred to above consists of the sum total of the profits after allowing for the payment of interest on capital, minus the sum total of the losses. Had the railway finances been conducted like those of an ordinary business enterprise, fresh capital would have been introduced to make up for the losses; while, if the profits had been kept as a reserve fund, interest would have been earned upon them. This method of book-keeping would have shown a net loss on the working of the Belgian railways of £2,840,000. The turnover from profit to loss

[11]The remarks which have been made above regarding the railway rates in Belgium refer to those charged upon the main railways. In the tables given in the Appendix, however, there is a column showing the rates upon the light railways. It will be noted that they are often much lower than those on the main railways.

is considerable, because the losses were chiefly incurred in the early years, and so compound interest upon them amounts to a large figure. Looked at from the business standpoint, however, it seems clear that there has been this actual loss on the railways.

Another point must be mentioned in connection with railway finances, namely, that although excellent regulations have been laid down, defining the forms of expenditure to be charged to the revenue and capital accounts respectively, these are not adhered to in practice. Much expenditure is charged to the capital account which in a prudently managed business would be paid out of revenue. If the official rules in this matter had been strictly enforced, there is no doubt that on many occasions instead of recording a profit the railways would have recorded a loss. Facts have been brought to the writer's notice, in the course of a detailed examination of the subject, which would illustrate this statement; but perhaps enough has been said to show that the official figures are based upon a system of book-keeping which it would be difficult to justify, from the standpoint of an ordinary business man. There is the less need to go more deeply into these financial matters, which at the best are complicated and difficult to state simply, because, from the social point of view, it is not of vital importance whether there has been a small margin of profit or of loss on the national railway enterprise. Even if it be proved that the past seventy-three years' working has resulted in a loss, the policy of the State might be justified by the facilities it has offered. It would, of course, be much more satisfactory in every respect if the accounts were so kept as to show accurately the financial position of the undertaking.

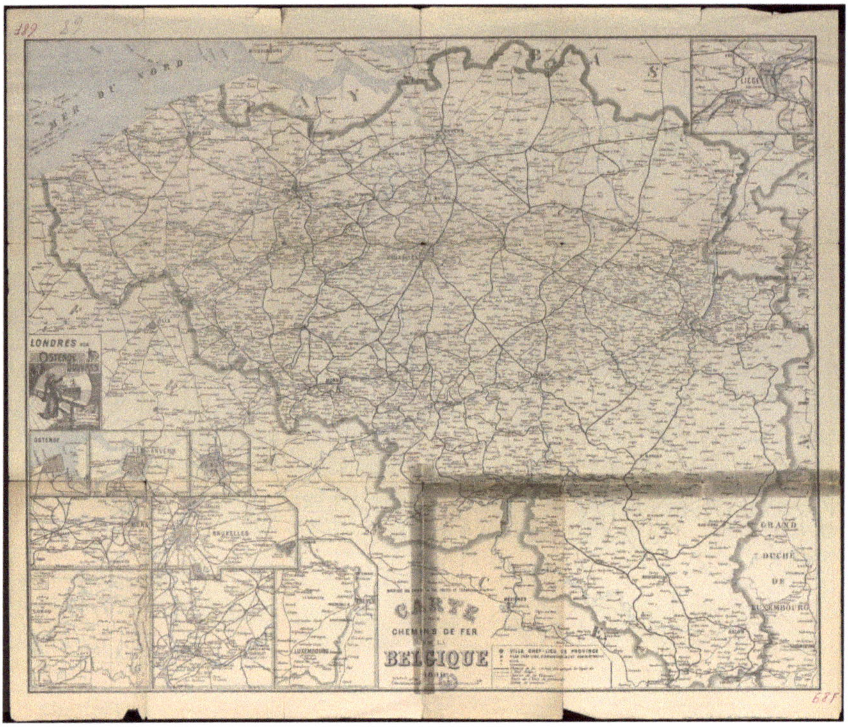

Fig. 7.1 "Railway map of Belgium" 1899.
Map showing the national and local railways. Bruxelles: Ministère des chemins fer, postes et télégraphes.
Source of data: © Liège University Digital Library

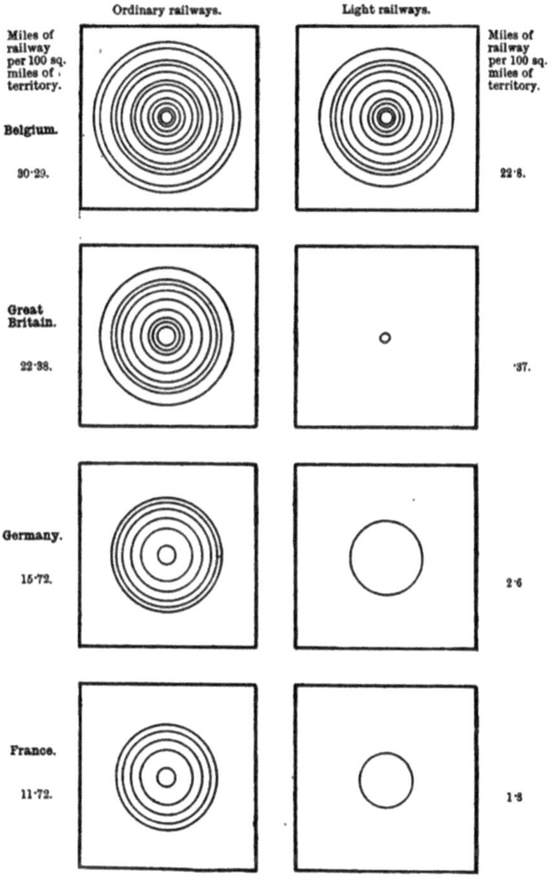

Fig. 7.2 "The Comparative Mileage of Railways for a given Area of Territory in Various European Countries".
Source: Rowntree, B. S. (1910). *Land and labour. Lessons from Belgium* (p. 302). MacMillan and Co.
"In 1885, when the present legislation was enacted, there were practically no light railways in Belgium [...]. At the end of 1908, there were 160 lines running or under construction, having a total length of 2586 miles, equal to 22.8 mile. for every 100 square miles of the national area. In addition, arrangements were in progress for the laying down of 149 other lines with a total length of 1272 miles. When these are completed, Belgium will have 3859 miles of line, or 33.7 miles for every 100 square miles of her total area. The contrast between these figures and those of other European countries is most striking, especially when it is remembered that Belgium has, in addition her light railways, a more complete system of ordinary railways than any other country in the world." (ibid., pp. 300–301)

Fig. 7.3 Ferraris map excerpt—1771–78 (Evergem area, Flanders).
Belgium as a "country of small holdings" is visible in the map realised more than one century earlier by the Count Joseph de Ferraris. Also the gardens described by Rowntree can be visualised in this map.
"[…] it is well known that Belgium has gone further than any other country in supplying her working-class population with gardens, which are usually well cultivated, and planted with a great variety of crops, providing a welcome addition to the family dietary, and often a substantial contribution to the family income. The possession of these gardens, so highly prized by the working-class population, is rendered possible by the system of cheap railway tickets, which enable men to live at some distance from their work in districts where land is comparatively cheap." (ibid., p. 108)

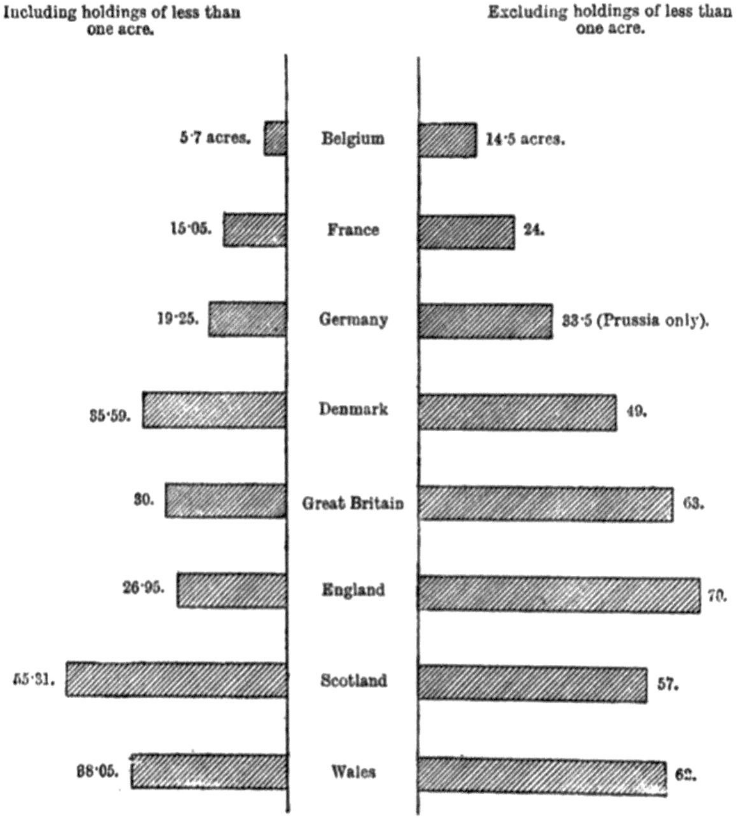

Fig. 7.4 "Average Size of Agricultural Holdings in Different Countries".
Source: ibid., p. 106.
"The average size of the holdings in Belgium is smaller than in any other country of Europe, over 94% of the total number being less than 25 acres in extent, while two-thirds are under 2½ acres. The number of holdings increased by 45% between 1846 and 1895 (from 572,550 to 829,625), the population of the country increasing in the same period in almost exactly the same ratio (48%). [...] Thus we see that there is a steady and continuous movement towards a still further reduction in the average size of a holding [...]" (ibid., p. 107)

Chapter 8
The City-Region (1915)

Patrick Geddes

Introduction

Chiara Cavalieri
Université Catholique de Louvain, Louvain-la-Neuve, Belgium
chiara.cavalieri@uclouvain.be

In 1915 Patrick Geddes (1854–1932) coined the term "City-Region" while discovering a new urban dimension in industrial Britain. By 1891 over 75% of the population of England and Wales (and 65% of Scotland's) was urban-dwelling, while industry, trade and transport employed over two thirds of the national workforce[1]*. As a consequence, the rapid growth of population and economic activity was deeply transforming the geography of the whole country, provoking massive socio-economic and spatial changes, converting what used to be "a province covered with houses" into an urban "conurbation". This "conurbation"—applied to what Geddes called City-Region, was the spatial consequence of industrial revolution: it consisted of cities, large towns, and agricultural areas that, through population growth and physical expansion—enabled by motorized transportation and electric power—merged to form one continuous urban and industrially developed area.*

The British City-Region, while radically—and for the first time in history—imposing a new scale of observation, was either a monocentric (i.e. Greater London) either a polycentric region (Clyde-Forth), where new and old, small urban centres were growing along the main transportation axes. At the same time,

[1]Lawton, R., Pooley, C. G. (1992). *Britain 1740–1950. A historical geography* (pp. 163, 195). Edward Arnold.

M. Barcelloni Corte and P. Viganò (eds.), *The Horizontal Metropolis*,
https://doi.org/10.1007/978-3-030-56398-1_8

the whole transportation network had significantly increased in terms of physical size and intensity of transport.

The City-Region—and its project—is envisioned by Geddes not as an expanding urban pattern that absorbs the resources of its natural environment, but rather as the result of the sum of the elements that had participated in its slow construction, through time: a city that still preserved the original relationship between dwelling, activities and natural environment. This pioneer approach, that intertwines the geography of resources with that of consumption, reflects Geddes' unique and hybrid attitude: he was in turn naturalist, geographer, re-former, social evolutionist and ultimately city planner.

Geddes had a multidisciplinary gaze while observing and portraying the evolution *of cities. On one hand, influenced by a spreading Darwinism, he attempted at building a retrospective vision of urban origins; on the other by an in-depth observation of contemporary social and spatial practices, he proposed a transformation trajectory.*

This multi-gazed approach is one of the main legacies of Geddes' method: the Geddesian survey *is a multi-disciplinary procedure of observation and data collection that ranges from natural to social science.*[2]

Within this evolutionary framework of redefining the past, analysing the present for ultimately planning the future of city-regions, Geddes introduces the concept of Valley Section as "the basis of any survey"[3]:

> *The Valley Section makes vivid to us the range of climate, with its corresponding vegetation and animal life. Then too the forests, the pastoral slopes, the mirror hills, the plains, their uniting rivers; all things are here. This is no longer a mere school book, with its image of a "country" as a coloured space on a flat map, with only "boundaries" and "capital", and so on; it is first of all the essential sectional outline of a geographer's "region", ready to be studied.*[4]

What emerges from this graphic categorization is a reconstruction of man's original relationship with his natural environment, where physical conditions and human activities become the basis for a more complex social organization. An organisation that marks the shift from what Geddes defines a paleotechnic civilization to a neotechnic one. A neotechnic civilisation that, in his vision, should evolve *while at the same time regaining its original boundary with natural environment. The City-Region, as described by the Valley Section, is hence a region ultimately defined by its environmental boundaries.*

It is then not surprising if Geddes' work became a primary reference throughout the tradition of American ecological planning, and that from the same notion of

[2]Ferraro, G. (1998). *Rieducazione alla Speranza. Patrick Geddes planner in India*, 1914–1924, p. 46. Jakabook.

[3]Geddes, P. (1925). *The third of the talks from my outlook tower. The valley plan of civilization.* Survey Graphic.

[4]Geddes, P. (1925). *The third of the talks from my outlook tower. The valley plan of civilization.* Survey Graphic, p. 289.

city-region arose further studies on regional geographies, i.e. the whole experience of the Regional Plan Association of America—RPAA.[5]

More in general, if later during the twentieth century, the study and the intervention on the physical environment[6] *became one of the mainstays of urbanism, it is certainly also thanks to the work of Geddes, to his capacity of intertwining site and history, survey and democratic participation into a new approach for planning the* evolution *of cities into city-regions.*

8.1 The Population Map and Conurbations (from: *Cities in Evolution*)[7]

The population map and its uses, London ("Greater London") as a spreading man reef.

Given a population-map, what has it to show us? Starting from the most generally known before proceeding towards the less familiar, observe first the mapping of London—here plainly shown, as it is properly known, as Greater London—with its vast population streaming out in all directions—east, west, north and south—flooding all the levels, flowing up the main Thames Valley and all the minor ones, filling them up, crowded and dark, and leaving only the intervening patches of high ground pale. Here then we have a fairly accurate picture of the growing of Greater London.

This octopus of London, polypus rather, is something curious exceedingly, a vast irregular growth without previous parallel in the world of life—perhaps likest to the spreadings of a great coral reef. Like this, it has a stony skeleton, and living polypes—call it, then, a "man-reef" if you will. Onward it grows, thinly at first, the pale tints spreading further and faster than the others, but the deeper tints of thicker population at every point steadily following on. Within lies a dark and crowded area; of which, however, the daily pulsating centre calls on us to seek some fresh comparison to higher than coralline life.

Here, at any rate, all will agree, is an approximation to the real aspect of Greater London as distinguished from Historic London. What matter to us, who look at it for the moment in this detached way from very far above, or even really to the actual citizens themselves to-day, those old boundaries of the counties, which were

[5]The first contact with Lewis Mumford was in Jerusalem (1919). See Carozzi, C. (1970). Preface. In P. Geddes (Ed.), *Città in Evoluzione*. Il Saggiatore. [first Italian edition of *Cities in Evolution*]. For further influence with RPAA see also Mumford, L. (1927). "Regionalism and Irregionalism". *Sociological Review*, a19, 277–288.

[6]De Carlo, G. (1965). Introduction, collection: Strutture e forma urbana. il Saggiatore [the same collection of *Cities in Evolution*].

[7]"Excerpt from: Geddes, P. (1915). The Population Map and Conurbations, first published in: Geddes, P. Cities in evolution, an introduction to the town planning movement and to the study of civics. London: Williams and Norgate".

once traced so painfully and are still so strictly maintained, from use and wont or for purposes other than practical ones? What really matter nowadays the divisions between innumerable constituent villages and minor boroughs whose historic names are here swallowed up, apparently for ever, like those microscopic plants, those tiny plants and animals, which a big spreading amœba so easily includes, so resistlessly devours?

Here for most practical purposes is obviously a vast new unity, long ago well described as "a province covered with houses." Indeed a house-province, spreading over, absorbing, a great part of south-east England. Even the outlying patches of dense population already essentially belong to it; some for practical purposes entirely, like Brighton. Instead of the old lines of division we have new lines of union: the very word "lines" nowadays most readily suggesting the railways, which are the throbbing arteries, the roaring pulses of the intensely living whole; or, again, suggesting the telegraph wires running beside them, so many nerves each carrying impulses of idea and action either way.

It interesting, it is necessary even, to make an historic survey of London—an embryology, as it were—of this colossal whole. We should, of course, look first into its two historic cities; we should count in its many boroughs as they grew up before being absorbed; we should take note of, however easily we forget, its innumerable absorbed old villages and hamlets, its ever new and ever spreading dormitory areas—loosely built and distant for the rich, nearer and more crowded for the middle class, and—where shall we seek or put the worker and the poor?

We see, we recognise these many corporate or at least associated units of the body politic, all growing more and more fully into one vast agglomerate, and this with its own larger corporate government, its County Council. Yet even this is already far outgrow; but in time, if the growth-process continues, as in every way obviously under present conditions it must, this governing body must overtake the spreading growth, and bring all that is really functional London into its province, with economy and advantage to the vast majority of all concerned.

Of course, in a general way all this is already known to the reader—to Londoners, greater or smaller; but does it not gain a new vividness with such a map before us, a new suggestiveness also? Do we not see, and more and more clearly as we study it, the need of a thorough revision of our traditional ideas and boundaries of country and town? As historians and topographers we cannot too faithfully conserve the record of all these absorbed elements; but as practical men governing, or being governed, we have practically done with them.

Let the Lord Mayor of London and his Corporation survive by all means, as historic monuments and for auld lang syne; let there be for the historic City, and for all the neighbouring boroughs—not merely Westminster, but every regional unit that can practically justify it, and as far as may be—local autonomy too. We are making no plea for over-centralisation; on the contrary, we are inclined to think that many ganglia may be needed to maintain the health of so vast and multi-radiate a body politic. But the essential thing is that common arrangements for life and health and efficiency be made in the main according to the present and the opening developments, and not maintained unduly upon the lines of history; otherwise we shall continue to have local friction, overlapping and wastage, arrests and encystments, congestions, paralysis even, instead of the general and local health and economy we surely all of us desire.

Look now at the map of London with any friend, or if possible with two—a Progressive and a Moderate. What real difference survives between them when they sit down like plain, open-minded citizens to look at the map? Do they not agree that both their parties would do well to sit down to it also, to survey the whole situation afresh? If so, our plea for City Survey is growing intelligible; and even its economy, its positive fruitfulness, would before long begin to appear.

As, however, our Progressive and Moderate friends continue these studies, and as the vastness of the problems of London thus increases upon them, they will admit that they are, separately or collectively, unable clearly to realise all that is going on in this vast man-reef, and still more to foresee what the morrow will bring forth. Still, one has this definite bit of knowledge and the other that—now of the part of London where he was brought up or lived as a young man or of the places where he works and lives now. So gradually we piece together in conversation a good deal of useful knowledge, it may be even of practical suggestion, here and there.

Need of inquiry into smaller cities and city-groups. But here the same growth-process appears, industrial towns and cities uniting into vast city-regions, "conurbations," which require the broadest of surveys.

But as our two type-Londoners' studies go on, as with growing interest they would, they would soon come to new points of difficulty, to problems too vast readily to be grappled with; and one would ask another, "Cannot we learn something as to this from what they are doing in smaller places, in simpler cities than this tremendous London of ours? There is Birmingham, it might help us." The other might agree; and even remember that he had heard from an American friend of an active municipality in Glasgow.

Suppose they look them up in the atlas. Alas, these also have spread beyond the simple dots we learned to identify as schoolboys; and instead we see great and growing masses, each essentially like another London. Let us try Lancashire, with its great cities; that will surely help us. There is Manchester with its great Liberal and Free Trade record; there is Liverpool, with its equally strong Conservatism; they surely must have threshed matters out between them. But behold, upon our map these, too, are fast becoming little more than historic expressions. The fact is that we have here another vast province almost covered with house-groups, swiftly spreading into one, and already connected up at many points, and sometimes by more than sufficient density of population along the main lines of communication.

Here, far more than even Lancashire commonly realises, is growing up again another Greater London as it were—a city-region of which Liverpool is the sea-port and Manchester market, now with its canal port also; while Oldham and the many other factory towns, more accurately called "factory districts," are the workshops. Even if this process be not in all respects so far advanced as in London, and as yet not organised in practice under any common government, is it not becoming fairly plain, a matter of reasonable foresight, that if growth and progress are to continue much longer as they have long been doing—in some respects of late faster than ever—the separate and detached towns, whose names we learned at school and still for local purposes employ, will become mainly of minor and district usefulness, postal and what not, like the practically unified cities and boroughs of London?

Hence, if we are to avoid the many mistakes and misfortunes of London through the past delay and present confusions of its organisation and government, is it not time to be thinking of, and even to be starting, a unified survey of urban Lancashire? This, as in the case of Greater London, we should consider at every point with the utmost respect to local history and even to administrative autonomy, yet also as part of a greater whole, already only too much consolidated at many points, and still growing together. Is it asked, "Of what use is all this?" Of many uses, but enough here if we cite two—Public Health and Town Planning. Only a word then of each; and first Public Health.

These great communities are already exercised, yet in most cases not nearly exercised enough, about their sanitation and their water supply. Moreover, if better crops of human population (as we are all becoming determined) are to be grown than the present one, the question of a fuller and a far more vital access of youth to the country and to country life and occupations must assume an incomparably greater importance, and correspondingly greater space than that which has yet been given it by municipalities even with the more exemplary of Parks Departments, bright patches though these show amid our vast labyrinths of streets.

Even in the town planning movement this enlarged way of looking at our enlarging cities is not nearly common enough. The architect is accustomed to single buildings, or to street plans at most; the city engineer is accustomed to streets or to street-quarters at most; and both are reluctant to enlarge their vision. They still speak as if any such wide outlook and foresight were "ahead of the times"—"might be useful fifty years hence"— and so on through a dozen variants of the grumbling protests which are a main symptom of the senile phase which fixity to environment may bring on at all ages.

But now, returning to Public Health, in each and every one of the congresses of Health and Sanitation which now meet so anxiously from year to year in one after another of these great cities, is it not obvious to every member of these, as regards the large cities around them, that they are late enough even if they begin forthwith? Their accesses to Nature and natural conditions have already been three-fourths destroyed; indeed more, so far as the working mother and her children are concerned—that is, the nation of tomorrow. The neighbouring great towns are rapidly linking up by tramways and streets no less than railways, while great open spaces which might have been not so long ago cheaply secured as unrivalled lungs of life are already all but irrecoverable.

Here are already solid arguments for our proposed survey, and they might be strengthened and amplified, were not our problem here and in this volume mainly the clearing of ideas before the shaping of policy.

To focus these developments, indeed transformations, of the geographic tradition of town and country in which we were brought up, and express them more sharply, we need some little extension of our vocabulary; for each new idea for which we have not yet a word deserves one. Some name then for these city-regions, these town aggregates, is wanted. Constellations we cannot call them; conglomerations is, alas, nearer the mark at present, but it may sound unappreciative; what of "Conurbations?" That perhaps may serve as the necessary word, as an expression of this new form of population-grouping, which is already, as it were sub-consciously, developing new forms of social grouping, and of definite government and administration by and by also.

Beside the vast conurbations of "Greater London" and "Lancaston" other colossal city-groups arising, here generalised as "West-Riding," "South Riding," "Midlandton," "Southwaleston," "Tyne-Wear-Tees" and "Clyde-Forth."

For our first Conurbation the name of Greater London is obviously already dominant beyond possibility of competition; but we need some name for the Lancashire region also, and for each similar one we may discover. Failing a better name, since we cannot sink Liverpool and other cities in a "Greater Manchester" or the like, let us christen the vast conurbation of the Lancashire millions as "Lancaston."

It is this "Survey of Lancaston" which its constituent cities and boroughs most need to realise; and this both in detail and in mass. Imagine it photographed from an aeroplane journey, as well as mapped street by street. In all these ways we can gradually accustom ourselves to visualise the region. What are its existing defects? and what its remaining possibilities? What natural reserves still remain to separate its growing villages and suburbs? What gardens and allotments are still possible, to sanify them?

Leaving Lancaston, we have but to cross the Pennines to see along the foot of their eastward slope another dark galaxy of towns. Huddersfield, Bradford and their neighbours constitute the world-metropolis of wool no less distinctly than does Lancaston that of cotton. What shall we call this province, this natural city-alliance? Why not, in an urban sense, as of old a rustic one, simply preserve the good name of West-Riding? Similarly for South-Riding, as we may call the conurbation centring round the steel and coal of Sheffield.

Note again the present expansion of Birmingham, which has of late legitimately succeeded in having its overflowing suburbs unified with itself, its extraordinary growth recognised, as now a city exceeding even Manchester or Glasgow. Invigorated by absorbing its outlying suburbs, Birmingham is already (1910) planning new extensions upon that bold and generous scale of civic design not so long ago characteristic of great cities; but lapsed, eclipsed, forgotten with the coming on of the Railway Age. Yet this present expansion is but a step in the old process.

This larger recognition of regional facts invokes the conception of a larger city-region—"Midlandton," as we may perhaps call it; and Greater and growing Birmingham is but the capital of this, though its exact limits may be hard to define. The recent union of the "Five Towns" (Municipal and Parliamentary Borough formed 1910) is thus not only a local event, but a regional pioneering, a noteworthy example of an incipient urban regrouping. And here let us hope that the Duke of Sutherland's generous gift of Trentham may similarly augur a period of better and closer relations of town and country throughout the land than have been those of yesterday.

Pass next to South Wales, where on its magnificent coalfield the same process of development is at work. And, speaking of coalfields, we may conveniently here call attention to the close coincidence of this great centre of population with its magnificent South Wales coalfield and thence note the parallelism of each great conurbation to its coalfield save in the case of London alone.

We plainly see the development of a Greater Cardiff, a veritable (South) Waleston, whose exact limits and relation to the metallurgic centre of Swansea are, of course, for its regional geographer to define.

Pass next northwards to the Tyne towns, with which we must plainly also take those of Wear and Tees, as constituting a new regional community, a natural province—Tyne-Wear-Tees we may perhaps call it. It is interesting to recall that our British Gallery at the Brussels Exhibition of 1910, unhappily burned down, was adorned with a well-painted perspective of this very region, shown with all its towns connected up by railways and roads, and presented as a bird's-eye view (or, as we may now-a-days say, an aeroplane view) from above the sea-coast. For does not this map clearly suggest that the economic and social unity of such new city-regions, such conurbations as are here described, is already becoming conscious to them?

The preparation and exhibition of such diagrammatic perspectives would be of no little service in making these ideas clear to all concerned, and in enabling the public and the rulers of each to realise the new situation, the new solidarity which are arising towards a fuller integration, a higher unity of the body politic. The great maps of railway systems which are at once a convenience and an adornment of German station-halls, have no little value and educational influence: so, and far more intensively, might enlargements of the conurbation-maps, which we are here discussing, bring before the public the needed conception of a local within a more general citizenship.

In conclusion, let us pass to Scotland. Here, again, the history and geography of popular notions, those of the school books on which we were brought up, and by which our children are still examined, are no longer adequate.

Glasgow, as everybody knows, is the main centre of activity and population in Scotland, far outnumbering and outweighing Edinburgh; it is the real capital in many respects. But Greater Glasgow—in the fullest sense, that in which we speak of Greater London—is something far vaster than the present name and burgh limits at all describe; it includes practically the Clyde ports and watering places, and runs far into Ayrshire, with inland burghs and villages not a few. It spreads far up the Clyde valley, indeed reaches its strenuous hands across the isthmus to Falkirk and Grangemouth, while its merchants have their villas at Stirling and beyond, as far as Bridge of Allan and even Dunblane. Again plainly, old, thinly populated provinces are on the way to be covered with houses.

Edinburgh has no doubt its marked regional individuality; and in its immediate growth is, more than is commonly realised, with Leith and minor towns and suburbs already approaching half a million: it is perhaps destined, with due development of its not inconsiderable adjacent coalfields, to double this within the century.

Though, from historic tradition and from present holiday associations, most people, even in Scotland, still think of the Scots as in the main a nation of hardy rustics, no population in the world is now so predominantly urban, and, as sanitary reformers know, none so ill-housed at that. More than half the population of Scotland is crowded upon this central isthmus; and, with the approaching construction of the Clyde and Forth Canal (which is so plainly a matter not only of Scottish, but even of national, imperial, and international policy), it is clear that we shall have a linking up of these two great cities and their minor neighbours of Scotland into a new conurbation— a bi-polar city-region indeed, which is more and more uniting into one vast bi-regional capital—Clyde-Forth as we may soon learn to call it.

Glasgow and Edinburgh are, of course, far remoter in type and spirit than their now-a-days small railway distance implies; and this difference, even contrast, is natural, inevitable, and so far permanent, for they are really the respective regional capitals of East and West Scotland, and contrasted in many ways—geographical and meteorological, racial and spiritual. To Glasgow indeed the contrast with Edinburgh may seem as great as that between Liverpool and York; while a still larger contrast might be made from the Edinburgh point of view, as that between the main cities of Sweden and of Norway, of both of which Scotland in many ways is a condensed miniature; say a Stockholm with Upsala for Edinburgh, and for Glasgow a greater Bergen and Christiania. Towns so widely distinct in nature and race, in traditions and in social functioning and structure do not easily recognise that even they are but the poles of a vast and growing conurbation: yet here, too, the growth-process is at work, and tends largely to submerge all differences beneath its rising tide. And, broadly speaking, the main limit of the modern city is that of the hour's journey or thereby, the maximum which busy men can face without too great deduction from their day's work; and hence it is above all with the constant extension and acceleration of the means of communication that each conurbation arises and extends.

> Thus a veritable New Heptarchy is arising, whose water supplies and coalfields, and kindred local affairs are the essentials of national existence; no longer negligible as the mere "parish pump" and "coal cellar" of metropolitan politics.

It is interesting now to return to the map and make our main conurbations clear, each upon its coalfield. Running downwards and leaving Clyde-Forth to Scotland, we have (1) Tyne-Wear-Tees, (2) Lancaston, (3) West-Riding, (4) South-Riding, (5) Midlandton, (6) Waleston, each a coalfield with its vast conurbation; while Greater London, without a coalfield, forms the seventh of our series. What is this but a new heptarchy, which has been growing up naturally, yet almost unconsciously to politicians, beneath our existing, our traditional political and administrative network; and plainly not merely to go on as at present, straining and cracking and bursting this old network, but soon surely to evolve some new form of organisation better able to cope with its problems than are the present distinct town and county councils. What are the new forms to be?

Leaving this sphinx-riddle for the present, and turning once more to the map, we recognise plainly enough that our political friend who was "not going back to the heptarchy" will have to go forward to it, indeed is already in it. Let him now observe closely, and in the very middle of our map, a great irregular white patch practically blank of population and separating Lancaston from South-Riding and West-Riding, which, indeed already are well-nigh run together. This white patch represents the heights of the Pennines, and consequently the water supply of these vast and growing populations on either side. Here, in fact, accurately speaking in synoptic vision, is their Parish Pump, one, however, no longer to be despised; but precisely the most important, the ultimate and determinant condition of population, and the inexorable limit of their growth.

Coal will still last a long time, and cotton might expand accordingly, but water is the prime necessity after air itself; and, unlike it, is limited in quantity. Food can be brought for almost any conceivable population as long as ships can sail the seas, and we have the wherewithal to buy; famine one can survive for months; total starvation even for weeks; but without water we can last barely three days.

Parish Pump indeed! the prime necessity of regional statesmanship, since even of bare survival. For life and health, for cleanliness and beauty, for manufactures too, what more need be said?

Return now to the question—What are the new social forms to be? It is not yet safe for us to speculate upon this until the needful Regional Survey is far more advanced. One suggestion, however, is practical enough; there should be, and that speedily and increasingly, amicable conference among all the representatives, rustic and urban, of the various cities and county regions concerned; and as a matter of fact, various beginnings of this are being forced into existence by the slicer pressure of their common interests. Such meetings will gradually increase in number, in usefulness, in co-operation, and by-and-by take more permanent form.

The old Borough Councils and County Councils can no longer separately cope with what are becoming so plainly yet larger Regional and Inter-Regional tasks, like those of water supply and sanitation for choice, but obviously others also. The growth of London and its County Council, its separate boroughs, is thus repeating itself; and its example merits study, alike for its suggestiveness and for its warnings. While conversely, to the Londoner such regional excursions may be suggestive. The contrasts of "London and the provinces," as Spending-town and Earning-town, again of Taxing-town and Paying-towns, and various others, also arise, and might lead him far.

It may not yet be time to press for political rearrangements: this might too readily come to mean premature disputes and frictions, not to speak of legal difficulties and expenses. But it is plainly time for the co-operation of the regional geographer with the hygienist, and of both with the concrete sociologist, the student of country and town, of village and city; and also for the furtherance of their labours, the discussion of them in detail, in friendly conferences representative of all the various groupings and interests concerned.

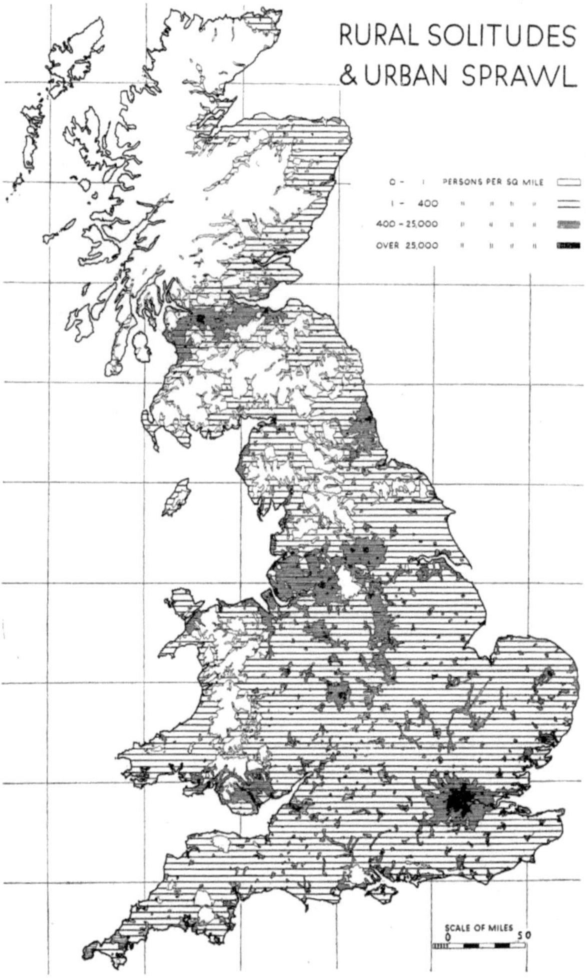

Fig. 8.1 "Rural solitudes and urban sprawl".
Source: Geddes, P. (1949). *Cities in evolution, an introduction to the town planning movement and to the study of civics* (p. 192). Williams and Norgate (New and revised edition)

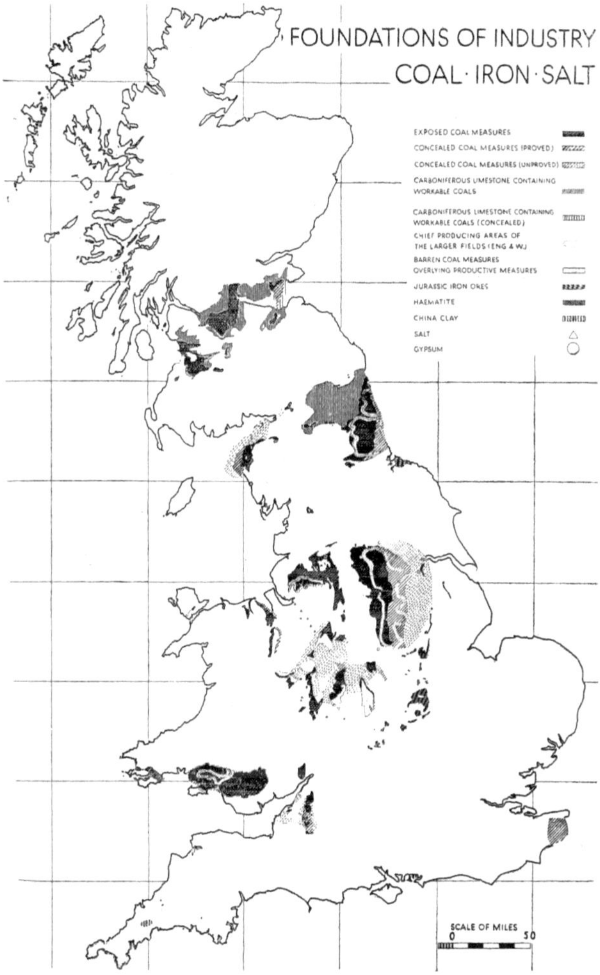

Fig. 8.2 "Foundations of industry Coal·iron·salt". Source: ibid., p. 193

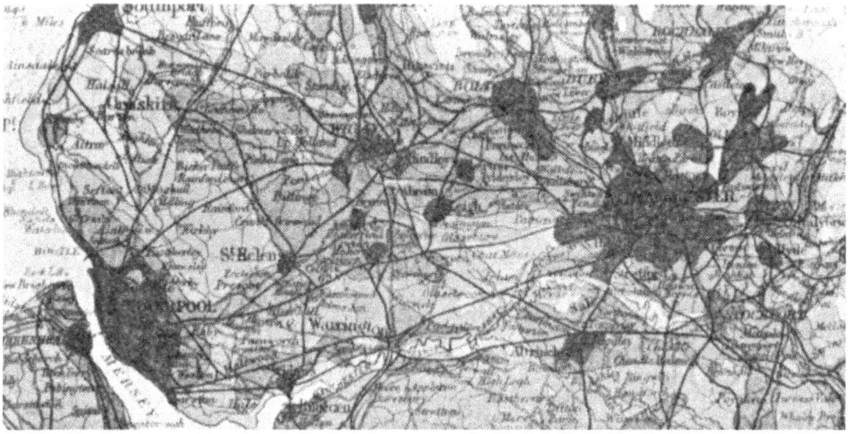

Fig. 8.3 "Lancashire towns agglomerating as «Lancaston»".
Source: Geddes, P. (1915), "The Population Map and Conurbations", published in: P. Geddes, *Cities in evolution, an introduction to the town planning movement and to the study of civics.* Williams and Norgate. p. 31.
"There is Manchester, with its great Liberal and Free Trade record, there is Liverpool, with its equally strong Conservatism; they surely must have threshed matters out between them. But behold, upon map these, too, are fast becoming little more than historic expressions. The fact is that we have here another vast province almost covered with house-groups, swiftly spreading into one, and already connected up at many points, and sometimes by more than sufficient density of population along the main lines of communication. Here, far more than even Lancashire commonly realizes, is growing up again another Great London as it were – a city-region of which Liverpool is the seaport and Manchester the market, now with its canal port also; while Oldham, and the many other factory towns, more accurately called "factory districts", are the workshops". (ibid., pp. 31-32)

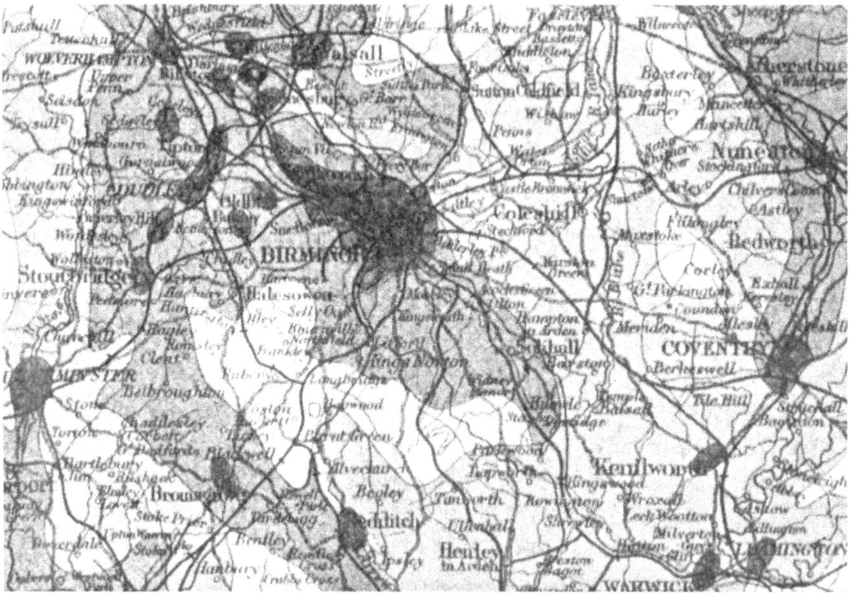

Fig. 8.4 "Midland towns agglomerating as «Midlandton»".
Source: ibid., p. 36.
«This larger recognition of regional facts involves the conception of a larger city-region- "Midlanton", as we may perhaps call it: and Greater and growing Birmingham is but the capital of this, though its exact limits may be hard to define. The recent union of "Five Tows" is thus not only a local event, but a regional pioneering, a noteworthy example of an incipient urban regrouping. And here let us hope that the Duke of Sutherland's generous gift of Trentham may similarly augur a period of better and closer relations of town and country throughout the land than have been those of yesterday." (ibid., pp. 36–37)

Chapter 9
The Land of Peasant Utopia (1920)

Alexander Vasilevich Chayanov

Introduction

Paola Viganò
EPFL, Lausanne, Switzerland
IUAV, Venice, Italy
paola.vigano@epfl.ch

During the first years after the Russian Revolution, many written works, posters and stories took on a utopian character, not always consistent with the new political and social condition, and often characterized by a strongly anti-urban approach.[1] *The opening to the future, which was characteristic of the time, then, immediately suspended and closed in the 1930s, was fed on the translation, among others, of the works of H. G. Wells*[2] *(see Chap. 5). The cities, at the centre of a harsh criticism, are stigmatized as places of perdition and conflict, atomization and lack of community by movements such as the socialist muzhiks and Scythian poets. In the various utopias, the cities are testimonies of the past, abandoned relics, museums of themselves, or underutilized. In this context full of experiments and visions, one has* The Journey of My Brother Alexei to the Land of Peasant Utopia, *published in 1920 by the State Publishing House, by an eminent Soviet agricultural economist, Alexander Chayanov (1888–1939), who set in 1984 a short story of*

[1]Stites, R. (1989). *Revolutionary dreams: Utopian vision and experimental life in the Russian revolution*. Oxford University Press.

[2]De Magistris, A., Scotti, A. (Eds.). (2018). *Utopiae finis? Percorsi tra utopismi e progetto*. Accademia University Press.

M. Barcelloni Corte and P. Viganò (eds.), *The Horizontal Metropolis*,
https://doi.org/10.1007/978-3-030-56398-1_9

which he wrote the first part, without the second ever seeing light. The writing is very dense and recalls elements belonging to Russian populism and anarchism.[3] *Kropotkin is, in that context, an influential presence with the themes of the decentralization of settlements (see Chap.* 4*) later taken up by Okhitovich (see Chap.* 8*) in the years of the first Soviet five-year plan.*

Alexander Chayanov's contribution to the peasant question goes much further. During the 1920s he developed an economic micro-theory, The Theory of Peasant Economy,[4] *starting from the role of the Russian peasant family who lived off its work without hired hands, and from this analysis challenged classic economic theory built around the capitalist entrepreneur. In the absence of wages, the other factors traditionally taken into consideration (rent, net gain, the interest on capital) also become meaningless indicators. For Chayanov, the peasant family, not only with its agricultural as well as artisanal and exchange related production, constitutes the unit of a different economic structure that, more generally, can be extended to the family economy. For this economy, a different economic theory is needed.*

Based on extensive field research carried out in Russia after the 1861 Land Reform in which many young urban intellectuals participated, Chayanov observes the production capacity defined in terms of family self-exploitation and within the limits decided by each family unit in order to maintain the affluence achieved. In terms now widely used, we could say that he highlights its resilience. Defending the hypothesis of greater competitiveness of the peasant family versus the large-scale capitalist farm, Chayanov elaborates the concept of "labour-consumer balance between the satisfaction of family needs and the drudgery (or irksomeness) of labour".[5] *In this way Chayanov assumes a position critical both of classical western and Marxist economic theory that considered the model of small family farming obsolete and destined to disappear. Different types of economies are in turn possible according to Chayanov and each of them needs adequate theories, this is the theme at the centre of another of his writings:* On the Theory of Non-Capitalist Economic Systems, *published in German in 1924.*

A great agricultural economist, a man of culture and a refined writer under various pseudonyms, in his travels outside of Russia Chayanov visited Lombardy[6]

[3]De Magistris and Scotti (2018), cit.

[4]Chayanov A.V. (1966). *The Theory of Peasant Economy*, ed. by Thorner D., Kerblay, B. and R.E. F. Smith, Homewood. Richard D. Irwing Inc., for The American Economic Association. The book contains the translation of two writings: *Peasant Farm Organization*, a publication of the Agricultural Economics Scientific Research Institute of Moscow, 1925, which extends *The Theory of Peasant Economy* published in 1923 in German and the short essay *On the Theory of Non-Capitalist Economic Systems*, also published in German in 1924.

[5]Chayanov A.V. (1966). *The Theory of Peasant Economy...* From the Introduction by Daniel Thorner, XV.

[6]Chayanov A.V. (1966). *The Theory of Peasant Economy...* From "A.V. Chayanov: Life, Career, Works", by Basile Kerblay, XXVI.

in 1908 for its irrigation system, while he went to Belgium to study the cooperative system that Seebohm Rowntree describes at length in Land and Labour *(see Chap. 7).*

9.1 The Journey of My Brother Alexei to the Land of Peasant Utopia[7]

Chapter 1

In which the well-disposed reader becomes acquainted with the triumph of socialism and Alexei Kremnev, the hero of our story.

It was long past midnight when the owner of labour book No. 317413, once, in the bourgeois world, called Alexei Vasil'evich Kremnev, left the stuffy, grossly overcrowded great auditorium of the Polytechnical Museum.

The misty haze of an autumn night blanketed the sleeping streets. The infrequent electric lights seemed lost in the far distance of intersecting alleyways. A wind stirred the yellowed leaves on the trees of the boulevard and in the gloom the walls of Kitaigorod loomed fabulously large and white.

Kremnev turned into Nicholas Street. In the misty haze it seemed to have taken on its former features. Wrapping himself up fruitlessly in his raincoat against the penetrating damp of the night, Kremnev looked nostalgically at the church of St. Vladimir and St. Panteleimon's chapel. He recalled how, when he was a first-year law student many years ago, with a faltering heart he bought Flerovskii's *ABC of the Social Sciences*[8] just here, on the right, from Nikolaev, the second-hand bookseller; how, three years later, he started his icon collection by finding a Novgorod Saviour at Elisei Silin's; and those many and long hours when, with the proselyte's burning eyes, he burrowed through the treasures of the manuscript and book stacks of Shibanov's secondhand shop, where now you could make out, under the dull light of a street lamp, the brief inscription 'Glavbum'.[9]

Banishing these guilty recollections, Alexei turned towards the Iberian Gate, went past the first House of the Soviets and plunged into the darkness of Moscow alleyways.

Words, phrases, and fragments of phrases just heard at the meeting in the Polytechnical Museum flashed morbidly through his mind:

---

[7]Source: Chayanov, A. (1920). Путешествие моего брата Алексея в страну крестьянской утопии. Moscow: State Publishing House. Source used in this publication: Chayanov, A. (1976). "The journey of my brother Alexei to the land of peasant Utopia". *The Journal of Peasant Studies*, 4(1), 63–108.

[8]V. V. Flerovskii, pseudonym of V. V. Bervi (1829–1918).

[9]*i.e.* an acronym, modelled on many actual Soviet examples, for the Chief Paper Board.

> By destroying the family hearth we are dealing the final blow to the bourgeois system!
>
> Our decree forbidding taking meals at home eliminates from our lives the joyful poison of the bourgeois family and firmly establishes the socialist principle for ever.
>
> The cosiness of family life leads to possessiveness; the petty proprietor's joys conceal the seeds of capitalism.

His tired head ached and was now, as usual, reasoning, but without conscious thought, and registering without drawing conclusions; and his legs mechanically moved towards the half-destroyed family hearth fated in a week's time to be destroyed completely in accordance with the decree of 27 October 1921, which had just been published and elucidated.

Chapter 2

Telling of the influence of Herzen on the inflamed imagination of a Soviet official.

Alexei buttered a large piece of bread, a blessed gift from the god-protected Sukharevka, poured himself a glass of boiling coffee and sat down in his working chair.[10]

The town could be seen through the panes of the large window; below, in the misty nocturnal haze, rows of street lamps stretched out in milky patches of light. Here and there, the dark bulks of houses, windows still alight, glowed a dull yellow.

> So, it's done.' Alexei thought, peering into the Moscow night. 'Old Morris, kind Thomas, Bellamy, Blatchford and you other good, dear Utopians.[11] Your solitary dreams are now common beliefs, your greatest. audacities have become an official programme and an everyday common-place! In the fourth year of the revolution, socialism can consider it holds undivided sway over the globe. Are you satisfied, Utopian pioneers?

And Kremnev looked at the portrait of Fourier hanging above one of his book-cases.

But for himself, old socialist, important Soviet worker in charge of a section of the World Economic Council that he was, there was something not altogether right about this realisation; there was a confused nostalgia for the past; some cobweb of bourgeois psychology still clouded his socialist consciousness.

He strolled up and down the carpet of his study, his glance sliding over the covers of the books; he suddenly noticed a row of small volumes on a half-forgotten shelf. The names of Chernyshevskii, Herzen and Plekhanov on the leather spines of the solid bindings caught his eye. Smiling a smile of childhood recognition, he took down Pavlenkov's edition of Herzen.

---

[10]The Sukharevka was a market where virtually anything could be acquired, at a price.

[11]Chayanov here evidently refers to William Morris, Sir Thomas More, the American Edward Bellamy (1850–98) and Robert Blatchford (1851–1943), editor of *The Clarion*.

It was two o'clock. The clock struck with a long-drawn hiss and was silent once more.

Good, noble, childishly naïve words revealed themselves to Kremnev's eyes. He was engrossed and troubled by what he read as one is troubled by recollections of one's first youthful love, one's first youthful vow.

His mind seemed freed from the hypnosis of Soviet daily life; new, unhackneyed, thoughts stirred in his mind; he found it possible to think in new ways.

Excitedly, Kremnev read a prophetic page he had long forgotten:

'Weak, puny, stupid generations,' Herzen wrote, 'will somehow last until the eruption, until a flood of lava smothers them under a stony pall and confines them to the oblivion of history. And then? Then spring will come, young life will burgeon on their gravestones; the barbarism of infancy, full of unmatured but healthy strength, will replace the barbarism of senility, a wild, new power will burst forth in the youthful breast of the young nations; it will be the start of a new cycle of events and of the third volume of universal history.

Its basic tenor can be grasped now. It will be an era of social ideas. Socialism in all its aspects will grow to the utmost limits, to absurdity. Then again, a cry of refusal will break from the titanic breast of the revolutionary minority, and there will be once more a mortal struggle in which socialism, in the position of today's conservatism, will be defeated by another, unknown revolution still to come'.[12]

> 'A new rising. But where is it? And in the name of what ideals?' he wondered. 'Alas, it has always been the weakness of liberal doctrine that it was incapable of creating ideologies and had no Utopias'.

He smiled regretfully. 'O, you Milyukovs and Novgorodtsevs, Kuskovs and Makarovs, what Utopia will you paint on your banners?[13] What, save the obscurantism of capitalist reaction, have you to offer in place of socialism? I agree... ours is by no means a socialist paradise... But what will you give in its stead?'.

The Herzen volume suddenly snapped itself shut with a crack; a bundle of loose octavo and folio sheets fell off a shelf.

Kremnev shuddered.

A suffocating odour of sulphur filled the room. The hands of the large wall clock spun faster and faster until they disappeared in a whirl. The pages of the tear-off calendar began to detach themselves noisily and spiralled upwards, filling the room with paper vortices. The walls became oddly distorted and trembled.

Kremnev's head swam and there was a cold sweat on his brow. He shuddered and in a panic of terror dashed for the dining-room door. It banged shut behind him with the crackle of breaking wood. He vainly felt for the electric light switch; it was

[12]The quotation is from Herzen's *From the Other Shore*, Chap. VII.

[13]P.N. Milyukov (1859–1943), a historian, was a member of the Kadet Party. P. I. Novgorodtsev (1866–1924) was a lawyer and historian E. D. Kuskova (1869–1958) was a member of the Kadet and later of the Socialist Revolutionary parties, a publicist and wife of S. N. Prokopovich (see note 11). She was expelled from Russia in 1922. N. P. Makarov (1886–[1980, editors' note]), an agrarian economist, was a member of the League of Agrarian Reform.

not in its old place. Groping about in the darkness, he repeatedly struck unfamiliar objects. He felt dizzy and befuddled, as if he were sea-sick.

Exhausted by his efforts, Alexei lowered himself onto a sofa which had not been there before, and lost consciousness.

Chapter 3

Depicting Kremnev's appearance in the land of Utopia and his pleasant conversation with a Utopian Moscow girl on the history of twentieth century painting.

Kremnev was woken by a silvery bell.

'Hello, yes, it's me', a woman's voice was saying, 'Yes, he's arrived... evidently last night... still sleeping... He was very tired, fell asleep without undressing... Very well. I'll call you.'

The voice ceased and the rustling of skirts indicated that its owner had left the room.

Kremnev propped himself up on the sofa and rubbed his eyes in amazement.

He was in a large yellow room flooded by the morning sun. Its appointments where of a strange, unfamiliar style: furniture made of red wood with greenish-yellow upholstery, yellow curtains, half drawn over the windows, and a table with wonderful metal fittings. Light female steps were heard in the neighbouring room. A door creaked and there was silence once again.

Trying to understand his position, Kremnev leapt up and quickly walked up to the window.

Thick autumn clouds were sailing across the blue sky. A little below them, just above the earth, some small and large airplanes of bizarre design darted about, the revolving metal parts gleaming in the sun.

The town lay spread out below... undoubtedly, this was Moscow.

To the left, the bulk of the Kremlin towers reared up, on the right, the Sukharevka shone red, and, far off, Kadashi rose proudly against the sky.

A view he had known for many, many years.

Yet how much it had all changed! The piles of stone which had once crowded the horizon were gone; whole architectural complexes had disappeared; Nirenzei's house was no longer in its place... Instead, there were gardens everywhere... Sprawling clumps of trees enveloped the whole space almost up to the Kremlin itself, dotted with solitary islands of architectural complexes. Boulevards intersected the sea of greenery, now turning to yellow. Streams of pedestrians, motor cars and carriages poured along them in a living river. Everything breathed a certain distinctive freshness, a cheerful confidence.

Undoubtedly this was Moscow, but a new, transformed and brighter one.

'Have I really become the hero of a Utopian novel?' Kremnev exclaimed. 'A pretty stupid situation, I must say!'

He began to look around him to get his bearings, hoping to find some reference point that would enable him to identify the new world around him.

> What is waiting for me beyond these walls? The blessed kingdom of socialism, now consolidated and enlightened? The marvellous anarchy of Prince Kropotkin? Capitalism restored? Or perhaps some new, hitherto unknown social system?

On the evidence of the view from the window, one thing was clear; people enjoyed quite a high standard of living and culture and lived in common. But this was not enough to understand the nature of his surroundings.

Alexei eagerly began to examine the objects around him, but they told him very little.

For the most part, they were articles of everyday life, notable only for the careful finish, a certain emphatic precision and luxury of execution and the peculiar style of their forms, reminiscent both of Russian antiquity and of the ornaments of Nineveh. A strongly russified Babylonian style, in a word.

Kremnev's attention was drawn to a large picture hanging over the heavily upholstered sofa on which he had woken up.

At first glance it might be confidently identified as a typical work of Pieter Brueghel the Elder. The same composition with its high horizon, the same bright jewel tints, the same foreshortened figures, but... the canvas represented people in coloured frock-coats, ladies with umbrellas and motor-cars, and the subject was undoubtedly something like the take-off of aeroplanes. Several reproductions lying on a neighbouring table were in similar style.

Kremnev went up to a large desk made from something like dense cork and hopefully began to look over the books that lay scattered there. These were the fifth volume of *The Practice of Socialism* by V. Sher, *The Renaissance of the Crinoline, a study of contemporary fashion*, two volumes of Ryazanov's *From Communism to Idealism*, the 38th edition of E. Kuskova's memoirs, a magnificent edition of the *Bronze Horseman*, a pamphlet *The Transformation of V-energy*,[14]... Finally, his hand trembling with excitement, he picked up a recent newspaper.

Impatiently, Kremnev spread out the small sheet. It was dated, 23.00 h, 5 September 1984. He had leapt forward sixty years.

There was now no doubt that Krennev had woken up in a land of the future; he plunged into reading the news-sheet.

'Peasantry', 'The past era of urban culture', 'State collectivism of sad memory'... 'This was in capitalist times, i.e. almost in pre-history...', 'The Anglo-French isolated system'—all these phrases and dozens of others penetrated Kremnev's brain, filling his mind with astonishment and a great desire to know.

---

[14]V. V. Sher (1884–?) was a Communist administrator.

D. B. Ryazanov (pseudonym of Goldenbakh) (1870–1938) was a Social-Democrat, later a Bolshevik, and was for a time Director of the Marx-Engels Institute.

E. D. Kushova (1869–1958) was a member of die Kadet and later of the Socialist Revolutionary parties, a publicist and wife of S. N. Prokopovch (see note 8). She was expelled from Soviet Russia in 1922.

The telephone interrupted his thoughts; steps were heard in the neighbouring room. The door opened and a young girl entered, together with a flood of sunlight.

'Ah, you are up...' she said gaily. 'I slept through your arrival yesterday.'

The telephone rang again.

'Excuse me, this should be my brother worrying about you... hello... yes, he is up now... I don't really know... I'll ask him at once... You speak Russian, Mister... Charlie... Mr Mann, unless I am mistaken.'

'But of course, of course,' Alexei surprised himself and exclaimed very loudly.

He does—and with a Moscow accent... Alright... I'll hand you over.

A confused Kremnev was handed something recalling an old-fashioned telephone set, heard a greeting offered in a soft bass, a promise to pick him up in three hours' time, and an assurance that his sister would take care of everything; putting down the receiver, he was aware very, very clearly that he was being taken for somebody else named Charlie Mann.

The girl had left the room. With the determination of despair Alexei rushed to the desk, hoping to find among the papers and bundles of telegrams some clue to the mystery surrounding him.

Luck was with him; the first letter he took up was signed by Charlie Mann who, in a few sentences, expressed his wish to visit Russia and become acquainted with its agricultural engineering installations.

Chapter 4

Continuing the third and separated from it only in order to avoid excessively lengthy chapters.

The door opened, and his young hostess came in carrying a tray with steaming breakfast cups above her head.

Alexei was enchanted by this Utopian woman, by her almost classical head, perfectly set on a strong neck, her broad shoulders and full breasts, which raised her shirt collar with every breath.

The momentary silence of first acquaintance soon gave way to lively conversation. Kremnev, avoiding the role of narrator, diverted the conversation into the area of art, assuming that this would not embarrass a girl who lived in rooms hung with fine examples of painting.

The girl, whose name was Paraskeva, spoke with the warm enthusiasm of youth about her favourite masters: Brueghel the Elder, Van Gogh, dear old Rybnikov and marvellous Ladonov. An impassioned admirer of neorealism, she sought in art for the secret of things, something either of god or the devil, but at all events beyond human capacities.

Recognising the supreme value of all that exists, she demanded of the artist a degree of harmony with the creator of the universe, prized the magic power in a picture, that Promethean spark which creates a new reality and all in all was closest to the realism of the Dutch Old Masters.

Kremnev gathered from her that after the era of the great revolution, when futurist painting was utterly disruptive of old traditions, there ensued a period of baroque futurism, of tame and charming futurism.

Then, as a reaction, like a sunny day after a storm, a thirst for craftsmanship came to the fore; the Bologna school came into fashion, the primitivists were somehow immediately forgotten, and hardly anybody visited the museum galleries containing the work of Memling, Fra' Beato, Botticelli and Cranach. However, while losing nothing of its excellence, craftsmanship, with the passage of time, gradually acquired a decorative tendency and created monumental canvases and frescoes of the period of the Varvarin conspiracy; the period of still-lifes and the blue palette passed like a rage, then the twelfth-century Suzdal frescoes became the yardstick of the world's ideas and, finally, realism came to rule, with Pieter Brueghel as its idol.

Two hours had slipped by, and Alexei could not make up his mind whether he should listen to the deep contralto of his companion or watch the heavy plaits entwined on her head.

Her wide-open, attentive eyes and a mole on her neck spoke to him of the superiority of neorealism better than any arguments.

Chapter 5

Excessively long, but essential to acquaint Kremnev with Moscow in 1984.

> 'I'll take you through the whole town', Paraskeva's brother, Nikifor Alexeevich Minin, said, settling Kremnev in the car, 'and you shall see our present-day Moscow'.

They drove off.

The town seemed to be a continuous park amidst which architectural groups rose up to right and left, looking like lost little skittle pins.

Sometimes an unexpected turn of the avenue disclosed to Kremnev's eyes the outlines of familiar buildings, mostly erected in the seventeenth and eighteenth centuries.

Beyond the dense crowns of the yellowing maples, the domes of Barashi could be glimpsed; the limes fell back to disclose the magnificent silhouette of Rastrelli's building, Kremnev's school in his 'teens. In a word, they were driving along the Utopian Pokrovka.

'How many inhabitants are there in your Moscow?' Kremnev asked his companion.

It's not so easy to answer that. If we consider the area of the town at the period of the great revolution and take the population constantly spending the night here, it now amounts to, perhaps 100,000 people; but forty years ago, immediately after the great decree on the destruction of towns, there were no more than 30,000. On the other hand, in daytime, if you reckon all the visitors and hotel guests, it might be over five million.

The car slowed down. The avenue was narrowing; the architectural blocks approached one another closer and closer, old fashioned town streets began to appear. Thousands of cars and horse-drawn carriages, several abreast, sped along in a solid stream towards the town centre; a dense crowd of pedestrians filled the broad pavements. The almost complete absence of black was surprising: the bright blue, red, navy and yellow of the usually plain men's jackets and smocks mingled with the multi-coloured women's dresses, which were something like sarafans with crinolines, but in a great variety of styles.

Newsboys, flower sellers, peddlers of toddy and of cigars scurried about in the crowd. Over its heads and above the stream of vehicles, waving banners, with slogans and masts bedecked with flags, flashed in the sun.

Boys darted about, almost under the carriage wheels, selling some kind of news-sheet and crying for all they were worth: 'The decider' Vanya of Vologda against Ter-Markelyanets! Two jacks and one dog!'.

There were lively arguments and exchanges in the crowd, and much was said about 'equalisers' and 'dogs'.

Kremnev raised his eyes in astonishment to his companion, who smiled and said:

The national sport! Today is the last day of the international knuckle-bone championship. The Tiflis dibs champion is challenging the man from Vologda for the knuckle-bone championship. . . But Vanya won't go down, and this evening in Theatre Square he will be the winner for the fifth time.

The car continued to slow down; it passed Lyubyanka Square, where both the Kitai-gorod wall and the Vitali cherubs had been preserved, and was going down past the First Printer.[15] A sea of heads, an explosion of bright flags flaming in the sun, a many-tiered platform rising nearly to the Bolshoi Theatre's roof, and the roars of the crowd filled Theatre Square. The knuckle-bone match was in full swing.

Without a doubt, on a thousand gun barrels, amicably supporting one another, stood Lenin, Kerenskii and Milyukov.[16]

The car turned sharply to the left and they passed close by the monument's plinth.

Kremnev had time to make out some of the figures on the bas-relief: Rykov, Konovalov and Prokopovich picturesquely grouped round an anvil; Sereda and

---

[15]I.P. Vitali was the sculptor who in 1835, produced the bronze figures for a fountain in Theatre Square. The First Printer was Ivan Fedorov, who printed books in Russia for the first time in 1564; there is a statue to him by the Kitai-gorod wall.

[16]These were the leaders of the Bolshevik, the Provisional Government and the Kadet Party.

Maslov engaged in sowing seed.[17] He could not hold back an exclamation of amazement. His companion replied through his teeth, without removing his smoking pipe: 'A memorial to the architects of the great revolution.'

The car swung to the left, the lady with the dog to the right; another turn, and the car dived into an underground tube, rushed along at a mad pace for some moments through a brightly lit subterranean tunnel, emerged on the banks of the Moskva and stopped by a terrace filled with tables.

'Let's have a *coca* and fruit-juice for the road.' Minin said, climbing out the car.

Kremnev glanced around; before him rose the bulk of a bridge so exactly reproducing the seventeenth-century Stone Bridge that it seemed to have come from Picart's engraving.[18] Behind it, the Kremlin loomed in its full splendour, embraced on all sides by the golden autumn woods, its gold domes flaming.

A waiter in traditional white trousers and shirt brought them some drink resembling an egg-flip mixed with candied peel, and our travellers were lost in contemplative silence for a while.

> 'Forgive me,' Kremnev began after a time, 'but as a foreigner I don't understand the organisation of your town and I can't quite conceive the history of its dispersal.'

'Initially, Moscow's reconstruction was determined by political factors', replied his companion. 'In 1934, when power was firmly in the hands of the peasant parties, the Mitrofanov government, persuaded by many years' experience of the danger to a democratic regime from huge conglomerations of urban population, decided on a revolutionary measure. At the Congress of Soviets they carried through the decree which, of course, is also known to you in Washington, on the abolition of towns with more than 20,000 inhabitants.

'Obviously, the decree was hardest of all to carry out in Moscow, with its population of over four million in the thirties. But the leaders' obstinate determination and the technical capacity of the engineer corps were able to cope with the task in ten years.

The railway workshops and the goods yards were moved back to the line of the fifth ring road. The railway men of the twenty-two radial lines, with their families, were dispersed down the line no closer than that fifth zone, that is Ramenskoe, Kubinka, Klin and similar stations. The factories were gradually evacuated to new railway junctions throughout Russia.

[17]A. I. Rykov (1881–1938) was a Bolshevik, twice a member of the Supreme Council of the National Economy, who became Prime Minister after Lenin. A. I. Konovalov was Kerenskii's deputy in the Provisional Government, and later a White emigré. S. N. Prokopovich (1871–1955) was an economist, a Social Democratic member of the Provisional Government, later banished from the USSR. S. P. Sereda (1871–1933) was Bolshevik Commissar of Arable Farming from April 1918 to December 1921 and S. L. Maslov (1867–1946), a Socialist-Revolutionary, had been the Cast Minister of Agriculture in the Provisional Government.

[18]Probably Pierre Picart, a Dutch engraver who worked in Russia in the seventeenth and early eighteenth century.

By 1937 the Moscow streets began to empty. After Varvarin's conspiracy, the work naturally intensified; the engineer corps set about planning the new Moscow; the city's skyscrapers were destroyed by the hundred; dynamite had often to be used. My father recalls that in 1939 the boldest of our leaders, wandering through the ruins of the town, were ready to acknowledge themselves vandals, Moscow was such a picture of devastation. But the destroyers had before them Zheltovskii's plans and the work continued stubbornly.[19] To reassure the population and European opinion, one sector was completely finished in 1940; that astonished them and calmed feelings; and by 1944 everything came to look as it does now.'

Minin took from his pocket a small plan of the town and spread it out.

> But now the peasant regime has grown so strong that this decree we regard as sacred is no longer observed with the former puritanical severity. Moscow's population is growing so much that, in order to observe the letter of the law, our councillors count as Moscow only the ancient White Town, that is the area within the perimeter of the pre-revolutionary boulevards.

Kremnev carefully examined the map, then looked up.

> 'Excuse me,' he said, 'but this is some kind of sophistry; all this, around the White Town, that's almost a town. And, altogether, I don't see how you could have agrarianised the country painlessly and what possible part your pathetic pigmy-towns could play in the economy.'

I can't easily answer your question in a couple of words. You see, formerly the town was self-sufficient, the countryside was no more than its hinterland. Now, if you like, there are no towns at all, there are only nodal points of the nexus of social relations. Each of our towns is simply on assembly point, the central area of an *uezd*. It's not a place for living, but a place for celebrations, gatherings and some other matters. A point, but not a social entity.'

Minin raised his glass, drained it at a gulp and continued.

> Take Moscow: with its 100,000 population, it has hotels for four million people; the *uezd* towns, with populations of 10,000, have hotels for 100,000, and they are hardly ever empty. Communications are such that every peasant can reach his town in an hour and a half, and visits it frequently.
>
> But it's time we were on our way. We have to make a pretty big detour and pick up Katerina at Arkhangel'skoe.

The car set off again, turning towards the Prechistenskii Boulevard. Kremnev looked round with surprise; instead of the golden church of Christ the Saviour, shining like a Tula samovar, he caught sight of gigantic ruins, ivy-covered and evidently carefully preserved.

[19]I. V. Zheltovskii (1867–1959) was an architect involved in a plan for the intended reconstruction of Moscow after 1918.

Chapter 6

In which the reader will conclude that after 80 years they have not forgotten in Arkhangel'skoe how to make vanilla cheese-cake for tea.

The old statue of Pushkin stood out among the spreading limes of the Tver' Boulevard.

Erected on the spot where Napoleon had once hanged those alleged to have set fire to Moscow, the monument had been a silent witness to the fearful events of Russia's history.

Pushkin in effigy had seen the barricades of 1905, the nocturnal meetings and Bolshevik guns of 1917, the trenches of the peasant guard of 1932 and Varvarin's mortars of 1937; and there he still stood, still placidly contemplative, awaiting further developments.

Once, and only once, had he tried to intervene in the elemental turmoil of political passions, when he reminded the crowd at his feet of his tale of the fisherman and the fish. But they had not listened...

The car turned into the Grand Avenues to the west. The quiet and dusty Tver' and Post-Horse Streets had once stretched here. Their monotonous series of buildings had been replaced by the luxuriant limes of the Western Park, and only the domes of the cathedral and the white walls of the Shanyayskii University rose from the thickets, like islands amidst the waving green sea.[20]

Thousands of cars glided over the asphalt of the great West road. Newsboys and flower sellers dashed about the avenues, dotted with the yellow awnings of cafes and alive with a motley crowd; hundreds of aerocutters were so many small and large black specks against the immobile clouds, while heavy passenger air-flyers took off from the Western Aerodrome.

The car sped past the avenues of Petrovskii Park filled with the noise of children's voices, slid by the hothouses of the Silver Wood, turned sharply left and darted like an arrow down the Zvenigorod Highway.

They still did not seem to be out of the town. The same fine avenues extended to right and left, they glimpsed many white two-storey houses and sometimes whole architectural groups, only instead of flowers between the rows of mulberry and apple trees, there would be strips of vegetable gardens, rich pastures and strips of harvested corn.

> 'Well,' Kremnev turned to his companion, 'your decree on the abolition of urban settlements has evidently been observed only on paper. The Moscow suburbs have stretched far beyond All Saints'.

'I am sorry, Mister Charlie, but we are no longer in the town, this is typical north Russian countryside.' And he explained to the astonished Kremnev that the density

[20]The Moscow City People's University was founded by a bequest from General A. L. Shanyayskii (1837–1905) in 1908. In 1912, Chayanov had published a brief account of it under the title *Istoriya Miusskoi ploshchadi*. It was abolished in 1918.

of peasant population in the Moscow *guberniya* had become so high that the countryside had taken on an aspect unusual for rural settlements. 'Now the whole area for hundreds of miles around Moscow is a continuous agricultural settlement, intersected by rectangles of common forest, strips of co-operative pastures and huge climatic parks.

In areas of farmstead settlement, where the family allotment is no more than 8–10 acres, you will find peasant houses almost side by side for dozens of miles. They are only screened from one another by closely planted rows of mulberries or fruit-trees, which have now become common. Actually, it is time to give up the old-fashioned division between town and country, because all we have is a more or less dense distribution of the same agricultural population.'

> 'You see those groups of buildings,' Minin pointed far to the left, 'that stand out a little because of their size? They are the "townships" as they are now usually called. The local school, a library, a hall for entertainments and dances, and other community facilities. A small social centre. Today's towns are just the same social centres of the same rural life, only bigger. And now we have arrived.'

The forest opened out to reveal, in the distance, the graceful silhouette of the Arkhangel'skoe palace.

The car made a sharp turn and, rumbling over the gravelled drive, passed under the wide gate crowned with a trumpeting archangel to pull up by the orangery wing, scattering as it did so a whole flock of young girls who had been playing hoop-la.

White, pink and blue dresses surrounded the newcomers and a girl of about seventeen threw herself with a cry of joy into the arms of Alexei's Companion.

> This is Mister Charlie Mann. My sister, Katerina.

A moment later the guests were seated on the lawn of Arkhangel'skoe park, amongst statuary busts of ancient philosophers, by a purring samovar set on a table; on the linen tablecloth, mountains of rosy cheese-cakes were piled up.

Alexei was gorged with the cakes, seductive, rich, vanilla cheese-cakes, with fragrant tea; he was bestrewn with flowers and questions on American morals and customs and whether they knew how to write poetry in America. Afraid of getting caught out, he counter-attacked, asking them two questions for every one that was put to him.

Tucking into cheese-cake after cheese-cake, he learnt that Arkhangel'skoe belonged to the Brotherhood of Saints Flor and Lavr, a kind of lay monastic order, whose numbers were recruited from talented young boys and girls excelling in the arts and sciences.

Down the suites of the old palace rooms and the lime-tree avenues of the park, once celebrated for Pushkin's visits and the brilliant, gallant life of Boris Nikolaevich Yusupov, with his Voltaireianism and his colossal library devoted to the French revolution and gastronomy, there now echoed the voices of this young band of bearers of Prometheus' creative flame, sharing life's labours and joys.

The Brotherhood had a couple of dozen magnificent, huge estates scattered over Russia and Asia, equipped with libraries, laboratories, picture galleries, and represented as far as could be seen one of the most powerful creative forces in the land. Alexei was astounded by the strictness of the almost monastic rule and the radiant, ringing joy which permeated all around; the trees, the statues, their hosts' faces and even the threads of the autumn cobwebs floating in the sun.

But all this was insignificant compared with the piercing eyes and melodious voice of Paraskeva's sister. Alexei was positively crazy about Utopian women.

Chapter 7

Convincing those so inclined that the family is the family, and ever shall be.

> 'Hurry, my friends, hurry', Minin urged his passengers, stacking Katerina's bundles and bags in the car, 'a general rain is due to start at 9 o'clock today and the meteorophores will be raising tornadoes in an hour'.

Although this tirade should have prompted Kremnev to surprise and Interrogation, it did not do so, since he was wholly absorbed by the process of wrapping up Paraskeva's sister in scarves.

When, however, the car was noiselessly gliding along the New Jerusalem Chausee, and fields, with thousands of peasants working hurriedly to gather in the last stacks of oats, flashed by on both sides, he could restrain himself no longer and asked his companions:

> Why the devil do you expend so much human labour on the fields? Surely your technology, which easily controls the weather, is capable of mechanising agricultural labour to free work hands for more skilled occupations?
>
> 'That's it, there's your American talking!' Minin exclaimed. 'No, dear Mister Charlie, you won't get far against the law of diminishing returns. Our harvests of more than three tons an acre are achieved by practically looking after each ear of grain individually. Agriculture has never been as manual as now. And at our population densities, this is no fad, but sheer necessity. That's why!'

He fell silent and increased the speed. The wind whistled past and Katerina's scarves fluttered over the car. Alexei watched her eyelashes, the lips glimpsed through folds of the scarf and felt he had known her since time began... A tender smile filled his heart with joy and comfort.

It was growing dark, and clouds had piled up in the sky, when the car approached the little houses on the steep banks of the Lama.

The extensive Minin family occupied several little houses built in the simple style of the sixteenth century; the surrounding palisade gave the estate the aspect of a small, ancient town. The barking of dogs and the hum of voices met them as they drove up to the gates. Some hefty lad seized Katerina in his arms, two little girls and a boy flung themselves on the parcels of provisions from Moscow, a girl in her teens was demanding some letter, and a grey-haired old man, who turned out to be Alexei Alexandrovich Minin, the head of the clan, took his namesake under his care and went to allocate him his room, wondering at the purity of his speech and the cut of his American clothes which vividly reminded him of the fashions of his distant childhood.

In about ten minutes, washed and combed, and feeling entirely confused, Alexei entered the dining room. At the flower-covered table a lively argument was in progress, and he had barely crossed the threshold when he was immediately chosen to arbitrate as a 'completely impartial' person. Two flat dishes were submitted to his expert judgment: one decorated with crayfish and black grapes, the other bearing a composition of a lemon, red grapes, and a cut glass goblet of wine. The two contenders, Meg and Natasha, demanded in the ringing tones of their fifteen-year-old voices a decision on whose still-life was 'more Dutch'.

Not without difficulty, Alexei solved his predicament by identifying one composition as an unrecognised authentic Jacob Pooter and the other a copy after Willem Kolff, and was rewarded with applause and a huge slice of cream cake, invented, as he was informed, by the professor of cookery herself, the absent Paraskeva.

Little Antoshka tried to learn from the American whether it were true that sperm whales could be caught with hook and line in Hudson Bay, but was immediately despatched to bed. An elderly lady pouring herself a third glass of tea, inquired whether Alexei had children and marvelled that his wife had allowed him to fly across the Atlantic. Much abashed by Alexei's assurance of the absence of any signs of spouse, she was preparing to continue the investigation, but at that point someone's hands tied a scarf over his eyes and he realised, or rather felt, that Katerina was standing behind him.

'Blindman's buff, blindman's buff', called the children, dragging him into the hall, and he was obliged to do a lot of chasing before he caught Katerina in his arms.

Alexei Minin appeared to restore order and, freeing Kremnev from his captivity, set him by the fireplace and spoke as follows:

> Tonight, when you are straight from your journey, I should not like to burden you with business talk. But all the same, tell me, what is an isolated American's first impression of our native land?

Kremnev launched into assurances of his surprise and delight, but the strains of a harpsichord terminated their conversation. Katerina had roped in her brother as accompanist and was singing Alexandrov's romance to Derzhavin's words:

Golden sterlet from the Sheksna,

Cream and borshch wait our desire

Wine and punch in the decanter

Tempt with sparks of ice and fire.

There followed 'The Peacock', and the duet 'The young couple's new home', and Kremnev felt she was singing for him, that she did not want him to attend to anyone else.

The 'general' rain due from 9 till 2 in the morning was pouring down outside in dense streams. The room became even cosier, the peace of the family contentment was warmed by the dying fire. Aunt Vasilisa was telling Natasha's fortune from the cards; the young people were planning how best to show the American round Yaropolets and White Kolp'. However, Alexei Minin categorically announced that he had bespoken Mister Charlie for the whole morning and that it was time for all to go to bed.

Kremnev asked Meg for her textbook on world history to read before going to sleep and, accompanied by Katerina and the hellish rain, made his way to the wing allotted as his quarters.

Chapter 8

Historical.

Katerina made Kremnev's bed, and putting some gingerbread biscuits and dates on the table, looked at him fixedly and suddenly asked:

Are they all like you in America?

Alexei was dumbfounded, and the equally confused young lady ran off, banging the door; through the steamed up window panes the retreating light of her lamp flickered away.

Kremnev was alone.

It took him a long time to recover from the impressions of that extraordinary day, all the marvels of which were nevertheless surpassed by the bewitching form of Paraskeva's sister.

Pulling himself together, Kremnev undressed and opened the history textbook.

At first he could understand nothing; it expounded at length the history of Yaropol'e *volost'*, then the history of Volokolamsk, and that of the 'Moscow *guberniya*, and only the final pages of the book dealt with the history of Russia and the world.

With mounting excitement, Kremnev swallowed page after page, gobbling up historical events together with Katerina's gingerbreads.

Coming to the events of his own era, Kremnev learnt that the world unity of the socialist system did not last long; the prevailing concord was very quickly disrupted by centrifugal social forces. No socialist dogma could expunge the ideas of military revenge from the German soul, and over the trifling matter of sharing out the Saar coal, the German trade unions compelled Radek, their president, to mobilise the metal workers and coal miners and occupy the Saar basin by armed force until the problem could be solved by the Congress of the World Economic Council.[21]

[21]K. B. Radek (pseudonym of Sobelson) (1885–1939) was a member of various European Social-Democratic parties before joining the Russian Communist Party.

Europe again broke down into its elements. The structure of world unity collapsed and a new, bloody war started, during which in France the aged Hervé succeeded in carrying out a social revolution and establishing an oligarchy of senior officials of the Soviets.[22] After six months of bloodshed, peace was restored by the joint efforts of America and the Scandinavian Union, but at the price of dividing the world into five closed economic systems—German, Anglo-French, American-Australian, Japan–China and Russian. Each of these isolated systems was allocated pieces of territory in all climatic zones, sufficient to ensure their economic existence, and there-after, while preserving a community of culture, they developed altogether different political and economic ways of life.

In Anglo-France the oligarchy of Soviet bureaucrats degenerated very quickly into a capitalist system. America returned to parliamentarianism and partly denationalised its production, though basically maintaining its state-run agriculture. Politically, Sino-Japan rapidly reverted to monarchism, but preserved peculiar forms of socialism in the economy. Only Germany continued the regime of the 1920s totally uncorrupted.

The history of Russia was as follows. While the system of Soviets was piously maintained, it was found impossible to complete the nationalisation of agriculture.

The peasantry, representing a massive social force, proved hard to communise, and within five or six years of the end of the civil war, peasant groups began to exert an impressive influence both in the local soviets and also in the All-Russia Central Executive Committee.

Their strength was considerably weakened by the policy of appeasement of the five Socialist Revolutionary parties, which frequently undermined the influence of purely class-based peasant associations.

For ten years no single tendency had a stable majority at the Congresses of Soviets; in practice, power belonged to the two communist factions, always able at moments of crisis to come to an arrangement and bring masses of workers out into the streets in impressive demonstrations.

But the conflict which arose between them over the decree on the compulsory introduction of eugenics created a situation from which the right-wing communists emerged victorious, but at the price of forming a coalition government and changing the constitution to give parity of voting power to peasants and townsmen. The re-elected soviets produced a new Congress of Soviets with an absolute preponderance of purely class-based peasant groupings, and from 1932 there had been a permanent peasant majority in the All-Russia Central Executive Committee and at congresses, and the system had been slowly evolving towards an increasingly peasant regime.

However, the dualist policy of the socialist revolutionary intelligentsia and the tactics of street demonstrations and uprisings had more than once shaken the foundations of the Soviet constitution, obliging the peasant leaders to retain the coalition formula in the composition of the Council of People's Commissars; a

---

[22]Gustave Hervé (1871–1944) was a French socialist.

procedure encouraged by the numerous attempts at a reactionary coup by certain urban groups. In 1934, after a rising aimed at establishing an intellectual oligarchy on the French model, and supported for tactical reasons by the metal and textile workers, Mitrofanov organised the first purely peasant-class Council of People's Commissars and at the Congress of Soviets carried through the decree on the destructions of towns.

Varvarin's rising of 1937 was the last manifestation of the political role of the towns before they dissolved in the sea of peasantry.

The General Land Plan was adopted and implemented in the 1940s; the meteorophores, a network of magnetic power stations, using A. A. Minin's method for controlling the weather, were set up. The sixties were marked by violent religious disturbances and an attempt by the Church to seize lay power in Rostov district. His eyelids were drooping and his exhausted brain could take no more.

Kremnev put the light out and closed his eyes. But Katerina haunted him for a long time, and it was late into the night before he fell asleep.

Chapter 9

Which young lady readers may skip, but which is recommended for the particular attention of members of the Communist Party.

The sole embellishments of Alexei Alexandrevich Minin's ample study were bookshelves gleaming with the dull gilt of leather bindings and a few Vladimir-Suzdal' icons.

A portrait of his father, a well-known professor in Voronezh and later in Constantinople, completed the room's appointments in deep indigo tones throughout.

> 'I must acquaint you', his hospitable host began, 'with the realities of life in these parts, for without such knowledge you will not be able to understand the significance of our engineering installations or the very possibility of their existence. But, Mister Charlie, I really am at a loss where to begin. You are, as it were, a visitor from the other world, and it is difficult for me to judge what aspect of our life is particularly new and unexpected to you'.
>
> 'I should like to learn', Kremnev said, 'about the new social principles on which Russian life was built after the peasant revolution of the thirties; without these, I think, it would be difficult to understand all the rest'.

His companion was slow to answer, as if considering how to tell his story.

'You want to know about those new principles', he began, 'which peasant power introduced into our social and economic life. In fact, we had no need of any *new* principles; our task was to consolidate the *old*, centuries-old, principles on which from time immemorial the peasant economy had been based.

Our only aim was to assert these great ancient principles, to enhance their cultural value, to transform them spiritually and to endow their embodiment with a social and technical organisation which would enable them not only to display that peculiarly passive resilience which has characterised them for ages, but also to have active strength, elasticity and, if you like, striking power.

Our economic system, like that of ancient Rus', is founded on the individual peasant farm. We considered it, and still do so, the ideal model of economic activity. In it, man confronts nature; in it, labour comes into creative contact with all the forces of the cosmos to produce new forms of existence. Every workman is a creator, each manifestation of his individuality represents the art of work.

I need not tell you that living and working conditions in the countryside are the healthiest, that the peasant farmer's life has the most variety etc., or any of the other self-evident facts. This is man's natural condition, from which he was exiled by the demon of capitalism.

However, in order to establish a twentieth-century nation on the basis of the peasant farm and the peasant way of life, it was essential for us to solve two basic organisational problems.

The *economic* problem was to create a national economic system which would be based on the peasant farm and retain its leading role, and also economic machinery which in its functioning would not be technically inferior to any other conceivable mechanism and would sustain itself automatically without recourse to state compulsion of a non-economic nature.

The *social*, or, if you like, the cultural problem, was how to organize the social existence of the broad masses so as to preserve, in conditions of scattered rural settlement, the highest forms of culture, which had long been the monopoly of urban civilisation, and to ensure that cultural progress in all spheres of mental and spiritual life would be at least equal to that under any other regime.

And mark, Mister Charlie, that we not only had to solve both these problems, but also had to think deeply about the means for solving them. For us not only *what* we wanted to achieve was important, but also *how* it might be achieved.

The period of state collectivism, when ideologists of the working class were realising their ideals on earth by the methods of enlightened absolutism, had brought Russian society into such a condition of anarchic reaction that no new regime could be introduced by edict or decree sanctioned by the force of the bayonet.

Besides, any idea of a monopoly in the sphere of social creation was alien to the very spirit of our ideologists.

Our leaders, who were not proponents of a monist outlook, thought or action, had, for the most part, a mentality capable of accommodating a pluralistic view of the world, and so believed that life is worthwhile when it permits the full realisation of all the possibilities, all the new departures, contained in it.

In brief, we had to resolve these problems in such a way as to leave to any initiative, any creative effort, the chance of competing with us. We endeavoured to conquer the world by the inner strength of our cause and our organisation, by the technical superiority of our organising principle, not by smashing in the face of anyone who thought otherwise.

Apart from that, we have always recognised that the State and its apparatus are by no means the sole manifestation of social life, and so, in our reforms, we have for the most part relied on social methods of solving problems, not measures of state coercion.

Still, we were never blindly addicted to principle; and when outside violence threatened our cause, and expediency made us remember that state power was in our hands, our machine guns worked no worse than those of the Bolsheviks.

Of the two problems I have outlined, the economic one presented us with no particular difficulties.

You know, of course, that in the socialist period of history the peasant farm was considered as something inferior, as that proto-matter from which the 'higher forms of large-scale collective economy' should crystallise out. Hence the old idea of grain and meat factories. It is now clear to us that this view has not so much a logical, as a genetic origin. Socialism was conceived as the antithesis of capitalism; born in the dungeons of the German capitalist factories, nurtured in the minds of an urban proletariat haunted by forced labour, by generations that had lost the habit of any individual creative work or thought, it could conceive of an ideal system only as the negation of the system it knew.

Hirelings themselves, the workers, in constructing their ideology, made, servitude an article of faith of the future system, and created an economy in which all were performers and only a few individuals possessed the right to creative activity.

But forgive me, Mister Charlie, I digress. Well then, the socialists conceived of the peasantry as proto-matter, since they had economic experience only in the field of manufacturing industries and could reason only in terms of their own organic experience.

But to us it was perfectly clear that, in social terms, industrial capitalism was merely a pathological, monstrous condition by which manufacturing industry, owing to its peculiarities, had been affected, and by no means a developmental stage of the economy as a whole.

Thanks to its fundamentally healthy nature, agriculture had avoided the bitter cup of capitalism and we had no need to direct our developmental process into that channel. Particularly, since the German socialists' collectivist ideal itself, in which the working masses were conceived as the executors of agricultural work in accordance with State directives, seemed to us socially very imperfect compared with a system of working peasant farmers, a system in which labour is not separated from creative management, in which the freedom of individual initiative allows each human being to develop his full spiritual potential, while enabling him also when necessary to make use of the whole might of the collective large-scale economy and of public and state organisations.

Already at the beginning of the twentieth century the peasants had collectivised and organised in large co-operative enterprises all those branches of their productive activity where big economic units scored over small ones; in its present form this gives us a very stable and technically advanced structure.

This is the buttress of our economy. It was much more difficult to deal with the manufacturing industries. There it would have been foolish, of course, to expect to restore family production.

For most sectors, artisan activity and cottage industry were out of the question in the overall majority of branches of production, given the present level of factory techniques. But here, too, peasant initiative got us out of our difficulties; peasant co-operatives with their guaranteed and very large market, nipped in the bud any chance of competition for most products.

Admittedly, we aided them a little in this, by breaking the backs of capitalist factories with swingeing taxation which was not applied to co-operative enterprises.

However, we still have private initiative of capitalist type; in those areas where collectively managed enterprises are ineffective, and in those cases where an organising genius can overcome, thanks to advanced technology, the effects of our Draconian taxation. We don't even try to kill it, for we consider it essential to preserve for our comrade co-operators a degree of threat from permanent competition, and thus save them from technical stagnation. We know that our present-day capitalists, too, have certain shark-like propensities, but then we all know that there are sharks in the sea so that the other fish should not doze.

But this residual capitalism is very tame and so, indeed, is the co-operative industry, though more inclined to rebel, for our labour laws protect the worker from exploitation even better than the laws of the dictatorship of the proletariat under which a colossal share of surplus value was absorbed by the herds of officials in the Chief Administrations and the Administrative Centres.

Well, apart from that, we got rid of all the economic enterprises, only keeping for the State monopolies in forestry, petroleum and coal; for having control of fuel, we control the whole of manufacturing industry.

If I add, that our trade is predominantly in the hands of co-operatives and that state finance is based on the taxation of profits of enterprises employing hired labour and on indirect taxes, that will complete for you the general picture of our economy'.

'Excuse me, I did not mishear you,' Kremnev inquired, 'you did say your state finances are based on indirect taxes?'

'Quite right,' Alexei Minin smiled, 'you are surprised by such an "old-fashioned" method; it jars in comparison with your American income tax system. But I assure you our indirect taxes are just as progressive as your income brackets. We have enough knowledge of the composition and mechanics of the consumption of all our social strata to arrange the incidence of taxes mainly not on necessities, but on luxury goods, and, besides, the differences between average incomes are not so great. The advantages of indirect taxation is that it does not cost the taxpayer a minute of his time. Our system is altogether so arranged that you may live for years in, say, Volokolamsk *uezd* and not once be reminded of the existence of the State as an oppressive power.

That is not to say that our State organisation is weak. Far from it. It is simply that we stick to methods of operation in which the State refrains from grasping the citizen by the scruff of the neck.

In former times it was very naïvely supposed that the only way to manage the economy was by giving orders, subordinating, nationalising, forbidding, commanding and issuing warrants, in a word, fulfilling the national economic plan by means of pliant performers.

We have always assumed, and can now prove by forty years' experience, that these barbarian trappings, burdensome both to ruler and ruled, are about as necessary now as Zeus's thunderbolts to maintain modern morality. We have long abandoned such methods, just as catapults, battering rams, the semaphore telegraph and the Kremlin walls were abandoned in their own day.

We have much more subtle and effective means of indirect influence, and can always create such conditions for any branch of the economy as to make it meet our requirements.

Later, in a number of practical examples, I shall try to demonstrate to you the strength of our power in the economic sphere.

But now, in concluding my outline of the economy, let me draw your attention to two organisational problems which are particularly important for an understanding of our system.

The first of these is the problem of stimulating economic activity. If you think back to the era of state collectivism and its concomitant reduction of the economy's productive forces, and reflect on the origins of this phenomenon, you will see that its principal causes lay by no means in the state economic plan.

Let us give their due to the organisational skills of Yu. Larin and V. Milyutin: their projects were very well conceived and elaborated in detail.[23] But it is not enough to elaborate, one must also implement; economic policy is the art of implementation par excellence, not the art of planning.

You not only have to design a machine, you also have to find the materials suitable for its construction and the motive power to drive it. You cannot build the Eiffel Tower of straw, nor will two workers put a rotary engine in motion by hand.

When we look at the pre-socialist world, we see that its complex machinery was driven by the power of human greed, by hunger; each component, from the banker to the lowest labourer, drew personal profit from his effort in economic activity, and this interest stimulated his work. Everyone as part of the economic machinery was a motor that kept it in motion.

The Communist system put all participants in economic life on a state wage and so removed all incentive from their work. The fact of work existed, of course, but there was no work effort, for there was nothing to justify it. The lack of incentive affected not only the producers, but also the organisers of production; they, like all officials, were concerned to perfect the economic process, they were concerned with the precise and smooth working of the economic apparatus, but not with the results of its work. To them the appearance of the matter was more important than its practical results.

[23]Yu. Larin (pseudonym of M. A. Lur'e) (1882–1932) was an economist. V. P. Milyutin was the first Bolshevik Commissar of Arable Farming in November–December 1917, then a member of the Supreme Council of the National Economy.

When we took over the management of the economy, we immediately put in motion all the mechanisms which stimulate private economic activity—piece rates, bonuses for managers and premium prices for those products of peasant farming which it was essential to develop; for example, mulberry farming in the north.

In restoring private economic incentives we had, of course, to be prepared to tackle inequalities in the distribution of the national income.

In this sphere, the main part of the work was already accomplished, since three-quarters of economic activity in regard to industry and trade had been taken over by co-operatives; nevertheless, we were constantly faced with the problem of democratising the national income.

We first turned our attention to reducing the share of unearned income—the principal measures we took were taxing rent in agriculture and the abolition of joint-stock companies and of private finance houses.

I am using the old economic terminology, Mister Charlie, so that you understand what I am talking about, it is still used in your country, but here... I wonder whether today's young people know them at all. That was our solution to the economic problem.

The social problem we found much more complex and difficult: how to preserve and develop culture while abolishing towns and high unearned incomes.'

'But now they are calling us to dinner', Alexei's partner stopped his story as he saw through the window that Katerina, with evident pleasure and determination, was striking an iron gong hanging in the midst of the extensive courtyard.

Chapter 10

In which the fair at White Kolp' is described and the author's complete agreement with Anatole France's dictum that a story without love is like fat without mustard is explained.

We know, from the surviving 'Book of Expenditure of the Patriarch's Office', that at the beginning of the eighteenth century the Most Holy Patriarch Adrian was daily supplied at table with 'porridge, fresh pike in brine, white sturgeon soup, a dish of stellate sturgeon, cabbage soup and underbelly of fish, fish joints and horseradish, a choice cut of white sturgeon, a pasty with meat' and at least a further twenty dishes in breath-taking quantities and of excellent quality. Comparing this feast of ancient times with the Utopian table in the hospitable house of the Minins, we have to admit that the Patriarch's fare was a little more sumptuous. But only a little... Paraskeva had arrived from Moscow and at her bidding such quantities of open pasties and pies, stuffed carp and carp in sour cream and other delicacies appeared on the table that its legs would surely have bent had they been but a little less sturdy; while the socialist politician Kremnev fully expected all the participants in the meal to be dead by the evening from over-indulgence. However, the national dishes prepared for the American's edification vanished rapidly and without trace to the accompaniment of a rising chorus of praise for Paraskeva, who modestly re-addressed it to 'The Russian Cuisine' compiled by Mr. Levshin in 1818.[24]

After a nap in the hay-loft in the traditional Orthodox manner, the young people carried Kremnev off to the fair in White Kolp'.

As Kremnev and his companions walked by the banks of the Lama, the shadows of clouds drifted over the mown meadows, flowering rowan bushes made yellow patches along the road, and in the dense autumn air spiders' webs floated about.

Katerina walked along with head held high; a gust of wind outlined clearly the contours of her body against the blue depths of sky beyond the river. Meg and Natasha were picking flowers. There was a smell of autumn wormwood in the air.

There's the main road!

They turned onto the highway lined with weeping birches and in the distance they could see the domes of White Kolp church.

Carts, bright-painted like trays and packed full of lads and lasses cracking nuts, overtook the travellers. The strains of a popular song rang out over the road.

Little pigeon on the roof-top

Soon you will be pigeon pie;

Come and help me, O my darlings,

Choose the one with whom I'll lie.

Kremnev was surprised by the almost complete absence of any difference between his companions and those who met or overtook them. The same dress and the same Moscow manner of speech and expression. Gaily, and with evident enjoyment, Paraskeva jokingly rebuffed the pleasantries of the passing lads, while Katerina simply hopped into one of the carts, kissed all the girls sitting in it, and took away a bagful of nuts from a surprised youth as she popped a chunk of banana into his mouth.

The fair was in full swing.

Mountains of Tula gingerbreads were piled on a stall, both the crisp variety and that with candied fruit; there were Tver mints in the shape of fishes and of generals and there were the juicy, multi-coloured Kolomna fruit-drops.

The passing of centuries had changed nothing in rustic delights, and only a careful observer would have noticed the considerable quantities of preserved pineapple, the bunches of bananas and the extraordinary abundance of good chocolate.

As in the good old days, little boys whistled on gilded clay cockerels, just as they had done in the days of Tsar Ivan Vasilevich and in Novgorod the Great. A double accordion played a fast polka.

In a word, everything was fine.

---

[24]V. A. Levshin (1746–1826), a prolific translator and author, had issued a six-part 'Cook's Calendar' in 1808.

Katerina, entrusted with the education of Mister Charlie, led him into a large white tent, and without comment pronounced: 'There!'.

The walls of the tent were hung with paintings of both old and modern schools. Kremnev joyfully recognised some old friends: Venetsianov, Konchalovskii, Rybnikov's 'St. Hieronimus', the Novgorod 'Elijah' from the Ostroukhov collection, as well as hundreds of new, unfamiliar paintings and sculptures which vividly put him in mind of yesterday's conversation with Paraskeva.[25]

He stopped before 'The Christ child' of Giampietrino which had held him spellbound in the Rumyantsev Museum and, at the risk of destroying his incognito, exclaimed: 'But however did these get to White Kolp' fair?'.

Paraskeva hastened to explain that the booth was a touring exhibition from Volokolamsk Museum to which some Moscow paintings had been temporarily loaned.

The densely crowded visitors, looking attentively at the exhibits and exchanging comments, showed Kremnev that the representational arts had become firmly established in everyday peasant life and met with an informed understanding. Of the latter he was further convinced by the rapid rate of sale at the door of the 132nd edition of P. Muratov's *A History of Painting in a Hundred Pages* and the booklet *From Rokotov to Ladonov*,[26] the dustjacket of which told him that Paraskeva not only could talk about painting, but also wrote books.

In the next tent, peasant women clustered round samples of ancient Russian embroidery, and two lads were sizing up a Boulle cabinet.

Soon the exhibition began to empty; the sound of voices and the ringing of a bell heralded the start of the eurhythmics. This was followed by a knuckle-bone match, a hurdle race and other Yaropole *volost'* championships. Huge blue posters promised 'Hamlet' by Mr Shakespeare at seven o'clock, to be performed by a company from the local co-operative union.

But it was time to hurry home and call in at the bee-garden for some honey on the way. So, abandoning the celebrations, the company only found time to visit the panopticum exhibited by the cultural and educational department of the Guberniya Peasant Union.

Along the walls stood wax busts, portraying historical personages; a panoramic display acquainted the spectators with the major events of national and world history and with the wonders of tropical lands.

Animated exhibits showed Julius Caesar at the Rubicon, Napoleon on the walls of the Kremlin, the abdication and death of Nicholas II, Lenin speaking at the Congress of the Soviets, Sedov routing the typists' rebellion, the basso cantante Chaliapin and the bass Gaganov.

---

[25]A. G. Venetsianov (1780–1847) is regarded as the first Russian naturalist painter and the originator of an important genre school.

P. P. Konchalovskii (1876–1942) was a realist painter.

I. S. Ostroukhov was a late nineteenth-century art collector and patron. His collection was nationalised after 1917.

[26]F. S. Rokotov (1735 or 1736–1809) was a portrait painter.

'Look, but this is your portrait!' Katerina exclaimed.

Kremnev was rooted to the spot. A bust, reminiscent of photographic postcards, stood before him on canvas under a glass. Underneath, the inscription read: 'Alexei Vasil'evich Kremnev, member of the college of the World Economic Council, the suppressor of Russia's peasant movement. According to medical opinion, most probably suffered from persecution mania; degeneration is clearly evidenced by the asymmetry of the face and the structure of the skull.'

Alexei blushed deeply and did not dare look at his companions.

'It's marvellous! An amazing likeness; even the jacket is like yours, Mister Charlie,' cried Nikifor Minin.

Everybody became oddly embarrassed and they left the panopticum tent in silence.

They were in a hurry to get home, but Katerina dragged Kremnev off to the bee-garden for honey. The path cut across some cabbage patches. The intense splashes of their firm, almost blue, heads accentuated the blackness of the earth. Two women, strongly built and wearing white dresses with pink polka-dots, were cutting the ripe ones and throwing them into a two-wheeled cart.

For the first time in his Utopian journey, Alexei, shaken by the confrontation with his waxen double, fully realised the seriousness and hopelessness of his position.

The original sin of his pretended birth bound him hand and foot, while his real name evidently branded him as an outlaw in the land of peasant Utopia.

But this surrounding world of cabbage patches, blue distances and the red rowan clusters was no longer alien to him.

He felt a new and precious bond with it, an affinity even closer than to the socialist world he had left; and the cause of this closeness, Katerina flushed from the rapid walk, was at his side, spellbound, clinging to him imperceptibly.

They slackened their pace as they descended the steep slope of an old stream. Alexei touched her hand and their fingers intertwined.

The crowns of apple-trees, their boughs bent as in an old Japanese print and laden with fruit, rose in neat rows above the utterly black, freshly tilled earth. Large, red, fragrant apples and the trunks white with lime filled the air with the smell of fertility; the scent seemed to him to soak its way through the pores of his companion's bare hands and neck.

So began his Utopian love.

Chapter 11

Very similar to Chap. 10.

When Kremnev and his companion returned home, supper had long been waiting.

They were greeted coolly, and the company silently sat down to table. There was a feeling of anxiety in the house. The talk was of sinister developments in Germany, of the German Economic Council's demand for a revision of the Galician frontier. It seemed to Alexei that not only he, but also Katerina, felt somehow guilty.

There was also a certain coolness about Alexei Minin when, later in the evening, Alexei went to his study to continue the morning's conversation.

'In our talk today', the grey-haired patriarch began, 'I omitted another peculiarity of our economic system. In striving to democratise the national income it was natural for us to distribute the incoming revenue and it was equally natural to prevent the amassing of large fortunes.

For all its merits, this also had some disadvantages. First, the accumulation of capital was hindered. The distributed revenue was nearly all consumed, and our society's capacity for capital formation was, of course, especially after the abolition of private credit agencies, virtually nil.

So we had to make considerable efforts to get the peasant co-operatives and certain state bodies to take serious steps to create special social capital funds, and thus speed up the process of capital formation. Among these measures is our generous financial support for every kind of inventor and entrepreneur in new areas of economic activity.

Another consequence of the democratisation of national income was a considerable decline in art patronage and the number of people with no occupation, i.e. of the two substrata which to a great extent have nurtured art and philosophy.

But here, too, peasant initiative, with, it must be admitted, some encouragement from the centre, proved equal to the task.

In order to flourish, art needs a high degree of interest on the part of society and an active and generous demand for its works. We now have both; today in White Kolp' you saw the exhibition of paintings and the public's reaction. I should add that, at present, frescoes ordered for the rural building programme are counted in hundreds, if not thousands, of square yards; fine examples of painting can be found in the schools and public buildings of every *volost'*. There is also a considerable private demand.

Do you know, Mister Charlie, that there is a demand not only for the artists' works, but also for the artists themselves? I know of several cases of a *volost'* or *uezd* paying considerable sums on a long-term contract to an artist, a poet or a scientist, just for him to reside in their territory. Not unlike the Medicis and Gonzagas of the Italian Renaissance, you'll agree.

In addition, we give strong support to the Brotherhood of Flor and Lavr, of Olympios the icon-painter, and many others, with the organisation of which you are, I think, already familiar.

As you see, in speaking of the economic problem we have imperceptibly reached the social problem; and this we found more difficult and complicated.

We had to solve the problem of the individual and society. We had to build a human society in which the individual could feel completely unfettered, while society takes care of the common interest by methods invisible to the individual.

And we never allowed society to become an idol or made a fetish of the state.

Our ultimate aim was always the enrichment of human life, an integrated human personality; everything else was a means to that end. We consider society and the state as the most powerful and essential of these means, but never forget that they are no more than means.

We are particularly cautious about the state, which we use only when necessity dictates. The political experience of many centuries unfortunately shows that human nature almost invariably remains human nature; the improvement of manners proceeds with the speed of geological processes. Strong natures, possessed of the will to power, will always seek to make for themselves a full, integrated and meaningful life, but on the ruins of the lives of others. We can fully believe that the life of a Herodes Atticus, a Marcus Aurelius or a Vasilii Golitsyn was scarcely less rich or profound than the lives of our best contemporaries.[27] The only difference is that, then, such lives were lived by a few individuals, while today they are the lot of tens of thousands—and of millions, I hope, in the future. All social progress merely consists in extending the circle of those who drink at the fountain-head of culture and life. Nectar and ambrosia have ceased to be the food of the Olympians alone, they now enrich the homes of humble villagers.

Society has been unswervingly developing in the direction of this progress for the past two centuries and, of course, it has the right to defend itself. When some strong natures, or even whole groups of strong natures, hinder this progress, then society may defend itself, and the state is a well-tried tool for this purpose.

It is also not a bad instrument for a whole number of technical needs.

You'll ask how the state is organised here? As you know, the form of states develops not logically, but historically. This, partly, accounts for many of our existing institutions. The system, as you know, is a Soviet system, a system of peasant councils. On the one hand, it is a heritage of the socialist period of our history; on the other, it has many merits. It should be noted that, among the peasants, it long predates October 1917, when it essentially existed in the management of the cooperatives.

The basic principles of this system are probably known to you, and I shall not dwell on them.

I'll only say that we value in it the idea of the direct responsibility of all authorities to the social groups or organisations which they serve. Only the courts, the organs of state control and some bodies in the communications field are exempt from this rule and are run entirely by central authority.

We see a particular value in the division of legislative power; under this, questions of principle are decided by the Congress of Soviets, having been first considered at local level, and I stress considered, since the law forbids imperative mandates for the delegates. The actual mechanics of legislation are entrusted to the Central Executive Committee and, in some cases, to the Council of People's Commissars.

This method of administration involves the masses very closely in state activity and at the same time ensures flexibility of the legislative machinery.

---

[27]Vasili Golitsyn (1643–1714) was an outstanding Russian statesman who dealt mairay with foreign affairs.

But in any case we are no sticklers for formality in putting even such a mechanism into effect, and local variants are readily allowed; thus we have parliamentarianism in Yakutsk *oblast'*, while the monarchists of Uglich have set up a local prince, restricted, it is true, in his power, by the local Soviet of Deputies; on the other hand, in the Mongolo-Altai territory, a "governor-general" appointed by the central authority rules alone.'

'Excuse me,' Kremnev interrupted, 'the Congresses of Soviets, the Central Executive Committee and the local Soviets of Deputies—all these merely legitimise power. But on what is real power founded here?'

'Ah, my very dear Mister Charlie, our fellow citizens have almost forgotten about such concerns, for we have stripped the state of virtually all social and economic functions, and the ordinary man has hardly any contact with it.

And, generally, we consider the state to be an outdated mode of organising social life and nine-tenths of our work is done by social methods; it is they that are characteristic of our system: various societies, co-operatives, congresses, leagues, newspapers, other organs of public opinion, academies and, finally, clubs—that is the social fabric which constitutes the life of the nation.

And it is here, in managing it, that we meet exceedingly complicated organisational problems.

Human nature, alas, inclines to go back to the simple life; left to itself, with no social contact or external psychological stimulation, it gradually fades and dissipates its content. A man cast into the forest goes wild. His mind is gradually emptied the content.

Quite understandably, therefore, in fragmenting the towns which had been the spring-heads of culture for many centuries, we were much afraid that, scattered amidst the forests and fields, the rural population would putrify and lose its culture as it did in the Petersburg period of our history.

To combat this putrescence, we had to provide for social drainage.

Even greater fears attached to the problem of the future development of culture, of that creative activity which, again, had been due to the towns.

We were constantly dogged by the thought: are the higher forms of culture possible with a population scattered in the countryside?

The era of landlord culture of the 1820s, which produced the Decembrists and gave Pushkin to the world, told us that, technically, all this was possible.

It remained only to find sufficiently powerful technical means.

We made every effort to create a perfect communications network, found ways of making the population use it, at least to go to the local centres, and we flung into these centres all the elements of culture we possessed: the *uezd* and *volost'* theatre, the *uezd* museum and its *volost'* branches, people's universities, every kind of sporting activity, choral societies; everything, even the church and politics, was thrown into the countryside to raise the level of its culture.

We risked much, but for several decades we kept the countryside in a state of psychological tension. A special League for the organisation of Public Opinion set up dozens of mechanisms for stimulating and maintaining the social dynamism of the masses; I'll even confess that draft laws threatening peasant interests were specially introduced in legislative bodies, in order to arouse peasant social consciousness.

But perhaps the law on obligatory travel for young men and women, and the two-year conscription for military and labour service played the biggest part in bringing our fellow citizens to the fountain-head of culture.

The idea of journeying, borrowed from the medieval guilds, allowed a young man to see the world and expanded his horizons. He was polished still further during his military service. Quite honestly, we attached to it almost no strategic significance; in case of foreign aggression, we have means of defence more powerful than all the guns and cannon taken together; and if the Germans carry out their threats, they will soon discover that.

But the educational value of labour service as a moral discipline is immeasurable. Sports, rhythmic gymnastics, eurhythmics, factory work, marches, manœvres, navvying—all this helps to mould our citizens and, verily, militarism of this sort redeems many a sin of the old militarism. There remains the development of culture; I have already told you something of what has been done in this area.

The main idea which eased the solution of the problem for us was the concept of artificially selecting talented individuals and promoting the organisation of their lives.

In the past there was no scientific understanding of human life; no attempt was made to formulate a science either of its normal development or of its pathology; we knew nothing of the diseases in men's biographies, we had no notion of the diagnosis and treatment of unsuccessful lives.

Those with a feeble stock of potential energy often burnt out like candles and perished, crushed by circumstances; individuals with enormous strength failed to use one tenth of their energy. Today we understand the morphology and dynamics of human life; we know how all the forces inherent in a man can be developed. There are special associations, large and powerful, which have millions of people under their observation; you can rest assured that today not a single talent can be lost, not a single human potentiality will vanish into oblivion...'

Shaken, Kremnev leapt up.

> But this is horrible! This is the worst of all tyrannies! These associations of yours that resurrect the German anthroposophists and the French freemasons are the equal of any state terror. Why have a state, indeed, when your whole set-up is no more than a sophisticated oligarchy of a couple of dozen very clever and ambitious men.

'Calm yourself, Mister Charlie; firstly, any strong personality will feel not the slightest hint of our tyranny; and, second, you would have been right about thirty years ago — then our system was in oligarchy of gifted enthusiasts. Today we are able to say: "Now let Thy servant depart". The peasant masses have learnt to play an active part in shaping the country's public opinion; and if, morally, we are in power, it is only because, as the Germans say, "Und der Kaiser absolut, wenn er unser Wille tut".

Let any organisation, no matter how strong, try to go against the opinion of those who live and reason in the cottages of Yaropolets, Murinovo and thousands of other places, and it'll immediately lose its influence and moral authority.

Believe me, once the people's spiritual culture has reached a certain very high level, it maintains itself automatically and acquires internal stability. Our task is to ensure that each *volost'* should have its own creative cultural life, that the life of Korchevskii *uezd* should not differ qualitatively from that of Moscow *uezd*; having achieved that we, the enthusiasts of rural renaissance, we, the followers of the great prophet, A. Evdokimov, can calmly go to our graves.'

The old man's eyes burnt with the fires of youth; Kremnev was facing a fanatic.

Kremnev rose and turned to Minin with evident irritation.

All right, you say that the free human individual, the whole state, duty, society, all these are means, Then, in your opinion, is a social criterion for judging their own actions necessary or superfluous for your citizens?

For convenience in managing the state, and as a characteristic of the masses, it is desirable, but from an ethical viewpoint it is not essential.

And you preach such views openly?

I wish you would understand, my dear fellow,' the old man flared up, 'that if we don't have theft, it is not because everyone recognises it is wrong to steal, but because no thought of stealing can arise in our fellow citizens' heads. In our view, if you like, conscious ethics are immoral.'

All right, but what about you, aware of all this, you, the heads of spiritual life and of society? What are you? Augurs or fanatics to duty? What ideas inspired you to work to create this peasant paradise?

'Unhappy man!' Alexei Minin exclaimed, drawing himself up to his full height. 'What inspired us, and thousands like us, to work? Ask Scriabin what inspired him to compose "Prometheus", what made Rembrandt create his fabulous visions. It was the sparks of Prometheus' creative flame, Mister Charlie! You want to know what we are, augurs or fanatics of duty? Neither one, nor the other — we are artists.'

Chapter 12

Describing the considerable improvement in Moscow's museums and places of entertainment and cut short by an exceedingly unpleasant surprise.

Next morning, Kremnev became conscious of a still greater coolness on the part of the inhabitants of the White Kolp' townlet. There was a reluctance about Alexei Minin as he explained to him the structure of the meteorophore system.

According to his account, a link between the state of the weather and the intensity of magnetic lines of force had been observed as early as the nineteenth century. The passing cyclones and anti-cyclones always had their magnetic counterpart. What was not altogether clear was which was the determining factor: did the weather determine the state of the magnetic field, or the magnetic field determine the weather? Analysis confirmed the second hypothesis; a network of 4,500 magnetic power stations was set up and this allowed almost complete control of the magnetic field, and thus of the weather. Minin went on to describe the meteorophore, but noticing Alexei's weak mathematics, abruptly broke off his explanation...

At lunch, Kremnev felt that his position was insupportable and disaster approached; he was therefore immeasurably pleased when Paraskeva asked him to accompany her to Moscow for shopping and to attend a liturgical concert on the Moscow bells.

The light aerocutter set them down at the central aerodrome by 3 o'clock and, as a good hour remained before the concert, Paraskeva proposed that they take a look at the Moscow museums; she said they had now succeeded in doing what the great revolution had been powerless to achieve, and had rescued all their treasures of the spirit from the vaults and the museum routine.

Even the Historical Museum was finally dragged from under its bushel in 1970!

The new building of the Rumyantsev Museum occupied a whole huge block, from the Manège to the Znamenka, the front facing the Alexandrov Gardens. Down the long galleries, there opened before them the wonderful visions of Sandro Botticelli, Rubens, Velasquez and other old masters, Japanese enamels and Chinese ones he had not previously known. All these, as Paraskeva explained, were gifts of other lands, exchanged with museums in the West and the East for Novgorod and Suzdal' icons. Rapidly scanning through dozens of rooms, Alexei found himself lingering in the historic relics section. He was struck by Pushkin's room; better than all the dozens of books he had once read about him, it revealed to Alexei the great poet's soul. There was the Ushakov album, the sheets of autograph-book verses, the portraits of his loved ones, the little house at Nashchokino and hundreds more tokens of a life of greatness.

He was overcome by the rooms devoted to the times of the great revolution, where familiar faces and objects, somewhat becobwebbed by time, stared at him with emphatic defiance.

But they could tarry no longer; the first bell was to ring in half an hour.

When they came out into the street, dense crowds filled the squares, parks and gardens along the Moskva. From a programme handed to him, Alexei learnt that the Alexander Smagin Society, to celebrate the harvest festival, invited the Moscow *oblast'* peasants to hear the following programme to be performed on the Kremlin bells in co-operation with those of other Moscow churches.[28]

PROGRAMME

1. Sixteenth-century Rostov peals.
2. Rachmaninov's Mass.
3. Ioakim's peals (1731).

[28]Aleksandr Vasil'evich Smagin (1843— ?), was a peasant from the Perm region who became a famous bellringer and expert on Russian campanology.

4. Borisyak's chimes.
5. A rising Georgii with a falling peal.
6. Scriabin's *Prometheus*.
7. Moscow peals.

A moment later a deep stroke of the Polyelei bell boomed forth and rolled over the city, followed by the answering chords of Kadashi, St. Nicholas of the Great Cross and the Monastery of the Immaculate Conception; the Rostov peal enveloped the whole of Moscow.[29] The metallic sounds floating down over the heads of the silent crowd were like the wingbeat of some mysterious bird. The elemental swirl of the Rostov peal completed its round and gradually rose away into the clouds, as the bells of the Kremlin intoned the austere scales of Rachmaninov's Mass.

Alexei, overwhelmed and prostrated by this solemnisation of high art, felt someone take him by the arm.

Turning quickly, he saw Katerina signalling to him with a secret air to follow her... He tried to say something to her, but his voice was drowned without trace by the ringing.

In a moment they were entering the gigantic 'Julia and the Elephant' restaurant in whose rooms one could take refuge from the sound of the bells.

> 'I don't know who you are,' whispered an agitated Katerina, 'I only know you are not Charlie Mann.'

And, excited and confused, she told him that his bad English pronunciation and pure Russian speech, the details of his dress and his ignorance of mathematics had from the very first day aroused the family's suspicion, which had been steadily growing; that he was now definitely considered to be an anthroposophist aiding the German adventure; that he was in danger of arrest, or worse; that she did not believe this slander; that in the past two days she had come to know and love him; that he was an out-of-the-ordinary man, predatory and beautiful, like a wolf; that she had searched for him to warn him and begged him now to escape; that she was frightened she might put on his track the judicial authorities which were at this moment arresting Germans and anthroposophists; that war would be declared any minute; and, kissing him unexpectedly on the forehead, she disappeared just as unexpectedly.

Kremnev, who had spent years living underground in Russia under the autocracy, was nevertheless bewildered and mortified by his inextricable situation. He shuddered as he noticed the waiters watching him fixedly and with suspicion.

---

[29]The main Rostov bells were cast when Iona was metropolitan there, from 1652 to 1691. Polyelei, the 'much-anointed', was the name of one of these bells, weighing over sixteen tons, and cast in 1682 by Filipp Andreev. Chayanov's sixteenth-century peal is thus either an anachronism, or a mistake for 'seventeenth-century.' The Iona or Sysoi peal was the first of the three full Rostov peals; this peal is perhaps referred to in item 1 of the programme. The others were those of Ioakim and Georgii (see items 3 and 5 in the programme). Ioakim was archbishop at Rostov 1731–41; he had been preceded by Georgii Dashkov as archbishop, 1718–31. Chayanov appears to imply that the Rostov bells had been brought to the Moscow Kremlin by 1984.

Quickly he went out of the restaurant into the square. The bells no longer shook the sky and the crowds were dispersing in alarm. Boys were scattering news sheets. 'War, war' was heard on every side.

Kremnev had not gone a dozen paces before a heavy hand was clapped on his shoulder and a voice said: 'Stop, comrade, you are under arrest.'

Chapter 13

Acquainting Kremnev with the bad organisation of places of confinement in the land of Utopia and with certain forms of Utopian judicial procedure.

The spacious 'Hotel for Visitors from Ryazan'', temporarily converted into a prison was surrounded on all sides by patrols of the Peasant Guard in the picturesque costume of the musketmen of Alexei Mikhailovich's time.

When the commissar who had arrested Alexei brought him into the lobby and handed him over to the commandant, the latter took his arrest number, rang the hall-porter, and then said:

> We've made a bit of a miscalculation about rooms and I shall have to put you in a common dormitory for tonight. You seem to have no luggage. If you are from Moscow, let me have the address, and we'll send home for what you need.

Kremnev said that, unfortunately, he was a visitor, and they promised to get him everything from the hotel's stock.

The hotel's concert hall, adapted as a prison cell, resembled a railway terminus in the good old days. Men and women of every condition and age sat by their travelling bags and bundles in attitudes of boredom and with gloomy faces.

There were Germans in leather jackets and caps, thin and slender, with their Teutonic superiority and disdain for their surroundings. There were pale Russian ladies, young men with unseeing, colourless eyes, and agile individuals of uncertain oriental origin.

As Alexei later discovered, the Russian ladies and the young men were anthroposophists, unfortunates caught up in the German intrigue, who had succumbed to the great German idea.

The prison commandant came into the hall and again apologised to the assembly for their deprivation of liberty and the intolerable accommodation; he expressed the hope that all would be free in a couple of days, and promised to make up for the inconvenience by a good meal and various entertainments.

In fact, the dinner, or rather supper, was not long delayed, and afterwards the Germans crowded round the card-tables to gamble the evening away, while the rest of the public heard a miniature concert hastily organised by the commandant.

They slept on camp beds, without undressing. In the morning Alexei was interrogated. When asked who he was and why he pretended to be the American engineer Charlie Mann, he frankly told his whole story, and fearful that his tale would be met with ridicule, offered as proof his bust in the White Kolp' panopticum and the reliable evidence contained in the historic relic galleries of the Rumyantsev Museum.

To his great astonishment, his tale encountered neither opposition nor incomprehension, but was calmly noted down and he was told that in the evening he would be handed over for expert investigation.

The whole, exhaustingly long day Kremnev sat before the windows of the room he had been allocated and looked at the town.

The social sea was in turmoil; like Old Man Black Sea, rural Russia was bringing forth from its depths thirty-three heroic forces.

Along the roadway before his windows, columns of troops in close formation marched with the rapid step of the French chasseurs. Some young lady, in blue riding-dress and a general's plume, was mounted on a white horse to take the parade of the women's light cavalry. Alexei's heart raced when he recognised Katerina's well-known features at the head of one of the squadrons passing by. Soon infantry replaced the cavalry and throngs of civilians filled every space in sight.

The crowd listened to orators who addressed it from moving cars and clutched at the sheaves of telegram tape that were scattered into the streets.

Towards evening, Alexei was put in a closed car and taken to the Mokhovaya, where the committee of experts awaited him in the circular hall of the university senate.

> 'Tell us,' a grey-haired old man in gold-rimmed glasses began his questioning, 'what is *Oblikomzap*? If you are really a contemporary of the great revolution, you should be able to explain the meaning of this word.'

Kremnev replied with a smile that it meant the Oblast' Executive Committee of the Western Oblast', a body which existed for a time in St. Petersburg after the capital had been transferred to Moscow.

What kind of institution was the *tsekmonkul't*?

The Central Committee of Monopolised culture, established in 1921 for the compulsory utilisation of cultural forces.

> Tell us what were the reasons for the forcible establishment and the subsequent abolition of the Committees of Poor Peasants?

Kremnev replied quite adequately to this question, too.

He was shown a series of documents of the period and asked to comment on them; this he also did satisfactorily and, finally, he was obliged to explain, with difficulty and at length, the idea of the urbanisation of agriculture, in reply to a question on state farms.

The outcome was that professorial interrogators shook their heads long and regretfully and informed him that, though undoubtedly he was well-read in the revolutionary literature, and had evidently familiarised himself with the archives, he was entirely unrepresentative of the spirit of the period and, from ignorance, monstrously misinterpreted historical events; he could thus in no case be acknowledged to be their contemporary.

When Alexei was driven back to the prison, the streets again seethed with a crowd that roared loudly and triumphantly, like the sea.

Chapter 14

And the last of the first part, which demonstrates both that sometimes ploughshares may successfully be turned into swords and that Kremnev finally turned out to be in an exceedingly piteous situation.

The triumphant and resonant ringing of bells woke the unwilling residents of the 'Hotel for Visitors from Ryazan" and they were soon informed that, the war being over, they were all free, but those who wished could stay for morning coffee.

The prison at once became a busy hotel, thus recovering its primary nature.

When Kremnev was leaving, the commandant handed him a packet with the findings of the investigating committee; these stated that, in the absence of evidence of a crime, the citizen calling himself Alexei Kremnev should be freed along with the rest. The committee considered his account of his origins improbable, but, finding no grounds to impute a criminal intent to the self-styled citizen Kremnev's imposture, was closing the investigation instigated by Nikifor Minin.

Alexei decided to exercise his right to breakfast at state expense on the terrace of his former prison and, sitting down at a table, he buried himself in the official account of the ending of war in a news sheet flung to him by a paper seller.

Alexei learnt that, on 7th September, three armies of German conscripts, accompanied by swarms of aeroplanes, had invaded the Russian Peasant Republic, and within twenty-four hours, without meeting any signs of resistance, or indeed of life, had penetrated to distances of 30, and in places 70, miles into its territory.

At 3.15 a.m. on 8th September, in accordance with a predetermined plan, the frontier-zone meteorophores exerted the maximum intensity of lines of force on a cyclone of small diameter; in the course of half an hour the millions-strong armies, and tens of thousands of aeroplanes, were literally swept away by monstrous whirlwinds. A wind curtain was established on the frontier and the Tara aerosleds gave such assistance as was possible to the defeated hordes. In two hours, the Berlin government announced that it was ceasing hostilities and would pay any compensation required for damage it had caused.

The form chosen by the Russian Council of People's Commissars was some dozens of canvases by Botticelli, Domenico Veneziano and Holbein together with the Pergamon altar, and 1,000 Chinese coloured engravings of the T'ang period, as well as 1,000 pedigree stud bulls of the famous 'Nur für Deutschland' breed.

The resounding trumpets of the peasant army played fanfares and the sounds of Scriabin's *Prometheus*, which turned out to be the national anthem, shook the Moscow sky.

His coffee drunk, his roast-beef finished, Kremnev rose from his chair. Hunched and dejected by his experience, he slowly came down the steps of the terrace, walking alone, friendless and penniless, to face life in a Utopian country he hardly knew.

Fig. 9.1 South-West environs of Moscow. Kon'kovo Village in the center.
Source: Schubert map (excerpt) of Moscow province, 1860, reprinted by the Office of Military Topographers. Scale of the original map: 1 cm = 840 m

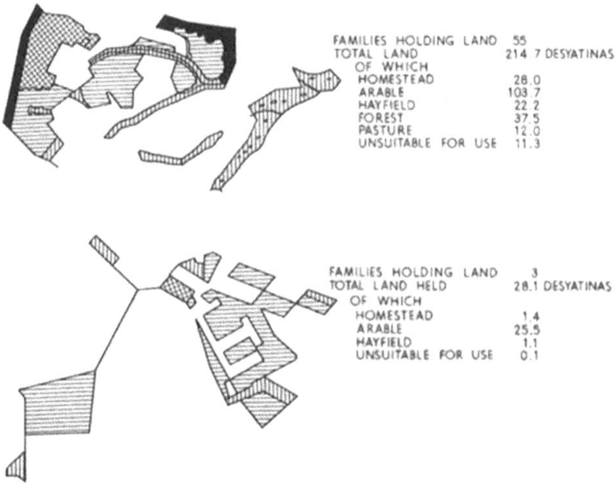

Fig. 9.2 "Plan of Kon'kovo Village, Moscow Uezd" and "Plan of Nikulin Village, Bronnesty Uezd".
Source: Chayanov, A. V. (Eds.) Thorner, D., Kerblay, B., & Smith, R. E. F. (1966). The theory of peasant economy. Richard D. Irwin, Inc., p. 175.
"In organizing peasant farms, one must almost always take account of their exceedingly poor layout. The excessive intermingling of the peasant household's strips, their length, and their fragmentation of arable and meadows frequently attain quite Homeric proportions. The fault for this lies partly with the excessively intermingled allocations made to peasant communes in the 1961 reform, and mainly with the communal egalitarian methods of repartitioning. The land was broken up into sections of equivalent quality, and a strip from each section was allocated to almost every member of the commune. (…) These plans show a somewhat hypertrophied land disorganization." (ibid., pp. 174–175)

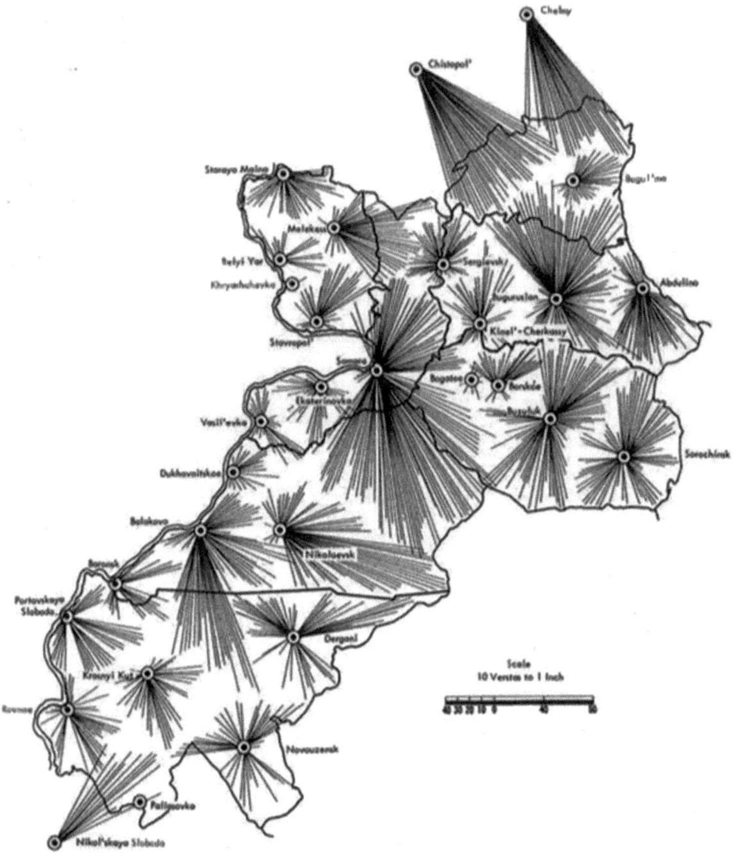

Fig. 9.3 "Trade catchment areas in Samara Guberniya before the war".
Source: ibid., p. 259. © Homewood: Richard D. Irwin, Inc.
"The system of the local rural bazaar at which the peasant sells his harvest and buys what he needs, and around which all the countryside's economic relations crystallize, has been very little studied. The bazaar is the primary cell of this general economic organism. (…) Observing local life shows that the bazaar site is a concentration of all local trading, cooperative, business, and even spiritual life for its catchment area, since the personal links of the area's inhabitants are united by the bazaar, where they invariably meet one another. In their turn, the bazars are attracted to a larger scale center of wholesale trade, and they construct a certain national economic whole from the scattered peasant farms through the firm links of their trading machine." (ibid., p. 258)

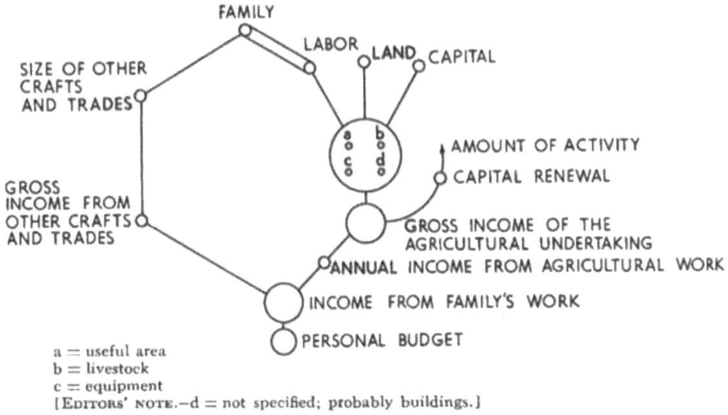

Fig. 9.4 Scheme of organization of the peasant economy around the family unit. Source: ibid., p. 102.
"By detailing the separate links, we obtain the scheme shown in the figure. We see that as a result of the mutual relationships of the factors we have studies the volume of agricultural, crafts and trades activity is established, the income from each of them is gathered in, and, combined, they give that synthetic single family labor income which, by comparison with its demands, also gives the basic economic equilibrium". (ibid.)

Chapter 10
The New Settlement Pattern (1930)

Mikhail Aleksandrovich Okhitovich

Introduction

Qinyi Zhang
EPFL, Lausanne, Switzerland
qinyi.zhang@epfl.ch

Urbanists and Disurbanists, two major opposing schools of utopian urban planning, emerged in the Soviet context in the twenties. They were rooted in the historical anti-urban attitude in Russian society, and in the utopian tradition—especially the one of the Garden Cities—in Europe and Russia, and within Marxist urban schemes.[1]

Both schools strongly criticized large-scale cities and advocated for the elimination of the urban–rural separation. While the Urbanists, led by the brothers Sabsovich (Viktor, Leonid and Vesnin) and Golosov (Panteleymon and Ilya), proposed medium-size cities of 50,000–75,000 inhabitants, with the elimination of traditional villages,[2] *Disurbanists, led by Moisei Okhitovich, Mikhail Barshch and Moisei Ginsburg, imagined a rather radical "non-urban redistribution of population"*[3] *through the entire Russian territory.*

[1]See Bater, J. (1980). *The soviet city: Ideal and reality*. Edward Arnold. and Stites, R. (1989). *Revolutionary dreams: Utopian vision and experimental life in the Russian revolution*. Oxford University Press.

[2]See French, R. (1995). *Plans, pragmatism and people: The legacy of soviet planning for today's cities*. UCL Press.

[3]See Stites, R. (1989). *Revolutionary dreams: Utopian vision and experimental life in the Russian revolution*. Oxford University Press.

M. Barcelloni Corte and P. Viganò (eds.), *The Horizontal Metropolis*,
https://doi.org/10.1007/978-3-030-56398-1_10

Okhitovich was a main figure within the Disurbanist movement, a Trotskyist, sociologist, town planner and architectural theorist, member of the OSA (Organization of Contemporary Architects) group and a prolific contributor to the Sovremennaia Arkhitektura *magazine. His ideas were known mostly through his design proposals brought forward together with his colleagues, such as the masterplan for Magnitogorsk (1930) or the "Green City Plan" for Moscow (1930). Only few texts of his, including the "Notes on the theory of settlement" proposed in this chapter have surfaced and have been translated.*

In his writing, the territory is never a background but a fundamental instrument for the reorganization of society. Beyond Kropotkin's idea of a mutual-aid among the different parts of the territory/society, as for instance large factories and small-scale workshops or urban centers and villages, Okhitovich's proposal takes the form of a horizontally organized network without center(s), where complex relations between architectural form and social dynamics are investigated in depth.

Utopian thinking, both by Urbanists and Disurbanists, gradually lost its strength and ceased during the 1930s. Okhitovich was executed in 1937.

10.1 Notes on the Theory of Settlement[4]

The reconstruction of industrial and agricultural production in the USSR sharply challenged all principles of building a dwelling as well as the social, geographical and productive approaches related to it. The most modest and simple statement on all of these issues immediately and painfully strikes the attention and tense nerves of huge masses. Ideas in conjunction with the masses reached an unprecedented scope. In the heated brawl of discussions, each participant matured rapidly, receiving an infinitely fruitful experience, developing attitudes towards competing proposals and clashing ideas. In the fire of ideological battles, formulations were sharpened, positions developed, new ones were created, old ones were abandoned … The timely practice with unexpected force felt the meaning of the seemingly abstract theory, concrete proposals began to be tested by the cold mind of the abstraction of historical judgments. Theoretical work rose from work offices, from the rustle of books and became actual practice. Much of the past work became relevant for the future. The future, which was painted by some before, amazed the spectator of the era, this future turned out to be buried in yesterday. This alliance of theory with practice is never to be abandoned. It's time to sum up the first results. It's time to prepare your own, revolutionary theory of resettlement. For Lenin said: "WITHOUT A REVOLUTIONARY THEORY THERE CAN BE NO REVOLUTIONARY MOVEMENT". The proposed "notes" aim to provide an outline for a serious development of the Marxist theory of the resettlement of

[4]Source: Okhitovich, M. (1930). Zametki po teorii rasseleniia. In *Sovremennaia arkhitektura, 1–2*, pp. 7–16.

people and the geographical distribution of productive forces in their interdependence. In our notes we deliberately eliminate the signs of the scientific apparatus as well as all elements of polemics. We give only a logical course of ideas, urging the reader to criticise them most severely.

The dwelling is always aligned with the place of production. A person always lives where he/she works. More precisely: a person ENDEAVORS to live where he/she works. The place of residence for a person is always their place of work, because labor turns an animal into a human. Labor is the only source of human development, or the only source of social existence, the only source of the HUMAN.

If the production is moving, then the place of habitation changes too—AFTER the production, FOLLOWING the production, TOGETHER WITH the change of the place of production; and the location of the dwelling of a human also changes. This is usual for nomads.

If a permanent place of work has been established, then a house is immediately built next to it. This is usual for the settled farmers. If a land plot is lost—the dwelling is lost too in search for a new place, and if a new land plot is found, a new dwelling is established, and the old one is abandoned.

If the place of work is abandoned, so would be the dwelling. Neither your poverty nor your avarice can stop this law of placing the dwelling near production.

It is not enough to say 'near production'.

The dwelling seeks to be as close to production as possible. The dwelling seeks to be located not even near nor around, but directly in production. It seeks to fit as close to production as possible, but of course in such a way as not to harm production itself but instead to supplement production.

For example, the dwelling of the peasant is not located on the plowed land. Here the production surrounds the dwelling.[5]

Both peasant's and craftsman's houses are places of habitation and production at the same time. Neither the peasant nor the craftsman go or ride to work.

The situation is more complicated in the case of manufacturing and industrial capitalist production. Of course, the Putilov worker does not live in Pskov. Of course, he will not live on the outskirts of Leningrad, nor in the center of the city. Even if a grand mansion is prepared for him in the center; he will always strive to live as close to the place of production as possible.

The need to use transport to come to work is a punishment for him. This is a misfortune[6] for him, as an inevitable result of the crowding of people around production sites. Housing which is far from work—it is not proper housing.

---

[5]While the craftsman's dwelling, on the contrary, itself surrounds the production, which is the machine.

[6]This is especially noticeable with the reduction of working time. The shorter is a working day, the longer the commute takes in proportion. That is why the transition to fixed working hours in agriculture will further distance the village from the city, since it should bring the producer closer to production even to greater extent.

A peasant who has received an allotment far from his village will inevitably move to another; if this allotment lies within the possession of the village, but far beyond other land plots, then the peasant, having obtained it, will settle there. If this one is an independent possession, located far from all the settlements, the peasant will simply start his own.

If the area of his work is vast, he will be torn apart. To keep up everywhere, in addition to his main, basic home in the place of his basic work, that is, the work that takes away most of his time, he also builds special 'zaimki' (land plots) for temporary, but sufficient, long-lasting other occupations, the movement to which and movement from which back to the dwelling would take him the time necessary for the main production.

Thus, for example, 'ZAIMKI' are made for forest, fur-bearing, hunting, fish, berry, mushroom, and other activities with the main occupation of agriculture.

The same is true, although for quite different reasons (which means—not the same thing) for the modern American farmer. The car, which is available to him, is not fast enough to deliver him to the production site in time. Here the role of the car is different—the car should (like a telephone) connect the various processes occurring simultaneously in different parts of the production territory, and it must turn individual private production efforts into one common product.

Thus, homelessness does not only mean the absence of the space that belongs to me. HOMELESSNESS also MEANS THAT I DO NOT HAVE AN AREA DIRECTLY IN PRODUCTION. According to objective reasons, independent of the will, the person settles in the place.

On the one hand, the forms of production strictly determines the remoteness of people from each other, the compulsory remoteness of people, up to the fact that people live in the wild, abandoned, idiotic rural life. On the other hand, they also create the proximity of people, the forced closeness of people, hence the tremendous calamities of CITY CROWDING.

The desire to eliminate homelessness (for what kind of house this is, when it is far from work) under certain circumstances leads to crowding. The desire to eliminate crowding under the same circumstances inevitably leads to homelessness. The reasons for isolation, the causes of crowding, the conditions for their occurrence, and therefore the conditions for their possible elimination must be sought in METHODS OF PRODUCTION ALLOCATION.

The moment of opposition between the city and the countryside is the most interesting and decisive in our thoughts about the sources of housing isolation, abandonment on the one hand, and causes for housing crowding and concentration on the other.

The city is the result of a social division of labor as well as the result of the separation of craft from agriculture, or otherwise the separation of the PROCESSES OF PROCESSING OF THE PRODUCT FROM THE PROCESSES OF ITS EXTRACTION.

The city, therefore, is not a product of some physical-geographical, i.e., natural conditions, but a consequence of PUBLIC, SOCIAL CHANGES. Raw materials are available everywhere, both as the product of nature and the product of agriculture. However, the city, industry or, more accurately, the manufacturing industry does not appear everywhere, evenly. It depends on the location of the elements of production, i.e., on geography, though not physical geography, but rather economic geography. To clarify:

In order for raw materials in the hands of a person to turn into craft products, fuel (or another source of energy: water, or wind) is necessary. The main thing is FUEL.

When the craft is separated from agriculture, such an upsurge of productive forces occurs, that if there is not enough local fuel reserves (which should be so), the fate of production must become dependent on the possibility of obtaining it from more remote areas. Thus, the lines of communication, not just as a means of exchange nor a means of trade,[7] but as a condition for production and a means of delivering fuels (mainly forestry), become the FIRST source of the development of a small industrial, meaning the URBAN settlement (hence the exchange of goods).

Transport is the FIRST condition for placing industry, separating it from agriculture, and becomes the FIRST cause of crowding on the land. Because of the uneven distribution of fuel on the surface of the earth, with the ubiquity of raw materials due to the need for fuel to be transported to the raw materials, there is uneven distribution of industry, unevenness and dispersal of people following the placement of industry, unevenness in the size of cities.

The centers of attraction of fuel became the centers of gravity of raw materials. The means of distribution fuel became also the means of distribution of raw materials (hence trade).

Thus, transport is reinforced as the reason for concentration of fuel, raw materials and housing.

Indeed, what is the city? It is a node of means of transportation: a node of roads and rivers.

Why is the city different from the village? There are intersections of roads, streets and rivers. The village, unlike the city, always has just ONE ROAD. If there is more than one road in the village, surely there has been a qualitative leap in its inner structure, and the craft has been separated from agriculture. It can still be called a village, not a city, but it cannot be rural anymore. This is the BEGINNING of the city, and the END of the village, its DEMISE.

Still, what causes the gravitation of people to one point? What causes people to pile their dwellings on each other? Time. Socially necessary time. Let's not forget that distance is measured by time.

Production is developed better in quantity, in quality and in pace where there is the easiest, fastest and most convenient way to receive fuel (and later raw materials and then semi-finished products) in order to be able to make products in the same

[7]The cause of the emergence of any trading center before all is manufacturing industry: craft, industry. The trading center is a point of exchange of products of industrial consumption, i.e. the city acquires raw materials for processing, the village—means of production. Raw material becomes also a means of personal consumption for a citizen, and goods are also a means of consumption for a farmer. In general it is necessary to mention that the means of communication arise before the exchange, but there is no exchange without means of communication.

given time but easier, faster and better. The craftsman is always (as well as the peasant) in pursuit of time. Time is a product, time is a commodity. That is why it is necessary for him to put housing on the crossroads of the ways delivering to him fuel (and later raw materials and semi-finished products).

Fuel! Transport of fuel! The revolution in the transfer of energy is the condition that allows us to hope for the removal and incredible isolation on the one hand and the extraordinary, hypertrophied crowding on the other.

Energy can be transferred to a distance by wire. The wire began to replace the road. Transportation of fuel has turned into TRANSPORTATION OF ENERGY.

It must be borne in mind that the longer the wire, the more expensive the energy is. Therefore, for the first time, imported coal is converted into energy in old industrial centers, and then the industry passes to new, local sources of energy because they do not require transportation of fuel. This causes some zahirenie (withering) of former centers, but creates not smaller and sometimes even larger industrial centers, large cities although in new areas.

Only the spread of the NETWORK of wires to very remote corners of the country, only the creation of local stations and especially the so-called circular transmission of energy (a common network of various stations) creates the conditions for a complete revolution in the forms of accommodation and resettlement, a revolution as human history has never seen.

The achievement is that if you do not have to transport fuel to Leningrad once the fuel in the form of electricity can be received at the place of receipt of raw materials, e.g. somewhere near the Magnetic ore from the Ural Mountains, there is no need to transport ore to Leningrad. Along with the ore, waste also can be transported, all production waste and other material that will eventually go into the waste.[8]

Moreover, in these conditions, the volume of freight reduces not only for fuel and raw materials, but also for finished products. Further, the transition to local raw materials, depending on the cost of transporting the finished product, will help to abolish the narrow industrial SPECIALIZATION of the districts, and their agricultural[9] MONOCULTURE.

All modern development of agriculture is going in the direction of reducing cargo transportation of a product from the place of manufacturing to the place of its

---

[8]It's different for flour, bread and other food products, and for some types of building materials (cement, concrete). It is more profitable to transport semi-finished products, because after processing it increases in volume and weight. This is one of the reasons for the backwardness of these industries in the past and fading present.

[9]Monoculture of Egypt (cotton) is the very product of the colonial system. The UK processed cotton on its own coal.

With the fall of the power of the English financial capital, Egypt will surely become the center of the textile industry on the basis of its own sources of energy.

Our Turksib (Turkestan–Siberia Railway), with its idea of turning Soviet Central Asia into a monocultural country, struggles with the monoculture of both Egypt and the cotton-growing south of the United States. With the further development of scientific farming, the expansion of cotton-growing to the Ivanovo-Voznesensk, Shuya might be possible.

processing and consumption. The selection and the whole technique of plant hybridization, the fertilizer technique, the translucence of seeds, the greenhouse technique and all the work of modern zootechnics are all focused on one task:

PRODUCE ANY PRODUCT IN ANY GEOGRAPHICAL REGION.

The consequences of changes in the way energy is transported, the consequences of replacing the fuel paths with metal wires, especially the circular transmission, and wireless transmission in the future cannot be calculated and taken into account carelessly.

Raw materials developed near the fuel,[10] raw materials were attracted to the fuel, to the crossroads of communications.

Now the energy reaches for the raw materials. Raw materials are everywhere—energy is stretched everywhere. Production is everywhere—housing is everywhere. The arrival of energy to local raw materials radically changes the position of agriculture. Agriculture is distributed not by centers, but across the entire land plot. How did this continuity of the agricultural production predetermine the great development of urban city centers?

The village, as the main engine of development of the city and a source of raw materials for the city (that is a commercial, not a communal village), developed based on its links with the market and the city, i.e., it stretched out with its raw materials to the fuel, i.e. to the city, to the intersection of the flows of raw materials with the flows of fuel. The fuel was located on the intersection of flows (but not necessarily where it was mined). So were raw materials.[11] Now energy is drawn to agricultural raw materials, NOT TO THE INTERSECTION OF FLOWS, but TO THE PLACE OF MANUFACTURING OF THESE RAW MATERIALS.

We reconstruct the village not only in the sense of its transition from small craft to large-scale, machine-made methods of production. More precisely, the transition of agriculture to the machine mode of production or the merging of industry with agriculture will mean that RAW MATERIALS WILL BE PROCESSED IN THE SAME PLACE as where it will be produced.

THE SMALLER[12] THE ENTERPRISE for processing agricultural products, the shorter the distance from the agricultural raw materials to processing, and the shorter the distance to deliver the finished product that is not intended for consumption in other districts.

So the greatest economic concentration of production leads to the greatest spatial decentralization of production. Lenin called this process DISPERSAL of industry. In this way will industry merge with agriculture.

---

[10]For if it did not have fuel, then it could not become raw material for the production of the product for the market, without fuel, therefore, it was impossible to separate industry from agriculture.

[11]A small town Lutetia had one radius of villages for its functioning, multimillion Paris different one.

[12]By territory.

We do not know yet what this process will lead to, in the technical sense. But already now it is not difficult to imagine a COMBINE, which would perform, say one more additional operation—the transformation of grain into flour.

Thus, production would return from the enclosed premises to open air. The new conditions for energy transport pose the question of energy sources differently. Will the means of energy transmission—by wire or wireless—matter? Will energy be distributed from some enormous centers—energy sources? Will local energy sources also develop?

The network will replace the centers. In this network, energy centers are not of great importance, unlike the opportunity to assemble large, small and the smallest sources of energy into one single power network. The transition from the centers of energy to the network turns the problem of the depletion of the world's energy reserves upside down, since the network deals with the collection of the smallest sources of energy, while the centers squander the largest ones.

How will these processes, as new forms of production location, affect the forms of settlement? How to put an end to both human isolation and housing overcrowding, these products of simple commodity society, infinitely amplified by large-scale capitalism? This is the question of:

RELATIONSHIP BETWEEN HOUSING CLUSTERS WITH PRODUCTION CLUSTERS.

Historically, the following four types of interaction between the clusters of dwellings in relation to the grouping of people in production are known.

First. CO-PRODUCTION CREATES COHABITATION. CO-PRODUCTION CREATES COMMUNAL HOUSING. This is confirmed by the example of primitive communism (shepherd cultures). In this culture, a dwelling was built for all participants of the production process. Researchers estimated 250–300 people in one dwelling.

Can this case be extended to other social formations? Communism, MODERN communism, should embrace at least hundreds of millions of people in a common production process. If joint labor led to cohabitation, it would be right to build one accommodation for several hundred million people. For a better verification of the of the situation, let's move on to the second question.

SEPARATE PRODUCTION LEADS TO COHABITATION.

The most independent, separate producer in history is the petty bourgeois. However, it is exactly the small-scale industry that created 'a conglomeration of gigantic masses in cities'. Take Paris, London, or Berlin. To guarantee accuracy, we will imagine them in the XIV–XV centuries, when there was no large-scale industry.

If we believe that joint collective labor leads to collective living, to a common house, then it turns out that the petty bourgeoisie would have to reduce the existing villages in size, while in fact it turned them, these small villages, into condensed formations in the form of cities.

Third. SEPARATE PRODUCTION LEADS TO SEPARATE ACCOMMODATION.

This happens with the village under the influence of the development of commodity relations. In that way capitalism affects the village. That very capitalism which in the course of its development ostensibly[13] creates overcrowding.

The second and third cases perfectly refute the assumption that the CAPITAL CONCENTRATION CREATES CONCENTRATION OF HOUSES. There is no capital yet, but the city is already there. There is no concentration of capital yet, but the city is already flourishing. Babylon itself would laugh at this idea.

But what about the village, which, contrary to these claims, is not crowded, but more than that, is being reduced in population due to the enlargement of production and concentration of capital?

Only the first and the third cases seem to confirm the dependence of cohabitation on joint labor and, consequently, separate living on separate work.

The fourth case. JOINT PRODUCTION LEADS TO SEPARATE LIVING.

The working barracks seem to confirm the accuracy of the law, derived from the experience of the first and third cases. We work together—we live together. Alas, this applies only to poor, underdeveloped capitalism, mainly in the era of manufacturing. In developed capitalism, the worker has a single–family dwelling, and even lives in a cottage whenever possible. We can condemn the 'petty bourgeois inclination' of the worker, attack his backwardness, 'opportunism', accuse him of biological bias, etc., but we do not know yet a single concrete worker who would exchange his apartment and, let alone the cottage for a cot in a dormitory. But we know plenty of cases of the opposite.

Here we turn to the question of what is the INTERNAL, SOCIAL FORM OF HOUSING.

Is the city a dwelling?—No. It is a form of settlement rather than a form of dwelling.

A house? A big house in the city—is it a dwelling? No. This is a vertical form of crowded dwellings, but not a dwelling.

A dwelling is a building that embraces the clan. A courtyard (but not a hut) is a dwelling. And the communal patriarchal village is no longer a dwelling; it is a form of resettlement.

A BARRACK, or a DORMITORY is a dwelling.

A KHUTOR (hamlet) is a dwelling.

A FARM is a dwelling.

A COTTAGE, an APARTMENT, a ROOM, a CORNER is a dwelling. But a city, a village, a house, or a hut is not a dwelling.[14]

---

[13]Indeed, ostensibly. Capitalism creates both crowding and isolation, one being the product of another. The greater the area covered by villages supplying the city with raw materials, the more it increases the overcrowding in city. Remember the example of Lutetia and Paris.

[14]Social form of housing does not have even a gram of materiality; social form of dwelling is not a thing, but rather a relationship.

The communist dwelling does not recognize any divisions on the urban and rural; divisions of the village into yards, divisions of the city into houses, divisions of the house into rooms.

This lack of divisions within the settlement and home signifies the absence of social division of labor. The patriarchal community already knows the first major division of labor in society: between a woman and a man. A man is engaged in agriculture, while the wife is engaged at home, hence her subordinate position in society. When there are no field works, the husband is at home engaged in craft, while the wife is again occupied with housekeeping. Why is one house of the clan, under these conditions, not divided into two halves—male and female, prior to the division of one dwelling into many, and then dwellings into yards?

The transition from cattle breeding to agriculture played a role in this. One dwelling transforms into a village[15] consisting of many dwellings. What it depends on? On the transition to a new mode of production, meaning a new way of accommodation.

The rural dwelling is rather an economic than an arithmetical unit. There are as many male workers as necessary to cultivate a particular land plot. What is the size of this plot?

Its size is measured in time sufficient to enable the cultivation during the favourable weather with given tools.

The increase in the number of employees will lead to a fragmentation of the land plot according to the number of new workers, and consequently, the creation of a new dwelling.

Why? Because this production method requires certain cooperation in production, since every extra person becomes not only an employee helping production, but at the same time a consumer, a waster of the common product. The number of employees must therefore be minimal.

This number is what is called the patriarchal family, which replaced the communist clan. In this case, why are the dwellings located not separately, but next to each other in one village? Because the division of labor in agriculture is not ultimate. Part of the territory such as meadows, pastures, forests, etc., remain in common use. Only the transition to a new mode of production and to a new division of labor leads to new forms of resettlement and to new forms of housing. Meadows, pastures, and forests unite people in villages (into one common grouping) as they did before, while farming splits them into separate yards (dwellings). But since the crafts become permanent and not seasonal occupation for part of the village, this new division of labor means the beginning of the city and the end of the village. This new division of labor also creates a new dwelling—a philistine dwelling, the home of an independent artisan. A new division of labor led to a new division and separation of the dwelling. The petty-bourgeois family, unlike the patriarchal

[15]It should be noted that the village here is a multitude of houses (patriarchal-peasant form of settlement), and in developed capitalism there is no actual villages in the rural areas but rather farms, khutors.

family, comprises only two generations, since the new division of labor obviously requires fewer people for the same operation.

The clan and the patriarchal family are replaced by a petty bourgeois family, then a capitalist family under capitalism in the city, but what about the village?

There is not only the separation of industry from agriculture, but the further division of labor in agriculture itself. Under capitalist agriculture, the producer is already detached from others, on the one hand, due to the strengthening of specialization, and on the other, by virtue of the separation of field crop cultivation from animal husbandry.

The community (hence the village) has disintegrated, and from the village the person moves to the farm. What will happen under the socialist method of settlement?

There will be no city or village,[16] there will be no boundaries between them, and they will unite as one.

And the dwelling? The dwelling is the product of a further technical division of labor, leading to further reduction in its size. After the clan dwelling, patriarchal dwelling spanning 4 generations, philistine dwelling for two generations, and capitalist dwelling for one, the technical division of labor under socialism will lead to a dwelling for one person—to an individual dwelling.

Why does this not happen under capitalism?

Because the wife and husband cannot yet make the final division of labor, as the capitalist is bound by the division of labor with the wage worker. They are connected by the unity of economic interests: inheritance of property and the unification of capital. Similarly, the proletarian family is bound by the common interests of the reproduction of its workforce and the hope of supporting its forces in old age by children.

Only the socialist relations will put society directly before the man-producer, and the man-producer will be directly faced only by public relations... For they destroy the division of labor between man and woman.

[It should be noted that the course of various forms of the family—patriarchal, philistine, petty-bourgeois, capitalist and proletarian—is consciously provided in abstract formulas.

In life, deviations are inevitable. The patriarchal family (4 generations) can reach in size the genus—6 generations; philistine (2 generations)—the patriarchal (4 generations); capitalist and proletarian (1 generation, or a couple)—the philistine (2 generations)]. Form of settlement: the city, or the village, the form affects the EXTERNAL FORM OF THE HOUSING, but not the internal one.

It does not matter, whether I live in a separate barrack or in a dormitory, located in a big house. The social character of the form of housing has not changed. Either way, the form of dwelling is primitive-communist.

---

[16]Since there will be no social division of labor into extracting and processing due to the merging of industry with agriculture, and not the opposite, which would mean only the mechanization of the extraction processes.

Similarly, it does not matter if I live in a separate apartment of a multi-family building, or my individual apartment is located outside the city and is therefore called a cottage. In both cases, the social form of the dwelling is same: family-philistine.

If I live in a separate room, it does not matter if it is located inside a huge house or not, because a room is a room, even if it was set up alone in the countryside, surrounded by greenery, silence, etc. Social character or rather a social form, the internal form of the dwelling has not changed.

This is the conclusion, if we break away from the obsessive physio-geographical criterion and will refer to socio-historical one…

In both cases, the fact of joint, collective production is obvious. But in the first case it leads to a common dwelling, and in the second case to separate apartments for families and even to separate rooms per person.

With low technology, joint labor also meant a common place of habitation; with highly developed, capitalist technology (division of labor), joint labor creates a SEPARATE HOUSING, but (here and there) JOINT LOCATION. Which is not the same.

Maybe, the commonality of location (house, quarter, city[17]) is created by commonality in production? No, not at all. That's the whole point. Work at Ford enterprises is a common work for one concentrated capital, but people live in different groups: some under Detroit, others in Ireland, the third in Constantinople, the fourth in Belgrade.

The same is true of any solid steamship business: some live on the shore, others on steamships. But all work in one enterprise, because if you stop the work of one of the groups, the other will stop.

The joint work of primitive peoples lead to joint housing. The joint work of modern peoples develops personality.

How, for what reasons does the personality develop in man? Due to the appearance of property. Property in its economic dimension, exploiting, overwhelming others, all other personalities except personalities-members of this exploiting collective, will disappear.

But the person will be born separately—not collectively. Eating, drinking, sleeping and dressing, etc., in one word CONSUMPTION will always be done separately.

In more developed society, individual consumption rights will increase. Let us not forget that socialism means the excess of products (the five-year plan is the aspiration to turn the shortage—the product of capitalism—into excess—that of socialism). Along with the private property, bourgeois and capitalist property, bourgeois and capitalistic personality will disappear; but personal property, personal consumption, personal initiative, personal level of development, personal hands, personal legs, personal head, brain will not disappear; moreover, these will

---

[17]The city is a form of settlement. It creates its own form of grouping: horizontal and vertical. Thus, the city is a condition for grouping of dwelling, its location, but not its form (grouping).

be accessible to everyone, not just the lucky few, the "privileged" ones, as it was before socialism.

Personality represents the need for a socialist mode of production, originating from the depths of the capitalist mode.

Personality is the product of a separate (at the moment, territorially, but not economically separate) production position. Personality is the product of the technical division of labor, not the social division. Not to recognize the collective, but at the same time to celebrate the PERSONALITY as Max Stirner does, means to admire the consequence, despising the cause. To praise the collective and ignore the personality means to praise the Russian language, but forbid using Russian words. This is what our modern Stirnerians are doing backwards, the supporters of the specific shade of Proudhonist communalism.

You can not oppose the individual to the collective and vice versa. The stronger the collective ties, the stronger the personalities comprising the collective, its components, and DISTINGUISHED BY IT. The stronger the personality, the stronger the collective to which it serves, and the stronger the class collective to which this personality serves…

Personality is not arithmetic, but a social unit.

In pre-socialist formations, the individual served as a unit in so far as it was an enslaving, exploiting, oppressive power. The others were not considered human. The task of socialism is not to perpetuate this attitude, but to change it.

It is necessary to destroy exploitation, to destroy its product—an aristocratic, capitalistic, philistine personality. But to destroy a personality in general would mean to destroy all division of labor between individuals. Our task is not at all to destroy the division of labor in general, as the ideals of the peasant community, represented by N. K. Mikhailovsky, demand.

Our task is to destroy the PUBLIC division of labor: the one existing between the entrepreneur and the wage worker, between a man and a woman, between the town and the country, and finally, between individual countries. For the evaluation of the division of labor between people, we consider the development of the productive forces of society as an inevitable, necessary and indispensable condition. But at the same time, we do not differ at all with N.K. Mikhailovsky[18] on the EVALUATION of the consequences of SPECIALISM, arising from the technical division of labor. With all determination, we condemn it. But while Mikhailovsky contraposes this specialism to the past with its low level of productive forces, we contrast it with the future, with specialization. Yes, even higher specialization. It is necessary to carry out the technical division of labor until there is an opportunity for an easy change of labor.

Then, finally, the social division of labor will be destroyed, the opposition between mental and physical labor, skilled and unskilled, responsible and simple, "women's" and "peasant", urban and rural activities, etc.

[18]Namely, N. K. Mikhailovsky on scientific socialism.

Only then a complete fusion of "labor" and "education" will become possible. While general education and physical education try to fill the failures of specialism, they try to be universal. With the division of labor between individuals, it is futile to fight with the product of this division of labor, personality; it is unnecessary and reactionary. Our historical task is only to eliminate the one-sidedness of the human personality, to transform the personality from the bourgeois limited with its professionalism, from capitalistically limited with its specialism to a comprehensively developed, socialist personality. We are for the division of labor, but we are also for a system of change of labor, which makes the division of labor possible.

The difficulty of mastering the essence of the relationship between the individual and the collective, due to the impossibility of solving it without studying the division of labor—the social and the technical division—is further aggravated by the fact that economic, industrial and territorial scales are forgotten.

Without the ability to sharply distinguish between these three things, one cannot understand anything about the agrarian economy, as well as the problem of industrial dispersion. The economic size of an enterprise in modern society is measured by the size of capital, but not by the number of workers employed in the enterprise—otherwise, a huge power plant could be named a small workshop, since just a few people work at the electric power station; and another craft workshop of a manufactory type that has hundreds of workers would have to be declared a large factory. The production size is also determined not by the number of people involved in the tools of production. The production size of the enterprise is characterized by the number of products produced by the worker of equal preparedness, equal effort, with the same amount of time. Territorial size can sometimes coincide with economic and production scales.

Especially it is so in the era of transition from small to large-scale production, for example, from small urban craft to manufacturing or from small-scale farming to cooperative, collectivization.

Under certain conditions of location, centers became a sign of the largeness of enterprises; in new, different conditions, the network will take their place.

It is the network that will become the condition, the territorial sign of size.

What is larger, one laundry for 25 thousand inhabitants of the city, mechanized "according to the latest technology," or 25 laundries, each for a thousand people, also mechanized with an even latest technology? Of course, 25 laundries is greater than one, although it usually thought otherwise. They are more necessary, since most of the settlements are less than 25 thousand people.[19] They will displace small laundries, for they are 25 times (in fact much more!) closer to the consumer. They will beat small equipment, because these laundry rooms are easy to produce in comparison with heavy equipment of large laundry ... But, of course, they are only possible when there is a network or at least local sources of energy.

Same with kitchen-factories, etc.

[19]The numbers are random.

The world will someday come to such an enlargement that there will be only one network of food, washing, housing, etc. But a human will never see such an enlargement of enterprises, that on the entire planet there will be built just one laundry or dining room, one center for washing, or food.

The network will win, the center will die out.[20]

In the future, the crowdedness of people in production will be unthinkable. On the contrary: around one group of automatic machines, which employs[21] one person, it will be possible to leave free space for greenery, light and air, and production will cease to be a source of occupational diseases, connected largely with the location of the enterprise. Such is the perspective of the inevitable result of the gradual displacement of man by the machine.

Reduction of the time required for production of the product caused overcrowding in these conditions (of obtaining energy and raw materials), creating an unprecedentedly high land rent. This same reduction in time in other conditions, instead of congestion, concentration, clusters, will lead to the MAXIMUM SCATTERING of industry, to MAXIMALLY-UNIFORM (i.e., without congestion on one side and deserts on the other, one that evokes the existence of the other…) resettlement of people.

It is interesting to compare our views with the views of the anarchist communists in the person of their ideologist P. Kropotkin.

According to Kropotkin, centers supply energy directly to the house of a local producer (in the future). In our view: There is no energy centers. There is a unified NETWORK of energy. There are no central sources; there are local, large, small and smallest, ubiquitous sources of energy. Each center is a periphery and each point of the periphery is a center.

The network should be a collector of the smallest energies, and not just a way of distributing energy from large sources.

P. Kropotkin frees a human from crowding by returning to small craft with the help of large-scale industry.

Scientific communism fights crowding and isolation through the new placement of agriculture and mining; by combining the manufacturing industry with them into one.

[20]Together with the center, the system of central water supply, sewerage, etc., will be wiped out, these foundations of a modern communal economy.

[21]It seems, that it is machine and not the person who actually works, produces. A person organizes the activity of machines or even the activity of people, who are organizing the activity of machines. In the first case, always near a group of machines; in the second, always on the go: by phone or by car.

TERRITORIAL solutions are close, almost identical: both Kropotkin and Lenin have the SCATTERING of production. But economically, one stands on the basis of small-craft, the other—a large-scale mode of production.[22]

And these former conditions for the location of production and settlement of people affected the very METHOD OF BUILDING PRODUCTION. Everyone talks about the backwardness of the construction industry. We must not talk, we must explain this backwardness.

...Capitalism is developing unevenly. Next to the greatest, sophisticated modern large-scale industry, there are infinitely backward:

(1) AGRICULTURE,
(2) construction technology,
(3) techniques of food processing, washing clothes, etc. (so-called housekeeping).

These three backward industries have not yet reached the level of other industries.

The main reason for the backwardness of construction equipment and agriculture is the same.

The decisive shift[23] in the technology of agricultural production is associated with the appearance of gasoline and technical transport in the countryside. For the first time, energy came to raw materials and construction materials came to village. For the first time, energy came to raw materials. As for the construction technology...

The modern direction of its development is connected with the same movement of energy to any local raw materials. The transfer of energy over increasingly long distances will end urban land rent (here the technical revolution will help the social); this last concrete, durable and heavy construction is an obstacle to revolution in construction technology.

In fact, why in the most developed centers of modern industry, where there is both raw materials and fuel, where energy was concentrated, why, in these seemingly most favorable conditions, construction production remained at the handicraft-medieval level?

Because of the high land rent. Because of the gravitation of industry to the crossroads of fuel and raw materials, semi-finished products.

---

[22]S. Motolyansky, in the journal 'Revolution and Culture' compares the ideas of William Morris ('News from Nowhere') with the ideas 'of our contemporary' M. Okhitovich. He stipulates, however, that W. Morris and 'our contemporary' reached similar conclusions based on **different** social motives. If S. Motolyansky were a Marxist, this statement would be not a stipulation for him but rather the basis of his judgment.

[23]The first revolution in agriculture happened with the replacement of plow and harrow with the machine, although still horse-powered.

External crowding led to the internal crowding. Everyone's rush to the place of production reduced the area of each place, and any expansion of the dwelling could occur only at the expense of a neighbor. Thus, the technical ideal of the builder of the city became a dwelling of minimal dimensions.[24]

Houses turned into boxes, receiving the shape of boxes. For the external form of buildings is changing depending not on the need or the internal social form of the dwellings, and not depending on the building material (clay, wood, stone were and are used to build round and rectangular houses),[25] but because of the DEPENDENCE OF THE EXTERNAL FORM ON THE LOCATION OF THE DWELLING, ON THE GROUPING OF DWELLINGS.

You cannot make a vaulted ceiling, for above you there is another dwelling, another room. And the need for residents (or for you) to walk on a flat floor inevitably makes you, living under them (or under yourself), having an even ceiling. You cannot change the shape of the dwelling in a horizontal direction, nor increase its size, for the external form of your home; the size of your home is limited by the arrangement, grouping of dwellings.

The road and the fence are the first reason for moving the dwelling to a "rectangular" shape from a "round" one. The ax made this transition from closed curves to straight lines of houses possible. Streets intersections and vertical grouping of dwellings are the reasons for the further development of rectangular houses.

The durability of buildings was the need for a manufactory mode of production and the inevitable consequence of crowding (in particular, vertical).

Changes in the technology of production, the need for the TECHNICAL RE-EQUIPMENT OF ENTERPRISES (pace, pace!) demand the destruction of old cities, old houses. They require the most short-lived buildings. Number of floors—this scourge of fragility—the last guard of the durability of buildings, must give way to floorless buildings.

A new way of production allocation has become the reason (which is already in effect today) that mankind would finally be able to move to a new method of production in the construction industry. Like every new revolutionary mode of production, it will come as the CHEAPEST…

The new way of building production will put an end to byt (the way of life), to the way of life in general. For construction should not perpetuate the relations of people (and byt, way of life is characterized by the permanence of relations), new construction should not displace them either. It must FOLLOW the person, the changes in the relations between people and the development of the individual

[24]Since the moment of the appearance of trade capital in the city, this principle comes in insurmountable clash with the intra-city transport. It makes the streets wider. In doing so, it pushes buildings closer together, and creates new demand for the street, since there is more people in the crowded buildings, leading to higher demand for transport.

For more details in the role of transport see our works in 'Journal of the Communist Academy' 35–36 (1930), 'Construction in Moscow' 1 (1930).

[25]Thus, it is not the material itself that matters, but the treatment of it.

person. FOLLOW him, and DO NOT overtake them and do not consolidate the old relationship.

Dwelling, like clothes, can be improved: by increasing it in size, in breadth, in height, increasing the size of windows, etc. Is this conceivable under the old technology?

No, only prefabricated houses that can be assembled, disassembled, reassembled will meet the actual needs of a particular constantly developing person.

While a new way of production allocation requires a NEW GROUPING OF HOUSING, a new way of building production makes this task feasible.

But as long as we have not yet destroyed the existing real city, until the urban industry is SCATTERED, we should consider about 40–50 years, in which we must lay down our projects of resettling the workers of these old cities. Buildings are not depreciated until the expiration of this (approximate) period, for energy is still obtained from the centers, not from the network. In this era, passenger transport will play a huge role—especially the car. The more the city develops, the greater is the need for intra-urban transport. The more the city develops, the smaller is the opportunity of using this transport. This is the result of crowding, accumulation, concentration. This is the manifestation of the contradiction between the urban form of allocation and the producer's desire for production.

The contradiction is so great that sometimes it is better for the modern worker to live further (geographically) in order to be closer (in time,[26] although this is possible only under the condition of USING THE PASSENGER TRANSPORT as a means of transportation to the place of production).

Thus, the passenger VEHICLE REPRESENTS THE NEED OF A TRANSITIONAL EPOCH from the city to its destruction and uniting with the village into one; the car is an inevitable companion of the spatially unscattered yet industry.

In the transitional era, any form of housing—whether the philistine one, the household commune, the proletarian or, finally, the socialist, individual dwelling—all of them will inevitably imprint coercion and closeness and remoteness, since the HOUSING MATERIAL is motionless, unchangeable for the continuation of the ENTIRE CENTURY, while the form of the dwelling is changed within half a decade, and sometimes even half a year.

In this way, the method of building production on the one hand and the planning method on the other becomes a means of resolving social issues.

[26]Therefore, while there is still a territorial concentration of production, the car is our condition of being closer to production, not the location or territory. But cars will not disappear even when the production is most **scattered**, while the social ties (family, friendship, communes, etc.) are preserved.

They will either interfere (forcedness is exactly in this) or facilitate the formation of a new social form of housing, will prevent new relationships between people or stimulate them. Under these conditions, the human is left with the choice to force the technology to serve society, serve the people, their needs; a man under these conditions will have to declare war, a relentless war on technology, old construction technology and its ideological representative—unprincipled imitation, boundless technicism. On the old technology, he will declare war. New machinery, he will harness in the chariot of history.

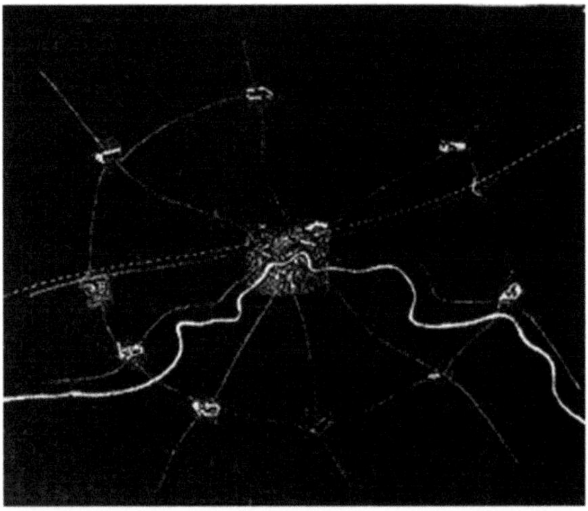

Fig. 10.1 "Disurbanist planning for urban areas (when the centre of the region is the city)". Source: Section on Socialist Settlement of the Construction Branch of Gosplan RSFSR, 1930, K. Afanasyev, M. Barsch, V. Vladimirov, M. Ginzburg, G. Sundblat, I. Milinis, A. Pasternak, G. Savinov, N. Sokolov (architects); M. Okhitovich (sociologist) and S. Orlovsky (engineer). Illustrations included in the article "Zametki po teorii rasseleniia", in: *S. A. Sovremennaia arkhitektura*, no. 6, 1927

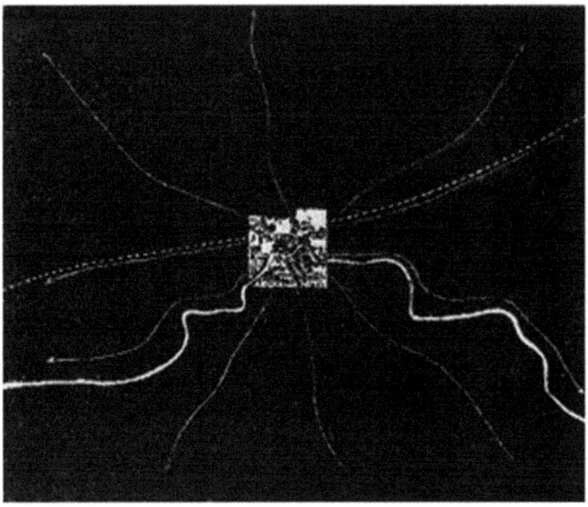

Fig. 10.2 "Decentralized planning for industrial areas (when the centre of the region is a large industrial company processing raw materials or semi-finished products)".
Source: ibid

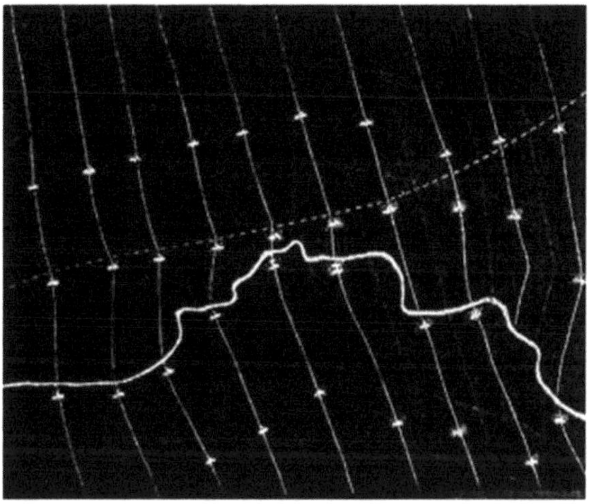

Fig. 10.3 "Acentric planning for agriculture or extractive industry areas (when the region's base is agriculture or mining)". Source: ibid

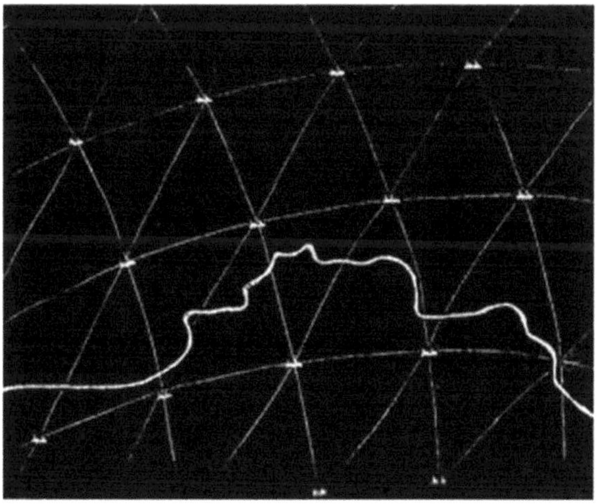

Fig. 10.4 "Dispersal planning for the consistent socialist city (when the distribution of work is based on the social relations of production; when an overall connection of industry and agriculture takes place; when the processing processes with the mining extraction processes take place as a whole)". Source: ibid

Chapter 11
Broadacre City (1932)

Frank Lloyd Wright

Introduction

Matthew Skjonsberg
ETHZ Future Cities Lab-Global, Zurich, Switzerland
skjonsberg@arch.ethz.ch

Proposing an indigenous model for North American settlements, Broadacre City (1932–58) embodied Frank Lloyd Wright's notion of 'ruralism' and his critique of modern industrial cities.

Broadacre City was both a planning statement and a socio-political scheme, explicitly championing "ruralism" as genuinely democratic and eschewing "urbanism" as a form of colonialism. Wright was sympathetic to Russian anarchist PA Kropotkin's notion of decentralized settlement—later popularized by Henry Ford—and believed that cities need no longer be centralized. Broadacre City illustrates a community built up over 'four generations of continuous inhabitation', sited along a river corridor and supported by a site-responsive regional network of transportation and communication infrastructure. The Jeffersonian grid is a secondary ordering system, supporting a primary armature of park systems woven through rural and urban settlements, adjoining protected forests and wilderness areas. Within this isotropic field, local governance administers a population of landowning citizen-farmers.

Wright's "The Art and Craft of the Machine" (1901), generally considered his first word on the city, cites Victor Hugo's idea of the printing press as democratizing knowledge—evoking Hugo's description of newsprint pages blowing in the

M. Barcelloni Corte and P. Viganò (eds.), *The Horizontal Metropolis*,
https://doi.org/10.1007/978-3-030-56398-1_11

wind like birds leaving the cathedral at dawn—asserting 'now knowledge…is free', and concluding: "Look out over the modern city at nightfall from the top of a great down-town office building, and you may see at a glance how organic the machine has become, how interwoven it is in the warp and woof of our civilization, its essential tool indeed, if not the very framework of civilization itself. Thus is the machine, the forerunner of democracy, into which the forces of art are to breathe the thrill of ideality – a soul".

This essay was initially presented March 1, 1901 at Jane Addams' Hull House in Chicago, a multi-faceted civic institution where Wright was actively involved. Most key Broadacre City themes were already there, but the central notion of decentralization was missing. In April Addams hosted Kropotkin who lectured on the subject of decentralization—and when Wright later takes up this theme he cites Kropotkin. Wright first explicitly addresses decentrality in his 1930 lecture on 'The City' at Princeton University, also first using the term 'ruralism', writing, "Ruralism as distinguished from Urbanism is American, and truly Democratic." Later that year, again echoing The Art and Craft of the Machine, he states: "Decentralization not only of industry but of the city itself is desirable and imminent…The greatest service sentient man is to receive from the machine…will be the death of urbanism! Hectic urbanism will be submerged in natural ruralism".

He clearly intends this as an explicitly polemic statement, elsewhere writing, "Ruralism as distinguished from 'Urbanisme' for future machine-age development is the business of the modem architect. Truly democratic business." His usage of French confirms that he is formulating his vision in direct opposition to European ideas, and is underlined by the fact that the term 'Broadacre City' was first used by Wright as the title for this article in March 1932, written as a critical response to Le Corbusier's article written for the same magazine just two-and-a-half months earlier. The magazine's editors described Wright's text as presenting 'a diametrically opposed program…which he sees as the logical development of the machine age.' Rather than "tear down the city and try to bring the green country in only to build the city up again on its old site," he proposed "the Broadacre City of free democracy: architecture and acreage seen as landscape." Wright illustrates the scheme with a design for a tower he had designed for New York in 1927—and eventually built in Oklahoma in 1956—but it seems he was also thinking of his 'non-competitive entry' to the City Club project in 1912, as the formal scheme of the Broadacre City models built from 1934 is proportionally identical to the 1912 project. Perhaps the closest thing we have to a built version of Broadacre City is found in Wright's home valley in Wisconsin, where from 1942 he designed a project for 'Regional Highways and Park Systems', and populated it with public services in the tradition of civic design—emphasizing ecology and community—the disciplinary predecessor to urban design—which has historically focused on economy and technology.

11.1 Broadacre City. An architect's Vision[1]

Recently in "The New York Time Magazine" the noted French architect Le Corbusier described the Green City which he has conceived as the solution of the modern urban problem-a substitute for the crowded centre. In the article that follows Frank Lloyd Wright, a pioneer American architect among moderns, presents another and diametrically opposed program – the Broadacre City, which he sees as the logical urban development of the machine age.

Let us approach the traffic problem not as a tinker or garage mechanic, nor childishly try to tear the city down to get the green country in and set the city up in it again on its old site-feudal towers a little further apart. Vested interest, once invested, cannot be divested except by agreement. And they will not agree.

With an architect's vision, let us see the natural law of organic growth at work as change: seeking the sequence to provide for consequences. Enough blind-alley non-sense has been talked about congestion by skyscraperites–words obscuring the simple issue. Of what practical use is the skyscraperites' expedient imagery? Super-space making for rent, to enable super-landlords to have and to hold the super-millions in super-concentration to make super-millions of superfluous millions? For organic reasons the traffic problem is insoluble for the future on any basis satisfactory to human life within any busy city.

True, the herd-instinct grows with the swarm in the erstwhile village streets. But the swarm is taking to wings—or to wheels, which is much the same because increased facilities of lateral movements are almost equivalent to flight—a fond human dream about to be realized.

By means of the motor-car and the inventions that are here with it, the horizon of the individual has immeasurably widened. A ride high into the air in any elevator today only shows the man how far he can go on the ground. And a view of the horizon gives him the desire to go. If he has the means, he goes, and his horizon widens as he goes. This physical release is at work upon his character.

His selfish interests might easily multiply and pile him up senselessly, in tiers of cells, ad infinitum, even after he got his release. These selfish interests may still turn the trick. He is still dazed by his new freedom. But, like a bird born in captivity, to whom the door of his cage has been opened, some time, soon, he will learn that he can fly. And when he learns that he is free, he is gone.

After all he himself is the city. The city is going where he goes and as he goes. When he goes he will be gone where he may enjoy all that the centralized city ever really gave him, plus the security, freedom and beauty of the ground that is his. That means he is going to the country with his machine by means of the machine, in larger sense, that is opening the way for him.

[1]Source: Wright, F. L. (1932). Broadacre city. An architect's vision. In *The New York Times Magazine*, March 20, 1932 (pp. 8–9).

The city is yet only about one-tenth the motor car city it will be if machine-made promises to the man are kept. Any dutiful devotion to the machine on man's part today should mean a motor car–comparative flight. The citizen and his increase either have a car or dream of having one or two or three which he already has.

If congestion is crucifixion now, what will congestion be within a few years when it is multiplied by "success" as many times as is inevitable?

Roughly calculate the mass of machines that machine-age "success" must mean to 6,000,000 people. More than several million private cars. Perhaps a quarter of a million trucks and as many delivery machines; a half million buses, displacing street-car tracks and unwholesome subways, hundreds of thousands of taxi cabs, meantime. With room enough for each and for incidental transients, to function at all, the mass would fill the busy city channels above the tenth story. Allowing for the crisscross or the gridiron, it would double and pile up over the skyscrapers themselves. Call this exaggeration and cut it in two–then cut that in two again, to be rash. There will be enough left at the rate to increase "success" will bring, to burry Manhattan out of sight with its own cars and transients.

The motor car has just begun on the city. Why deck or double-deck or triple-deck the city streets at a cost of billions of dollars, only to invite increase and meet inevitable defeat? Why not allow the citizenship to keep the billions that decking would cost to buy more motor cars and get more out of living in a more natural and fruitful life, as freedom dawns on the citizen? The utility of the city as a form of centralization vanishes by way of the machine that built it.

Democracy means just that freedom for the citizen–if the machine is working for the citizen. Monarchy was the ideal of centralization–the unit compelled to revolve around about a common centre. Democracy is the ideal of integration—many units free in themselves functioning together in freedom.

Monarchy has fallen. Our capitalistic system, if it persists as a form of centralization, stands to fall. Electrified mechanical forces that are building our modern world now are, by nature, turning upon centralization to destroy it. That means that the city is destroying itself.

Centralization, whether expressed as the city, the factory, the school or the farm now has the enormous power of the machine-age setting against it. It is in the nature of universal or ubiquitous mobilization that the city spreads out far away and thin. It is in the nature of flying that the city disappears. It is in the nature of universal electrification that the city is nowhere or it if everywhere.

Centralization by way of the city has had a big day and a long day. It is not dead yet. But it is no longer a necessity or a luxury. Mobilization of the human animal, volatilization of human thought, voice and vision make the city as troublesome an interference to life as "static" is troublesome to radio.

Once more humanity has overcome a physical disability with a mechanical appliance. The man may already get more out of his new release by way of increased mobility for lateral movement than ever came to him before in the history of his race. Imagine then, what is coming to him in the next twenty-five years.

Democracy, as the systematized integration of small individual units, is a practical and rational ideal of freedom: machine in hand. Integration over the whole surface of the nation is ideal and no less practical.

The Broadacre City is not merely the only democratic city. It is the only possible city, looking toward the future. Exaggerated vertical lanes of transport impinging upon congested, narrow horizontal lanes; tall channels, ruinous to privacy, to let light and air into offices or habitations; the shelf and the pigeon hole–all these are landlord expedients that it is high time to have done with. There is no life in them. There is only rent.

As for securing privacy by hermetically sealed and blinded buildings, hot air circulating between two glass surfaces opaque or transparent—that expedient means heating the inside and the outside impartially, 50–50, with no gratitude from the outside. And 1000 people to the hectare (2.47 acres) is looking not so far ahead. It is 980 too many.

It has been discovered that severe standardization is no bar at all to even greater freedom in self-expression than was ever known before. Always providing we mean by self-expression genuine individuality—and not personal idiosyncrasy. These two discoveries are the Magma Carta of the new liberty into which the citizen may now go by way of machinery, and go in his own machine: to join the building of the Broadacre City of the twentieth century that, all unknown to him, has already begun to be built where he belongs.

The present-day city, in the light of our new opportunities, has become a stricture in distribution and transport; a handicap in production; an imposition upon family life.

The present form of the automobile is crude and imitative compared to the varied forms it will soon take. The flying machine is only in a more or less experimental form, unwieldy in scale and a hostage to the elements. Tele-transmissions of sight and sound are not only experimental; they are in their infancy, as is the intelligence to which their operation is entrusted.

We are proud of the great network of highways–the road systems of the country we are just beginning to build. And yet it requires but little imagination to see in all the power of these new resources, machines and materials, new release of human activity within reach of every one–the basis not only of adventure and romance with nature, but a safer, saner, less anxious life for free people; a longer, happier life wherein the man is sure of a living for himself and his own; sure also of some healthful association with beauty. Not only is the city itself a structure, a handicap in production, but the contributing railroad is too limited in movement, too expensively clumsy and too slow in operation. The end of the day of the back-and forth haul, long or short, demanded by centralization, is in sight; the end, too, of mass transport by iron rail.

No great flight of imagination is needed to see the coming highways as great architecture, the telephone and telegraph poles down. Poles and wires are already useless. Signboards down, too, as gratuitous insults to intelligence. Fine bridges and noble viaducts, may be great architecture in great landscape, landscaping healing the cuts and fills of the constructions of the roadbeds.

Highway travel is growing more and more interesting because, becoming as safe as the train, and swifter, it is more desirable because it is more spontaneous. Mobile human transport is getting down within reach of every one at last: no waste time or waste motion to pay for.

Produce is fresh every hour because distribution is taking place where producer and consumer happen to stand or choose to live. No flight of immature imagination is needed to see these great highways with their tributaries and ramifications hooked up at intervals with safe systems of noiseless, compact air transport; to see the network of fine roads passing public-service stations that live up to all the name implies, passing roadside markets, integrated with groups of three, five and ten acre intensive farm units; diversified manufacturing units not far away; great automobile objectives as amusement resorts built on and with the natural features of environment; fine homes in parks and gardens that are small farms, too, all windings up the beautiful natural features of our great landscapes into the spacious broadacre cities of the near future: the greatest work of man for the liberation of mankind that has ever been realized.

Human environment on this soil of ours dedicated to human freedom may now be spacious and humane; life free as to individuality, yet integrated spontaneously, instantaneously as a whole with all life; the new cities affording each man and his family greater facility for culture, the amenities of life on a larger scale, human intercourse more liberal and healthy. Every home is a centre and a delightfully varied association with nature, the homemaker's birthright.

Not 1000 to the hectare. No, nor space sold by the square foot. But space belonging to those who use it, by the acre. There is space enough for this "distribution"—good for nothing else.

Let us say, for the sake of argument, that the objectionable mechanical features of travel are removed, as they are rapidly being removed; that the utilitarian object of concentration in centralized cities is entirely gone, as it is indeed, or is fast going, hastened by the new leverage of the new power of a new life—we call it the machine for lack of a more comprehensive term. Suppose the motorbus is as comfortable and inviting to us as the Pullman never was and together with the low-gravity truck-train it takes to the right of way the railroads; the average highway has four traffic lanes, the main arteries six or more.

These safe, broad traffic lanes—noble engineering in themselves—are thronged with life, moving not humdrum in standardized machines as we know them, but in many forms suited to many needs and individuals. Swift singles, doubles; safe quads and dignified sextets; safety and gayety now in the swift-moving fleet. Grade crossings eliminated Charm returning to the road.

Self-contained mechanical units, as varied and well designed, are taking off into the air from plane stations at intervals—noiseless, compact and safe. Beguiling resorts, famous for hospitality, cuisine, entertainment, occur where the beautiful features of the land occur. Extensive, complete markets, with produce fresh every hour, are integrated with the intensive small farm units. Other markets pertain to manufacture of classified foreign products. These are scattered along the highways, but each series integrated as links in a great chain. Service stations that are hostelries and rendezvous for comfort and entertainment at strategic points are slightly features of the highway itself.

The farm units that produce and the factories, too, are within a ten-mile radius of the extensive roadside markets, integrated with them—and a ten-mile radius means within ten minutes of each other. The factories might be within walking distance of the acre home units of the workers, but why, when ten miles is ten minutes?

Interchange of products is by continuous movement in beautiful environment each way, spaciousness determined by the man in his machine and no traffic problem because traffic is the norm of the new spacing or standard of measurement of the centring of all units.

As the citizen sits in his car, he may press a variety of buttons or turn an indicator and obtain any section he desires of the modern newspaper—the forest saved and millions of tons of waste paper eliminated. He picks up by sound and sight whatever he is interested in, learns by listening where the day's specialties are to be found; where events of interest are occurring or are occurring to occur, near or far away. All over the surface of the globe, in fact, if he pleases, he may listen in.

His home is now a centre to which all this comes and he is integrated with all other homes and forms of production or of distribution spontaneously and instantaneously. More, he has the means of easy, direct personal contact with all forms of production or distribution within a radius of 150 miles: two hours by wheeling or a half-hour by flying.

He can cross the continent, for conference or pleasure, within twenty-four hours. The home, you see, takes on a new significance. A fresh importance as the centre.

Business, too, takes on a different character. It is a form of freedom, not slavery. Anxiety is easier to bear. Individual initiative, in this new freedom, comes alive. Relieved of the inordinate, merciless pressure of centralization, architecture bursts into bloom. Hugo, the great modern of his time, said it might do so in the twentieth century. There is no prevailing style, but style is everywhere. The common sense of the age flowers into new forms born of new materials.

Better ideas of life take rational imaginative form. The garden as a park is normal to the "new" city. Play spaces for all the sports we have now and new ones to come. The integrated three, five and ten acre farm units are coveted things for beauty, and are a basis for a safe, rich life for the rearing of families: work to be done now with lessons to be learned. Man's relation to his animals takes on a better character in architecturally planned association, in surroundings where the nature of the association is recognized and provided for intelligently.

Spaciousness, transport, buildings, all of life intimate with the ground, appropriate to each other and to each and every one according to his need or the nature of his meed and love of life. Woods, streams, mountains, ranges of hills, the great plains–all are shrines, beauty to be preserved.

Imagination is our human divinity. It alone is what the human herd needs to distinguish and save it from the fate that has overtaken all other herds, human or animal. In the nature of the traffic problem lies the liberation, and that soon, of such human enslavement as the city now stands to exploit. Or else in its nature lies the further degradation of humanity.

In this collision and irreconcilable war of mechanical factors which we call the traffic problem, an element of human freedom is caught and held against its larger interest—held to serve only the will and act of more or less gratuitous forms of fortune that function mostly as rent, in some guise—the will that built and maintains the city. The struggle is on.

The traffic problem will be solved as architecture. But not by verticality. Narrow lanes of vertical transport impinging on inadequate horizontal alleys, narrow vertical channels destroying privacy to let light and air into dreary miles of shelving upon which human beings, by paying tribute, may perch or roost—this arrangement, however it may be rationalized or visualized, affords no measure of life to the man, such as was promised to him by our charter of freedom, the Declaration of Independence.

In this traffic problem our future as a nation is at stake. The problem will be solved as architecture, but the twentieth-century cities will be no scrap-pile or brick yard nor any standardized park with standardized colossi standing in it and scraping the sky. It will be solved by the horizontal line of the machine age, indefinitely extended as the great architectural highway and by the flat plane of the machine age expanded into the free acreage of the Broadacre City of free democracy: architecture and acreage seen as landscape. No power of man can long preserve any imposition reversing these new social forces when they are understood and put to work by man for man. No, not even if held by himself against himself.

11.2 The Broadacre City (from: *The Disappearing City*)[2]

We are concerned here in the consideration of the future city as a future for individuality in this organic sense: individuality being a fine integrity of the human race. Without such integrity there can be no real culture whatever what we call civilization may be.

We are going to call this city for the individual the Broadacre City because it is based upon a minimum of an acre to the family.

And, we are concerned for fear systems, schemes, and "styles" have already become so expedient as civilization that they may try to go on in Usonia as imitation culture and so will indefinitely postpone all hope of any great life for a growing people in any such city the United States may yet have.

To date our capitalism as individualism, our eclecticism as personality has, by way of taste, got in the way of integrity as individuality in the popular

[2]Excerpt from: Wright, F. L. (1932). *The disappearing city* (pp. 3–20). W. F. Payson.

understanding, and on account of that fundamental misunderstanding we, the prey of our culture-monger, stand in danger of losing our chance at this free life our charter of liberty originally held to us.

I see that free life in the Broadacre City.

As for freedom; we have prohibition because a few fools can't carry their liquor; Russia has communism because a few fools couldn't carry their power; we have a swollen privatism because a few fools can't carry their "success" and money must go on making money.

If instead of an organic architecture we have a style formula in architecture in America, it will be because too many fools have neither imagination nor the integrity called individuality. And we have our present overgrown cities because the many capitalistic fools are contented to be dangerous fools.

A fool ordinarily lacks significance except as a cipher has it. The fool is neither positive nor negative. But by way of adventitious wealth and mechanical leverage he and his satellites—the neuters—are the overgrown city and the dam across the stream flowing toward freedom.

It is only the individual developing in his own right (consciously or unconsciously) who will go, first, to the Broadacre City because it is the proper sense of the dignity and worth of the individual, as an individual, that is building that city. But after those with this sense the others will come trailing along into the communal-individuality that alone we can call Democracy.

But before anything of significance or consequence can happen in the culture of such a civilization as ours, no matter how that civilization came to be, individuality as a significance and integrity must be a healthy growth or at least growing healthy. And it must be a recognized quality of greatness.

In an organic modern architecture, all will gladly contribute this quality, as they may, in the spirit that built the majestic cathedrals of the middle-ages. That medieval spirit was nearest the communal, democratic spirit of anything we know. The common-spirit of a people disciplined by means and methods and materials, in common, will have—and with no recognized formula—great unity.

Already the centripetal city is itself an "ism" for ists and ites. Individuality has no longer a place in it more important than a burrow. Individuality is driven into nooks and corners or thwarted or aborted: frustrated by the mass-life only competing with, never completing, life.

So no healthy human-soul may longer grow or long survive in the vicarious life of the machine-made city because life, there, must be a surrender of true correlation of the human faculties to the expedient in some form; expedients imposed senselessly upon every soul in it to no purpose at all—except as they may be found to be some form of rent.

Voluntary self-sacrifice may be constructive. But to be condemned to the servile sacrifice of a voluntary life-long use of petty expedients to get by to, eventually, nowhere, is quite another matter. The human soul grows by what it gives as well as by what it feeds on. But the soul does not grow by what is exacted from it. Urban life having served its term is become a life-sentence of vicarious acts and the petty exaction of the expedient. A life out-moded. The big city is no longer modern.

"BROADACRE CITY": AN ARCHITECT'S VISION

Spread Wide and Integrated, It Will Solve the Traffic Problem and Make Life Richer, Says Frank Lloyd Wright

Recently in The New York Times Magazine the noted French architect Le Corbusier described the Green City which he has conceived as the solution of the modern urban problem—a substitute for the crowded centre. In the article that follows Frank Lloyd Wright, a pioneer American architect among moderns, presents another and diametrically opposed program—the Broadacre City, which he sees as the logical urban development of the machine age.

By FRANK LLOYD WRIGHT

LET us approach the traffic problem not as a tinker or garage mechanic, nor childishly try to tear the city down to get the green country in and set the city up in it again on its old site—feudal towers a little further apart. Vested interests, once invested, cannot be divested except by agreement. And they will not agree.

With an architect's vision, let us see the natural law of organic growth at work as change: seeking the sequence to provide for consequences. Enough blind-alley nonsense has been talked about congestion by skyscraperites—words obscuring the simple issue. Of what practical use is this skyscraperites' expedient imagery? Superspace making for rent, to enable super-landlords to have and to hold the super-millions in super-concentration to make super-millions of superfluous millions? For organic reasons the traffic problem is insoluble for the future on any basis satisfactory to human life within any busy city.

True, the herd-instinct grows with the swarm in the erstwhile village streets. But the swarm is taking to wings or to wheels, which is much the same because increased facilities of lateral movement are almost equivalent to flight—a fond human dream about to be realized.

By means of the motor car and the inventions that are here with it, the horizon of the individual has immeasurably widened. A ride high into the air in any elevator today only shows the man how far he can go on the ground. And a view of the horizon gives him the desire to go. If he has the means, he goes, and his horizon widens as he goes. This physical release is at work upon his character.

His selfish interests might easily multiply and pile him up senselessly, in tiers of cells, ad infinitum, even after he got his release. These selfish interests may still turn the trick. He is still dazed by his new freedom. But, like a bird born in captivity, to whom the door of his cage has been opened, some time, soon, he will learn that he can fly. And when he learns that he is free, he is gone.

After all he himself is the city. The city is going where he goes and as he goes. When he goes he will be gone where he may enjoy all that the centralized city ever really gave him, plus the security, freedom and beauty of the ground that is his. That means he is going to the country with his machine by means of the machine, in larger sense, that is opening the way for him.

* * *

THE city is yet only about one-tenth the motor car city it will be if machine-made promises to the man are kept. Any dutiful devotion to the machine on man's part today should mean a motor car-comparative flight. The citizen and his increase either have a car or dream of having one, envying the neighbor the one or two or three which he already has.

If congestion is crucifixion now, what will congestion be within a few years when it is multiplied by "success" as many times as is inevitable?

Roughly calculate the mass of machines that machine-age "success" must mean to 6,000,000 people. More than several million private cars. Perhaps a quarter of a million trucks and as many delivery machines; a half million buses, displacing street-car tracks and unwholesome subways; hundreds of thousands of taxicabs, meantime. With room enough for each and for incidental transients to function at all, the mass would fill the busy city channels above the tenth story. Allowing for the criss-cross on the gridiron, it would double and pile up over the skyscrapers themselves. Call this exaggeration and cut it in two—then cut that in two again, to be rash. There will be enough left at the rate of increase "success" will bring, to bury Manhattan out of sight with its own cars and transients.

The motor car has just begun on the city. Why deck or double-deck or triple-deck the city streets at a cost of billions of dollars, only to invite increase and meet inevitable defeat? Why not allow the citizenship to keep the billions that decking would cost to buy more motor cars and get more out of living in a more natural and fruitful life, as freedom dawns on the citizen? The utility of the city as a form of centralization vanishes by way of the machine that built it.

Democracy means just that freedom for the citizen—if the machine is working for the citizen. Monarchy was the ideal of centralization—the unit compelled to revolve around about a common centre. Democracy is the ideal of integration—many units free in themselves functioning together in freedom.

Monarchy has fallen. Our capitalistic system, if it persists as a form of centralization, stands to fall. Electrified mechanical forces that are building our modern world now are, by nature, turning upon centralization to destroy it. That means that the city is destroying itself.

Centralization, whether expressed as the city, the factory, the school or the farm, now has the enormous power of the machine-age setting dead against it. It is in the nature of universal or ubiquitous mobilization that the city spreads out far away and thin. It is in the nature of flying that the city disappears. It is in the nature of universal electrification that the city is nowhere or it is everywhere.

Centralization by way of the city has had a big day and a long day. It is not dead yet. But it is no longer a necessity or a luxury. Mobilization of the human animal, volatilization of human thought, voice and vision make the city as troublesome an interference to life as "static" is troublesome to radio.

Once more humanity has overcome a physical disability with a mechanical appliance. The man may already get more out of his new release by way of increased facility for lateral movement than ever came to him before in the history of his race. Imagine then, what is coming to him in the next twenty-five years.

* * *

DEMOCRACY, as the systematized integration of small individual units, is a practical and rational ideal of freedom: machine in hand. Integration over the whole surface of the nation is ideal and is no less practical.

The Broadacre City is not merely the only democratic city. It is the only possible city, looking toward the future. Exaggerated vertical lanes of transport impinging upon congested, narrow horizontal lanes; tall channels, ruinous to privacy, to let light and air into offices or habitations; the shelf and the pigeon hole—all these are landlord expedients that it is high time to have done with. There is no life in them. There is only rent.

As for securing privacy by hermetically sealed and blinded buildings, hot air circulating between two glass surfaces opaque or transparent—that expedient means heating the inside and the outside impartially, 50-50, with no gratitude from the outside. And 1,000 people to the hectare (2.47 acres) is looking not so far ahead. It is 980 too many.

It has been discovered that severe standardization is no bar at all to even greater freedom in self-expression than was ever known before. Always providing we mean by self-expression genuine individuality—and not personal idiosyncrasy. These two discoveries are the Magna Carta of the new liberty into which the citizen may now go by way of machinery, and go in his own machine: to join the building of the Broadacre City of the twentieth century that, all unknown to him, has already begun to be built where he belongs.

The present-day city, in the light of our new opportunities, has become a stricture in distribution and transport; a handicap in production; an imposition upon family life.

* * *

THE present form of the automobile is crude and imitative compared to the varied forms it will soon take. The flying machine is only in a more or less experimental form, unwieldy in scale and a hostage to the elements. Teletransmissions of sight and sound are not only experimental; they are in their infancy, as is the intelligence to which their operation is entrusted.

We are proud of the great network of highways—the road systems of the country we are just beginning to build. And yet it requires but little imagination to see in all the power of these new resources, machines and materials, new release of human activity within reach of every one—the basis not only of adventure and romance with nature, but of a safer, saner, less anxious life for a free people; a longer, happier life wherein the man is sure of a living for himself and his own; sure also of some healthful association with beauty. Not only is the city itself a stricture, a handicap in production, but the contributing railroad is too limited in movement, too expensively clumsy

Drawn by Frank Lloyd Wright.

The Future—A Design for an Apartment Building in the "Broadacre City."

Fig. 11.1 Wright's article "Broadacre City: An Architect's Vision" was published in *The New York Times Magazine* in 1932—the same year he founded his Taliesin Fellowship and authored the books "The Disappearing City" and "An Autobiography". The term "Broadacre City" was first used by Wright in this article, written as a critical response to one that Le Corbusier had written for the same magazine just two and a half months earlier. The editors described Wright's text as presenting "a diametrically opposed program [to that of Le Corbusier]…which he [Wright] sees as the logical development of the machine age." Source: The New York Times Magazine. © 2019, ProLitteris, Zurich. Credits: M. Skjonsberg

Fig. 11.2 One quarter section of four constituting the full Broadacre City model, here displayed in the Hillside Building at the Taliesin campus circa 1935–36. Behind are two other models, and two of at least a dozen panels each proclaiming key concepts and ambitions of this profoundly pedagogical project—as further attested to by the attribution "MODELS BUILT BY THE TALIESIN FELLOWSHIP, FRANK LLOYD WRIGHT ARCHITECT". As he described Broadacre City that year, "...the ground itself predetermines all features; the climate modifies them; available means limit them; function shapes them." ("Broadacre City: A New Community Plan," Architectural Record no. 77, April 1935). Source: Frank Lloyd Wright Foundation 3402.036. © 2019, ProLitteris, Zurich. Credits: M. Skjonsberg

Part II
Interpretative Concepts and Visions

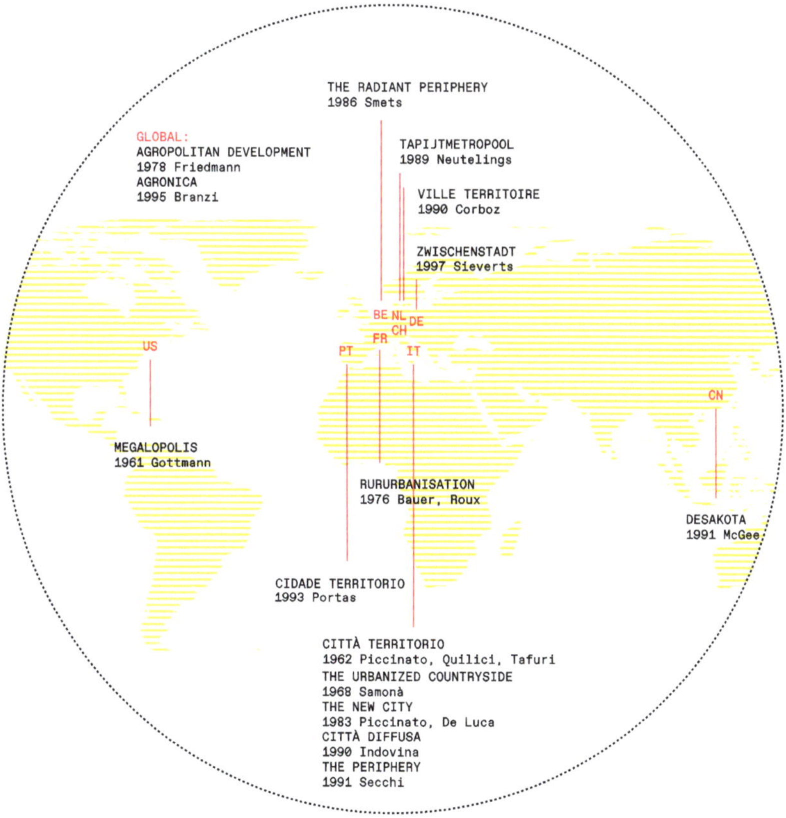

Chapter 12
Interpretative Concepts and Visions: Introduction

Paola Viganò

The second part of the Anthology reconstructs a more recent and empirical tradition of thought that emerged in the twentieth century between the 60s and the beginning of the 90s from the direct observation of different contexts and case-studies developed by a group of scholars confronted with radically changing urban conditions.

Through a selection of themes and research paths that take in the prophecies and clues of the first part, as well as the interpretative concepts and visions of the second part, this introduction works as a bridge between the two sections of the book. We can observe the high frequency of oxymorons to emphasize the unusual combination of opposite characters, sometimes dealing with unusual spatial ones, as in the "*campagna urbanizzata*" (Samonà, in the plan for the Province of Trento, 1968), or the "*città diffusa*" (Indovina 1990); or to underline the novel topology, as in the "*Zwischenstadt*", indicating positions in the territory that, simply due to being "in-between cities", develop new forms and characteristics of a novel urban landscape (Sieverts 1997).

Often, a degree of ambiguity is created by the "free" translation of the original titles as in *Zwischenstadt*, which in the English version becomes, quite strangely, "cities without cities", insisting on the presence of a new urban constituent where, in fact, no cities existed prior to that, instead than on the "in-between" character of the new urbanity. This is also the case of another "lost in translation" definition, the *Tapijt metropool*, where the same author of the formulation uses English terms which are not the translation of the Dutch ones he himself proposed. The *Patchwork metropolis* coined by Neutelings (Neutelings 1990) refers to a very structured organizational device (the patchwork), where rules of proximity and connection among pieces must be respected by each patch, which has to be coherent with the

P. Viganò (✉)
EPFL, Lausanne, Switzerland
e-mail: paola.vigano@epfl.ch

IUAV, Venice, Italy
e-mail: paola.vigano@iuav.it

M. Barcelloni Corte and P. Viganò (eds.), *The Horizontal Metropolis*,
https://doi.org/10.1007/978-3-030-56398-1_12

group, both in form and measure. Against this, the concept of carpet (*tapijt*) might and usually involves a common figure, or a regular motif that emerges during the carpet's construction. Translations and misunderstanding[1] are part of the genealogy of the Horizontal Metropolis, where misinterpretations and the nomadism of concepts through languages and regions have produced original and often fertile terminological confusion. However, a set of common and cross-over themes can be recognized.

The Industrial Revolution and the City Countryside Continuum as a Policy

As already described by Geddes, the structuring of the industrial city along the coal seams defines a new urban scale and, thanks to technological progress, the introduction of novel relationships between the city and production. Here the construction of a Horizontal Metropolis is a conscious investment in the definition of an alternative to the concentrated city as a solution to the problems of congestion as well as social and political conflict: Belgium is one of the most exemplary cases. The idea of a "radiant periphery" ("*banlieue radieuse*", Smets 1986), opposed to the Leorbusierian Radiant City, is anchored in Emile Vandervelde's *L'exode rural et le retour aux champs*, 1903, and Seebohm Rowntree's 1910 report on labour and land in Belgium. Beyond the physical and topological, the centrality of this space is highlighted as part of the political reformism from which Modern Urbanism draws its main roots. Smets acutely locates the "radiant periphery" at the core of the emancipatory program of Modern Urbanism. Its project no longer lies in the traditional urban space, although differently imagined and designed in Le Corbusier's "Radiant City". It lies in the less heroic space of the periphery where the scattered dreams of the masses can find a concrete realization (and this, of course, is not only the case of Belgium). The progressive faith in a better future finds in the "Radiant Periphery" its closest, democratic and more concrete realization.

The Urbanized Countryside: In Search of New Spatial Hybrids

The terms *campagnes urbanisées*, later used by Giuseppe Samonà in the 1960s to conceive a project of decentralised urbanization for the Trentino region, was used at the beginning of the twentieth century by Émile Vandervelde, the socialist exponent

[1]This text takes up and extends my text "The Horizontal Metropolis Deux ou trois choses que je sais d'elle", in N. Katsikis with F. Farinelli, A.Lahoud, P. Viganò, A. Wall, *Geo-Graphical Urbanism*. In L. Degli Esposti (Ed.) (2015).

and professor in Brussels, in connection with the "new" (at the time) system of transportation. Discussing the study produced by Constantin Pecqueur in 1838[2] at the launch of a vast railway plan in France (as an answer to the question of its impact formulated by the *Académie des sciences morales et politiques*), Vandervelde criticizes the negative separation of rural activities from industry, a situation that could be overcome by a wider accessibility. Following Pecqueur, this hybridization would have spread to the countryside the "matériel de la civilisation" (Vandervelde 1903) still peculiar of the city. The invention of new hybrid spaces would have been possible: farms, factories and town houses would have redesigned the old villages as a regular *composé*, while villas could infiltrate the urban space, bringing air and greenery. There is no opposition to the city or the big city: the idea is that urban civilization, notwithstanding its problems, brings better living conditions, and the redistribution of some of its characteristics to larger territories could only be beneficial. The "trains ouvriers", with their incredibly low-priced tickets, are in Vandervelde's opinion "the most unconsciously revolutionary" (*la plus inconsciemment révolutionnaire*) decision made by the Belgian government in the last thirty years since his book, *L'exode rural et le retour aux champs*, after the French-Prussian war (Vandervelde 1903: 203).

The emphasis on the new hybrid condition emerges again in the definition of *Rurbain* (Bauer and Roux 1976): the main characteristic of the new inhabited space is the unusual mix of urban and rural landscapes and activities. This combination might point towards a new urban-rurality or, more simply, indicates a future model of urban agriculture. Here what needs to be underlined is the coexistence, following Kropotkin's utopia, of different economies and practices with the dissolution of the opposing concepts of city-countryside. For Francesco Indovina, the trajectories of the *diffuse city* describe a territory of urban centres and scattered nuclei, within which the same population expresses an improvement in economic conditions through widespread home ownership, in continuity: one house after another. This, according to Indovina, is diffuse urbanization.

Other definitions, as *Ville-Paysage*, lie in the same vein of a novel integration between traditionally contrasted terms. In this case it is the translation and probable expansion of the German *Stadtlandschaft,* a concept proposed by the German geographer Passarge,[3] which constitutes a partial reconfiguration of the same. Indeed, it might also be simply translated as *Urban Landscape*, in this case being much closer to the "townscape" tradition of Raymond Unwin and later of Gordon Cullen than the invention of a new urban-landscape substance.

[2]Published in 1839 with the title: *L'Économie sociale: des intérêts du commerce, de l'industrie et de l'agriculture, et de la civilisation en général, sous l'influence des applications de la vapeur*, Paris: Desessart.

[3]Coined around 1920, by Siegfried Passarge (1867–1958) (*Die Grundlagen der Landschaftskunde*, Hamburg 1920/21), also author of *Stadtlandschaften der Erde* (Hamburg 1930), *Stadtlandschaft* is today a term reinterpreted keeping a distance from the context in which it was proposed. I always thank Thomas Sievert for the precious insight. See also: Mantziaras (2008).

A New Integration: Against the Division of Labour

The agrarian crisis of the 1860s, with the competition from goods produced by the rest of the world, depleted the European countryside and reinforced the attractiveness of the city. The stronger and stronger concentration in the metropolis pushed several authors, at the turn of the century, to revive the idea of power redistribution and territorial re-balance already proposed by Ruskin, Morris, Proudhon and others. The industrialization of agriculture and the implantation of new industries in the fields is Kropotkin's manifesto against the division of labour and Taylorism: "having no idea how and why the machinery performs its rhythmical movements" (Kropotkin 1974). Kropotkin favoured the idea of recovering the "most varied texture of soils and climates"(Kropotkin 1974:25) of any nation, a variety which can only exist within "a variety of occupations" (Kropotkin 1974:25). The theme of integration against division was crucial in a period in which, as Colin Ward (editor of the 1974 edition) underlined, "Britain was the workshop of the world" (Kropotkin 1974:41). At the beginning of the twentieth century, the decentralization of industrial production seemed possible also thanks to electricity and hydro power (the "white coal", *houille blanche*). It was the presence of dispersed energy production in the mountains that fed Rousseau's imagery while describing, as early as the end of the eighteenth century, Switzerland as a unique city with workshops distributed in the valleys close to creeks and rivers. In the Alps or Pyrenees, all corners close to a water course could provide a workshop, electric tramways could run along the valley and the remote alpine village could be equipped with electricity. Parallels can be made with the American factories drawing their power from New England's waterfalls as contrasted with the later steam power engines of factories in crowded European industrial cities. In fact, the isotropic availability of energy, due to the proximity to the source, as in the Alps, or its transportability and easy distribution (through an extended transmission network and a fine and connected distribution network) define, as in Okhitovich's vision, conditions of the isotropic implantation of industrial production. In the diffuse city of the Veneto Region, dispersed small and medium size industries now stand alongside the proto-industry close to the water and the big industry of Porto Marghera, for which large dams and transmission networks were created, also feeding the rest of the territory.

At the beginning of the twentieth century, an "industrial exodus" of industries leaving the cities was announced by Émile Vandervelde, opening up to a further hybridization of the landscape.

A Revolution in Land Use and the Role of Open Space

The idea of a revolution in land use is also the fundamental trait of the megalopolis described by Jean Gottmann in his research on the eastern-coast of the United-States, published in 1961, and is repeatedly considered a specific character inside the territories of dispersion (as in McGee's *desakota* system or the *città diffusa* described by Piccinato, Indovina, Secchi and others). This character, even at first glance, is what makes the European City Territory so different from the traditional North-American suburbs. Although the two forms have today more and more similarities, they obviously refer to different cultural, ideological and utopian backgrounds. The presence of the open space–soil, forest, wetlands, meadows, gardens… cultivated, natural and equipped, today more and more understood as a fundamental set of ecosystem services for all forms of life–represents an inversion of the traditional urban figure-ground relation: a "Reverse City" (Viganò 1999). The possibility of a productive and sustainable agriculture developed within an urban rural framework is today realistically considered, as well as the potential qualities and functionalities of rurban soils, with benefits and services offered by the extensive presence of open areas. These considerations and awareness transformed urban and regional design in the last decades.

The Extended Metropolitan Character

In the presentation of the research "Hypothesis on the central Veneto area. The diffuse city" (1987),[4] Francesco Indovina clarifies the link between research on the "nature of the central Veneto area", the inclusion of Venice among metropolitan municipalities and the consequent problem of the delineation of its boundaries. The phenomenon has taken on a "gigantesque" characteristic, almost a "change of status" (Indovina 1990:21). Today in the representations of the region the diffuse city has become part of the urban and is no longer an exception.

The research tackles the issue of the possibility of defining a city as an original spatial aggregate whose urban functioning must be proven. The term is "descriptive of a type of spatial organisation" that "presents some urban characteristics in the absence of others of such nature" (Indovina 1990:16). In this case too, an oxymoron is involved.

What was to be grasped was the novelty of the phenomenon. According to Indovina, the new conceptual elaboration was urgent as many parts of the country were changing in this direction and in order not to reinforce misconceptions due to the similarities, for example, with the American suburbs (Indovina 1990:21). Searching for evidence of urbanity is interesting, not among the positive aspects linked to the city, but rather among the negative ones: the diffuse city, like the city, is highly selective. The urban use of the territory is reserved for some brackets of the population. "A partial, inequitable and discriminatory use of the space typical of the city" is proposed (Indovina 1990:16). This is why the practices of space should be investigated, between hierarchy and loose networks.

[4]"Ipotesi sull'area centrale veneta. La città diffusa", research funded by the Mpi 60%, 1987.

While in the city "accessibility is measured in terms of the multiplicity of opportunities and closeness" (Indovina 1990:32), it is the greater propensity for movement and the greater accessibility in terms of time that cancels out the difference between the diffuse city and the consolidated city. The role of the car in the creation of an "intense catchment area" for private services is central; services capable of attracting a vast audience are found at metropolitan level. The horizontality (as in a city) and multi-directionality of relationships (movements within the diffuse city and outside of it) develop with the support of a territory of ancient urbanization. Nevertheless, the complexity of spatial hierarchies remains to be investigated. In 1990, the diffuse city does not yet seem to attract the "current power," it fails to conceptualize the leap in scale that occurs once the services to businesses and customers attractive on a vast scale outline a new type of metropolis. The diffuse city might perform a "balanced role and that of a balancing element, both in terms of (innovative) production and the social quality of life" (Indovina 1990:42). However, to achieve this, the diffuse city should become a project.

The Problem of a Project

The question of the "project", of designing a different type of space, made of different urban materials, for a different society and economy tightened by horizontal relations, where links and connections are ever looser, from all points of view, is more or less implicitly considered in each interpretation. This is clearly the case, in this second part, in the contributions of John Friedmann and Andrea Branzi—today presenting themselves as perhaps even more relevant due to the accumulation of new urban issues, regarding the ecological, socio-demographic and economic transition.

Through gashes in the thick canvas of common ideas of city and territories, visions and prophecies of the first part of the Anthology filter through and resurface in the second part. Here, each of the interpretations contains a project *in nuce*, or at least its possibility, within the potential that the reading of each situation has revealed. These are *situated* visions, rather than generalized theories, in which the difference, the being different from other cases and places, is forcefully asserted. Bernardo Secchi places the project and its problem at the centre of his research, and this does not mean losing sight of the social, economic and spatial complexity with which every aspiration of change has to grapple with. It is the form of the project to be revisited, its main figures—spatial figures and figures of discourse—the emergence of which marks new paradigms, concepts under formation, new types of spaces and practices. Horizontality, isotropy and the new urban materials, at the different scales, become the subject of description and of design research.

The six topics briefly outlined here have not yet been entirely developed, since. They still contain many unknown elements, and offer an agenda from which to discern the current problems raised by the pervasive rural urban environment, the increasing marginalisation of spaces and populations, the ecological crisis. Taken as a whole, interpretations, prophecies and implicit projects for a Horizontal Metropolis feature as a legacy to be critically interpreted, yet with a mind free of prejudices and reductive assumptions.

References

Bauer, G., &, Roux, J.-M. (1976). *La Rurbanisation ou la ville éparpillée.* «Espacements», Éditions du Seuil. pp. 1–25.

Indovina, F. (1990). *La città diffusa.* In F. Indovina, F. Matassoni, M. Savino, M. Sernini, M. Torres, & L. Vettoretto (Eds.), *La città diffusa.* Daest. pp. 21–43.

Kropotkin, P. (1974). *Fields, Factories and Workshops Tomorrow.* In C. Ward (Ed.). Harper & Row.

Mantziaras, P. (2008). *La ville-paysage—Rudolf Schwarz et la dissolution des villes.* Metis Presses.

Neutelings, W. J., Sulsters, W., van Wesernael, P., & Winkler, E. (1990). De transformatie van de Haagse zuidrand, Stedebouwkundige studie verricht in opdracht van de Dienst Stadsontwikkeling en Grondzaken van de gemeente 's G ravenhage, in het kader van het stadsplan—1988–89. In R. Geurtsen, P. Verschuren, & G. Kwakkenbos (Eds.), Stadsontwerp in 's-Gravenhage: stedebouwkundig ontwerpen. Delft: Faculteit der Bouwkunde, pp. 80–95.

Vandervelde, É. (1903). *L'exode rural et le retour aux champs* (p. 9). Felix Alcan.

Viganò, P. (1999). *La città elementare.* Skira.

Viganò, P. (2015). In L. Degli Esposti (Ed.). "The Horizontal Metropolis 'Deux ou trois choses que je sais d'elle". In *Milan Capital of the Modern.* New York: Actar Publisher.

Sieverts, T. (1997). *Zwischenstadt, zwischen Ort und Welt Raum und Zeit Stadt und Land.* Vieweg.

Smets, M. (1986). *La Belgique ou la banlieue radieuse.* In M. Culot, & C. Mierop (Eds.), *Paysage d'architectures* (Catalogue published for the exhibition *Paysages d'Architecture* organized by the Fondation pour L'Architecture and The Musée des archives d'architecture moderne June 21 to September 20, 1986 at La Fondation Pour L'Architecture), Bruxelles: Archives d'Architecture Moderne, pp. 33–35.

Chapter 13
Megalopolis (1961)

Jean Gottmann

Introduction

Chiara Cavalieri
UC Louvain, Louvain-la-Neuve, Belgium
chiara.cavalieri@uclouvain.be

Published in 1961 by the French geographer Jean Gottmann (1915–1994) Megalopolis *is the first attempt to legitimize the existence of a new urban form, that of a super metropolitan character, emerging along the northeastern coast of US. This legitimation represents the result of 20 years of research*[1]*: already in 1942 indeed, to the eyes of a foreigner travelling from New York to Washington, the most striking realisation was the frequency of large urban units scattered along Atlantic seaboard.*[2] *Later on, the new census of 1950 made it possible to visualise this "frequency' in the form of a map of the northeastern region, later published in the booklet* State Economic Areas. *According to Gottmann, that map showed the evidence of the existence of a "metropolitan economy', thus defining an object that was new both for dimensions and size: within an area 600 miles long and already in 1950, dwelt by about 30 million people.*[3] *Within this nebulous object that had the dimension of a nation and the characteristic of a metropolis, the merit of Jean Gottmann was that of capturing the existence of a unity rather than a dissolution of elements. A unity that was represented by new integration between concepts and trends that used to be separated and opposed: a new symbiosis integrating what*

[1]The research was funded by the Twentieth Century Fund.

[2] Gottmann, J. (1957). Megalopolis or the urbanization of the northeastern seaboard. *Economic Geography, 33*(3), 189.

[3]ibid, p. 189.

M. Barcelloni Corte and P. Viganò (eds.), *The Horizontal Metropolis*,
https://doi.org/10.1007/978-3-030-56398-1_13

used to be urban and rural. Its roots belong to the eighteenth century, when a vast number of small clusters had been installed in the region as triggers for colonial conquest.[4] *This system of small poles soon became stronger and more extensive, thanks to an impressive growth of industries, services and financial institutions. Gottmann describes the Megalopolis as the largest existing conurbation, as a device for spatial democracy, as a space where food, accessibility, and wealth are equally distributed among the population. And it was indeed for this character of novelty and unicity that Gottmann sought a new name and chose the one of* Megalopolis, *defined as "a very large city" but at the same time recalling that Greek city established in the Peloponnesus, that, despite the hope of its founders, never really grew much into being a city. Against this, the other side the north-eastern seaboard surpassed everything dreamers of the past may have visualised.*[5]

And it is exactly in this tension, that emerges from the choice of the title, between the enthusiasm of a continuous discovery and the curiosity of digging down through history, that the specificity of Gottmann's approach is somehow disclosed. The geographer indeed, thanks to both his French and American education, was provided with a historical attitude inherited from the French school of Vidal de La Blache, and with a wide eco-anthropological attention thanks to the lessons of the schools of Chicago. Somehow, such a specific and external eye was necessary in order to grasp the Megalopolis as the result of an early mushrooming of sea-heading towns[6] *along the coast, from Boston, to New York, Philadelphia and Baltimore, until Washington. A result that was at the same time an exceptional and pioneer object: exceptional because the Megalopolis featured a concentration of population, industrial and commercial facilities, financial wealth and cultural activities that were not comparable with any other situation worldwide; pioneer because, in its new organisation of inhabited space, more of such bodies were expected: according to Gottmann and his team, perhaps the situation most nearly approaching it within the following twenty years would be northwestern Europe, from Amsterdam to Paris. Furthermore, because of this extraordinary dimension, according to Gottmann, the Megalopolis had a crucial role in defining national trends: in his time one fifth of the American population lived in a Megalopolis, and that fifth was the healthiest, the best-paid and ultimately the best-educated. Surprisingly enough, this concentration of wealth was not grounded in particularly advantageous natural conditions*[7]*: there was nothing particular in the natural potential of the Megalopolis, unless we count its geographical location, which*

---

[4]Gottmann, J. (1961). *Megalopolis: The urbanized northeastern seaboard of the United States.* Twentieth Century Fund, Inc. © 1964 Twentieth Century Fund, Inc., by permission of The MIT Press. p. 216.

[5]Gottmann, J. (1957). Megalopolis or the urbanization of the northeastern seaboard. *Economic Geography*, *33*(3), 189, 190.

[6]ibid, p. 192.

[7]Gottmann, J. (1961). *Megalopolis: The urbanized northeastern seaboard of the United States.* Twentieth Century Fund, Inc. © 1964 Twentieth Century Fund, Inc., by permission of The MIT Press. p. 274.

enabled it to play its historical role as the landing ground for European immigrants. The land in itself was not as important as the way it was used: in contrast with theories of ecological planning grounded in the same years in the US, Gottmann pointed out the growing dependence of employment—and thus of urbanisation—on human organisation rather than the location of resources. Despite Gottmann's vision of growing urbanisation being certainly intertwined with the theories of Constantinos Dioxadis—Gottmann indeed became president of the World Society of Ekistics in 1971—his discourse on urbanisation and civilisation remains grounded on western society, therefore the whole question of eastern urbanisation—further brought to the debate by other authors—had not yet been taken into consideration.

13.1 The Main Street of the Nation (from: *Megalopolis: The urbanized Northeastern Seaboard of the United States*)[8]

The Northeastern seaboard of the United States is today the site of a remarkable development—an almost continuous stretch of urban and suburban areas from southern New Hampshire to northern Virginia and from the Atlantic shore to the Appalachian foothills. The processes of urbanization, rooted deep in the American past, have worked steadily here, endowing the region with unique ways of life and of land use. No other section of the United States has such a large concentration of population, with such a high average density, spread over such a large area. And no other section has a comparable role within the nation or a comparable importance in the world. Here has been developed a kind of supremacy, in politics, in economics, and possibly even in cultural activities, seldom before attained by an area of this size.

A Very Special Region: Megalopolis

This region has indeed a "personality" of its own, which for some three centuries past has been changing and evolving, constantly creating new problems for its inhabitants and exerting a deep influence on the general organization of society. The modern trends in its development and its present degree of crowding provide both examples and warnings for other less urbanized areas in America and abroad and call for a profound revision of many old concepts, such as the usually accepted distinctions between city and country. As a result new meanings must be given to some old terms, and some new terms must be created.

---

[8]Excerpt from: Gottmann, J. (1961). *Megalopolis: The urbanized northeastern seaboard of the United States*. Twentieth Century Fund, Inc. © 1964 Twentieth Century Fund, Inc., by permission of The MIT Press.

Great, then, is the importance and significance of this section of the United States and of the processes now at work within it. And yet it is difficult to single this area out from surrounding areas, for its limits cur across established historical divisions, such as New England and the Middle Atlantic states, and across political entities, since it includes some states entirely and others only partially. A special name is needed, therefore, to identify this special geographical area.

This particular type of region is new, but it is the result of age-old processes, such as the growth of cities, the division of labor within a civilized society, the development of world resources. The name applied to it should, therefore, be new as a place name but old as a symbol of the long tradition of human aspirations and endeavor underlying the situations and problems now found here. Hence the choice of the term *Megalopolis*, used in this study.

Some two thousand years before the first European settlers landed on the shores of the James River, Massachusetts Bay, and Manhattan Island, a group of ancient people, planning a new city-state in the Peloponnesus in Greece, called it *Megalopolis*, for they dreamed of a great future for it and hoped it would become the largest of the Greek cities. Their hopes did not materialize. Megalopolis still appears on modern maps of the Peloponnesus but it is just a small town nestling in a small river basin. Through the centuries the word *Megalopolis* has been used in many senses by various people, and it has even found its way into Webster's dictionary, which defines it as "a very large city." Its use, however, has not become so common that it could not be applied in a new sense, as a geographical place name for the unique cluster of metropolitan areas of the Northeastern seaboard of the United States. There, if anywhere in our times, the dream of those ancient Greeks has come true.

An Urbanized Area with a Nebulous Structure

As one follows the main highways or railroads between Boston and Washington, D. C., one hardly loses sight of built-up areas, tightly woven residential communities, or powerful concentrations of manufacturing plants. Flying this same route one discovers, on the other hand, that behind the ribbons of densely occupied land along the principal arteries of traffic, and in between the clusters of suburbs around the old urban centers, there still remain large areas covered with woods and brush alternating with some carefully cultivated patches of farmland. These green spaces, however, when inspected at closer range, appear stuffed with a loose but immense scattering of buildings, most of them residential but some of industrial character. That is, many of these sections that look rural actually function largely as suburbs in the orbit of some city's downtown. Even the farms, which occupy the larger tilled patches, are seldom worked by people whose only occupation and income are properly agricultural. And yet these farm areas produce large quantities of farm goods!

Thus the old distinctions between rural and urban do not apply here any more. Even a quick look at the vast area of Megalopolis reveals a revolution in land use. Most of the people living in the so-called rural areas, and still classified as "rural population" by recent censuses, have very little, if anything, to do with agriculture.

In terms of their interests and work they are what used to be classified as "city folks," but their way of life and the landscapes around their residences do not fit the old meaning of urban.

In this area, then, we must abandon the idea of the city as a tightly settled and organized unit in which people, activities, and riches are crowded into a very small area clearly separated from its nonurban surroundings. Every city in this region spreads out far and wide around its original nucleus; it grows admist an irregularly colloidal mixture of rural and suburban landscapes; it melts on broad fronts with other mixtures, of somewhat similar though different texture, belonging to the suburban neighborhoods of other cities. Such coalescence can be observed, for example, along the main lines of traffic that link New York City and Philadelphia. Here there are many communities that might be classified as belonging to more than one orbit. It is hard to say whether they are suburbs, or "satellites," of Philadelphia or New York, Newark, New Brunswick, or Trenton. The latter three cities themselves have been reduced to the role of suburbs of New York City in many respects, although Trenton belongs also to the orbit of Philadelphia. (Sec Fig. 1, the distribution of population density.)

The "standard metropolitan areas,"[9] first used by the U. S. Bureau of the Census in 1950, have clarified this confused situation somewhat but not entirely. For example, the New York—Northeastern New Jersey standard metropolitan area cuts across political boundaries to reveal the relationships of this vast region to the core city of New York. And yet the mechanical application of the term "standard metropolitan area" has resulted in the establishment of separate areas for Trenton, which is closely tied to both Philadelphia and New York, and for Bridgeport, which is for many practical purposes part of the New York area. Similar problems can be found in other parts of Megalopolis.[10]

---

[9]The U. S. Bureau of the Census defined a standard metropolitan area as "a county or group of contiguous counties which contains at least one city of 50,000 inhabitants or more. In addition to the county, or counties, containing such a city, or cities, contiguous counties are included in a standard metropolitan area if according to certain criteria they are essentially metropolitan in character and socially and economically integrated with the central city." In New England, "towns and cities, rather than counties, are the units used in defining standard metropolitan areas."

[10]For the 1960 Census the term "standard metropolitan area" was changed to "standard metropolitan statistical area." The definition was modified and a somewhat different set of criteria used which resulted in breaking down several of the formerly recognized larger metropolitan areas into smaller such units. The results thus achieved may be more precise in some respects but in the case of Megalopolis they may cause some confusion. The New York—Northeastern New Jersey standard metropolitan area of 1950 has been replaced by four standard metropolitan statistical areas: one for New York in New York State and three in New Jersey, those of Paterson-Clifton-Passaic, Jersey City, and Newark. The stricter definition of metropolitan integration of adjoining counties now excludes Somerset and Middlesex counties, formerly classified as metropolitan. As a result the percentage of the population of New Jersey residing in metropolitan areas fell from 89.9 in 1950 to 78.9 in 1960—a statistical trend surprising to those who know how much more metropolitan—or should we say Megalopolitan—the whole of New Jersey grew through the 1950's. To compensate for such an impression and for the separation between New York City and Northeastern New Jersey, a new term has been created and defined:

Thus an almost continuous system of deeply interwoven urban and suburban areas, with a total population of about 37 million people in 1960, has been erected along the Northeastern Atlantic seaboard. It straddles state boundaries, stretches across wide estuaries and bays, and encompasses many regional differences. In fact, the landscapes of Megalopolis offer such variety that the average observer may well doubt the unity of the region. And it may seem to him that the main urban nuclei of the seaboard are little related to one another. Six of its great cities would be great individual metropolises in their own right if they were located elsewhere. This region indeed reminds one of Aristotle's saying that cities such as Babylon had "the compass of a nation rather than a city."

Megalopolis—Main Street and Crossroads of the Nation

There are many other large metropolitan areas and even clusters of them in various parts of the United States, but none of them is yet comparable to Megalopolis in size of population, density of population, or density of activities, be these expressed in terms of transportation, communications, banking operations, or political conferences. Megalopolis provides the whole of America with so many essential services, of the sort a community used to obtain in its "downtown" section, that it may well deserve the nickname of "Main Street of the nation." And for three centuries it has performed this role, though the transcontinental march of settlement has developed along east–west axes perpendicular to this section of the Atlantic seaboard.

In recent times Megalopolis has had concentrated within it more of the Main Street type of functions than ever, and it does not yet seem prepared to relinquish any of them. Witness, for example, the impact of the Federal government in Washington, D. C., as it tightens up over many aspects of national life; the continued crowding of financial and managerial operations into Manhattan; New York's dominance of the national market for mass communication media, which resists all attempts at erosion; and the pre-eminent influence of the universities and cultural centers of Megalopolis on American thinking and policy-making. Megalopolis is also the country's chief façade toward the rest of the world. From it, as from the Main Street of a city, local people leave for distant travel, and to it arriving strangers come. For immigrants it has always served as the chief debarkation wharf. And just as passing visitors often see little of a city except a few blocks of its Main Street, so most foreign visitors see only a part of Megalopolis on their sojourns in the United States.

"Standard Consolidated Areas," of which there were two (recognized for 1960) in the country: the New York—Northeastern New Jersey area (which included Somerset and Middlesex counties in New Jersey), and the Chicago—Northwestern Indiana area. The recognition of these broader areas was interlined to stress "the special importance of even more inclusive metropolitan statistics" (see Executive Office of the President, Bureau of the Budget, *Standard Metropolitan Statistical Areas,* U.S. Government Printing Office, Washington, D.C., 1961). The metropolitan area of Philadelphia remained unchanged in both its Pennsylvania and New Jersey parts.

Just as a Main Street lives for and prospers because of the functions of the whole city, rather than because of any purely local advantages of its own, so is Megalopolis related to the whole United States and its rich resources. In general, Megalopolis itself was blessed only moderately by nature. It has no vast expanse of rich soils (there are some good soils but more poor ones), no special climatic advantages (its cyclonic climate is far from ideal), and no great mineral deposits (though there are some). In these respects it cannot compare with the generous natural potential of the Middle West or Texas or California. But it does excel in locational advantages—deep harbors of a drowned shoreline, on which its principal cities were early established, and a connecting-link relationship between the rich heart of the continent and the rest of the world. By hard work man has made the most of these locational resources, the most outstanding ones in an otherwise average natural endowment. As a result, early in its history Megalopolis became a dynamic hub of international relations, and it has maintained and constantly expanded that role to the present day. It is now the most active crossroads on earth, for people, ideas, and goods, extending its influence far beyond the national borders, and only as such a crossroads could it have achieved its present economic pre-eminence.

Megalopolis as a Laboratory of Urban Growth

Modern technology and social evolution provide increasing opportunity in urban pursuits on the one hand, and on the other steadily improving means of producing more agricultural goods with less manpower. The forces at work in our time, coupled with the growth in population, are, therefore, bound to channel a rising flow of people toward urban-type occupations and ways of life. As this tide reaches more and more cities they will burst out of old bounds to expand and scatter all over the landscape, taking new forms like those already observable throughout Megalopolis. This region serves thus as a laboratory in which we may study the new evolution reshaping both the meaning of our traditional vocabulary and the whole material structure of our way of life.

Tomorrow's society will be different from that in which we grew up, largely because it will be more urbanized. Nonagricultural ways of life will be followed by more and more people and will occupy much more space than they ever did, and such changes cannot develop without also deeply modifying agricultural life and production. So great are the consequences of the general evolution heralded by the present rise and complexity of Megalopolis that an analysis of this region's problems often gives one the feeling of looking at the dawn of a new stage in human civilization. The author has visited and studied various other regions of the world but has not experienced such a feeling anywhere else. Indeed, the area may be considered the cradle of a new order in the organization of inhabited space. This new order, however, is still far from orderly; here in its cradle it is all in flux and trouble, which does not facilitate the analyst's work. Nevertheless, a study of Megalopolis may shed some light on processes that are of great importance and interest.

A Study in Entangled Relationships

As the work of data-gathering and analysis progressed it became evident that the key to most of the questions involved in this study of Megalopolis lies in the interrelationships between the forces and processes at work within the area rather than in the trends of growth or the development of techniques. Thus the trend of population increase, easy to measure and perhaps to forecast approximately, provides less insight into the nature of the area than do the interrelations existing between the processes that caused the local population to grow, those that attracted certain kinds of people to Megalopolis, and those that supplied the swelling crowds with the means to live and work together there. Many of these processes are statistically measurable and some of them can be mapped, but the degree to which each of them stems from the others or determines them is a much more subtle matter, and is more basic to an understanding of what is going on and what can be done about it.

Most regional studies stay on the safer and more superficial grounds of statistical description and functional classifications. Had this report followed that pattern it would have been devoted mainly to summing up the abundant data available from the Censuses and other sources of general information about the various characteristics of Megalopolis. A description of natural conditions, such as topography, climate, hydrography, and vegetation, would have introduced a historical sketch to be followed by chapters on population, industries, trade, transportation and communications, the real estate market, other occupations, and descriptions of the main cities and of the general features of "rural areas." Such a report would have concluded with a description of present problems and forecasts of the future presented by means of graphs, based on the assumption that the trends of the past twenty to fifty years will continue for the next twenty years.

A mere compilation of such data would probably be of service to some people but it could hardly help those who need further insight into and understanding of the basic problems of the area. By attempting to find out more about the deeper processes and their entanglements, one may hope to achieve a more fundamental kind of knowledge, which can be applied to another area or projected into the future more safely, though not always more easily. This is why the present report is organized along a somewhat less classical outline, its goal being a more reasoned discussion and an objective analysis. For such complicated phenomena as the social and economic processes at work in Megalopolis there are, of course, numerous and interlocking determining factors. The author has endeavored to search for *all* these factors, keeping in mind their multiplicity and entanglements and avoiding any arbitrary choices among them.

Outline of This Report

Part One presents a sketch of the *dynamics of urbanization* and attempts to show, in terms of the region's history, why things have come to be as they are and where they are Although this section is largely descriptive it cannot avoid raising some new questions.

Part Two takes up what may be called the "*modern revolution in land use*." The new mixture of urban and rural must be dissected and each part related to the others in the newly developing system. Separation between place of work and place of residence creates within the area the system of daily "tidal" movements involved in commuting. Over these are superimposed other currents, some seasonal and some irregularly recurrent. These reflect relations between different parts of Megalopolis that stem from more complicated needs than the simple journey from home to work. These other needs grow more complicated and more general as average family income rises and both goods and activities that were once considered dispensable come to be regarded as necessary by large numbers of Megalopolitans. As Montesquieu observed two centuries ago, on the eve of the Industrial Revolution, "It is the nature of commerce to make the superfluous useful and the useful necessary." Perhaps it is not commerce but just human nature that produces this sequence. At any rate it has certainly been proven true of the consumption of goods, and now it seems to apply to the consumption of activities and space. The modern urban revolution, so apparent already in the affluent society of Megalopolis, devours time and space as well as food and industrial goods, and the fulfilling of these needs requires many types of movements.

These various tidal movements involve a reshaping of land use. Much agricultural land has been taken over by residential and industrial development. On the remaining farms a new specialized type of agriculture is developing, which requires less space than did the old system of farming. Woods have spread over much of the land abandoned by the farms, and this expansion of forests calls for new methods and concepts of forestry management, to provide for recreational and other suburban needs and for a better conservation of the landscape and of wildlife. Simultaneously the old city cores or "downtowns" are evolving toward decline or renewal, while uptowns, suburbs, and outer suburbia are becoming interlocked in a new and still constantly changing web of relationships. Regional integration is taking on forms unknown a generation or two ago, and the old system of local, state, and national authorities and jurisdictions, which has changed little, is poorly suited to present needs.

New *patterns of intense living* that have become normal in Megalopolis affect not only land use. They also exert a strong influence on the economic and social foundations of society, and Part Three endeavors to describe the problems thus created. The density of activities and of movement of all kinds is certainly the most extraordinary feature of Megalopolis, more characteristic even than the density of population and of skyscrapers. It has become a means of maintaining economic growth and stabilizing society; but how far can it go without destroying itself? For example, the growth of Megalopolis owes much to the automobile, but highway traffic jams are beginning to strangle city activities and to take the pleasure and efficiency out of driving a car. At the same time cars contribute to the ruination of other means of transportation, made more necessary than ever by the massive tidal currents of people and goods. The self-defeating effect of dense concentrations may be observed also in other fields than transportation. Many industries, for example, are now aiming at decentralization. The intense living of Megalopolis makes a great deal of waste inescapable, waste of space and time as well as of materials. For a long time such waste may have seemed justifiable, for, paradoxically, the crowding that caused it brought higher economic yields. Now this crowding seems at times to defeat its own aims. Why and how does such intense living grow and threaten itself? Answers to these queries build up a general picture of a dynamic and prosperous society, obviously responsible for maintaining the growth of large-scale urbanization but responsible also for the problems the process creates and for finding the badly needed solutions.

It is easier to accept responsibility for solutions than to provide them. The many millions of people who find themselves *neighbors in Megalopolis*, even though they live in different states and hundreds of miles from one another, are barely becoming aware of the imperatives of such a "neighborhood." Part Four attempts to point them out. Responsible public opinion is becoming conscious of the problems involved, and the struggle to find solutions has started. It is especially difficult because no one problem can be tackled without affecting the others. Transportation, land use, water supply, cultural activities, use and development of resources, government and politics—all are interrelated.

Today it is essential that solutions be found to save this area from decay and to reassure the nation and the world about the kind of life modern urbanization trends presage for the future. Megalopolis has been built and often reshaped by its people. These people are now wealthier, better educated, and better endowed with technological means than ever. They ought to be able to find ways of avoiding decline of the area.

For the Better or for the Worse?

The preceding paragraph may seem to imply an unwarranted optimism about society's ability to control itself. True, history records a long list of brilliant civilizations that have sunk under the pressure of internal decay and external jealousy. We remember their names: Babylon, Corinth, Sparta, Athens, Rome, and many others. In the shadowy vistas of ancient times they vanished into the distance like shipwrecked ships loaded with ambition and precious cargo. Can such a fate be looming in the offing for Megalopolis? Modern urban sprawl is viewed by many as a threat to progress and general welfare. What is happening in Megalopolis today has been described as a pathological phenomenon, a sickness, a cancer. Such views are held by distinguished and respectable citizens of the area. One may well be alarmed by their invectives, all the more so as one does not have to go far away from Megalopolis to hear expressions of distrust and jealousy inspired by the amazing concentration of wealth and power in the great seaboard cities. Are people both in and out of this extraordinary region united in condemning it?

Urban growth in general has been discussed and condemned on moral grounds for a long time. Such debate is expectable and desirable, but on the whole history has shown the condemnation to be unjust, as can be seen by a brief review of some of the consequences of crowding.

Contrasts between rich and poor, for example, are especially striking in the crowded communities of cities. These may exist in rural areas too, but there they are diluted by scattering and veiled in greenery. The growth of urban pursuits (industries, trade, services) sharpens the contrasts by condensing them into a smaller area. Rich and poor live within short distances of one another and mix together in the streets in a way that often arouses righteous indignation. It seems brutally amoral to witness destitution neighboring on elegant sophistication, poverty mixing with prosperity. And yet, alas, a growing city's environment can hardly escape such sights. For many centuries there was an enormous difference between the advancement possible in trade and industry on one the one hand and in farming on the other (though modern farm mechanization and subsidies to agriculture have substantially increased the profit possibilities of farming), and so to rise economically within the span of one lifetime has traditionally been easier in cities than in rural areas. The affluence of those people who have so risen draws to the city large groups of humbler people, who come there to profit by the local abundance of money and the volume of spending and to serve the wealthier. In contrast to the more conservative "open" country, the "closed-in" city offers a more dynamic environment, socially and economically.

In cities, too, other vicious aspects of economic growth and social life have always been more evident than in the country. As urban development was accelerated by the Industrial Revolution, some of the vicious aspects became increasingly obvious. Slums and mobs grew worse than ever, making the urban landscape ethically and aesthetically shocking to those who cared about the people. From his sojourns in an industrializing western Europe, and especially from Paris during the French Revolution, Thomas Jefferson brought back impressions that reinforced his normal Virginian opposition to great cities and the development of manufactures or large-scale commerce. As slums and mobs became more general in European cities in the first half of the nineteenth century there arose more awareness about the classes of society and social injustice. There was more discussion of these matters, and the early Socialist doctrines were largely inspired by them. Then came the teachings of such philosophers as Fourier and Proudhon in France and Engels and Karl Marx in Germany, opposing great urban concentration as much as great concentration of capital. Engels' writings on the slums and working conditions in the then fast-developing British cities, such as Manchester, are well known. Because urban conditions of living and working were largely at the root of nineteenth-century Socialist doctrines, Karl Marx stressed that his theories applied much more to the industrialized countries of western Europe, which had accumulated large amounts of capital, than to the rural, little-urbanized countries to the east. Twentieth-century events have proved him wrong on this score, however, for communism has conquered the mainly rural countries, and the forms of socialism that developed in the more urban and capitalistic countries of the West have turned away from Marxism.

Crowding of population within a small area creates shortages of various resources, and most of the crowded people are bound to suffer in some ways because of the shortages. To alleviate them, to make crowding more bearable and the population happier, ways and means of constantly better distribution must be found. Otherwise no lasting growth can develop, and the whole enterprise will soon be doomed. Front the struggle against such shortages have come some of mankind's most important advances. In the arid areas of the Middle East, for example, early civilization arose when people first congregated around the main springs and permanent rivers. As the settlement grew, the supply of both water and irrigable land became scarce. To insure survival of the people a proper distribution system had to be achieved, and rules and regulations had to be set up and accepted. Thus organized society, ruled by law, was born. Because authorities were needed to enforce law, political power arose, and people organized themselves to avoid more oppression than was necessary. Everywhere, the more crowded people have become in cities the more they have craved both security and freedom. Modern political life and its concepts of liberty, self-government, and democracy are the products of urban growth, the inheritance of cities in process of growth and development—places such as Jerusalem, Athens, Rome, Bruges, Florence, Paris, London, to mention only those that have been most studied by historians. And the same places, or similar urban centers, have contributed most of our scientific and technological developments, either because people there were struggling to solve pressing problems or because urban societies make possible a leisurely enough elite, some of whose members can devote themselves to disinterested research and a search for a better understanding of the universe.

Thus urban crowding and the slums and mobs characteristic of it may be considered growing pains in the endless process of civilization.

In the same way, the picture of Megalopolis is not as dark as the outspoken pessimists and frequent protests would seem to paint it. Crowded within its limits is an extremely distinguished population. It is, *on the average,* the richest, best educated, best housed, and best serviced group of similar size (i.e., in the 25-to-40-million-people range) in the world. The area is still a focus of attraction for successful or adventurous people from all over America and beyond. It is true that many of its sections have seen pretty rural landscapes replaced by ugly industrial agglomerations or drab and monstrous residential developments; it is true that in many parts of Megalopolis the air is not clean any more, the noise is disturbing day and night, the water is not as pure as one would wish, and transportation at times becomes a nightmare. Many of these problems reflect the revolutionary change that has taken place as cities have burst out of their narrow bounds to scatter over the "open" countryside. In some ways this suburban sprawl may have alleviated a crowding that had threatened to become unbearable, for residential densities of population per square mile have decreased. But new problems have arisen because of the new densities of activities and of traffic in the central cities and because the formerly rural areas or small towns have been unprepared to cope with the new demands made upon their resources. New programs are needed to conserve the natural beauty of the landscape and to assure the health, prosperity, and freedom of the people. In spite of these problems, however, available statistics demonstrate that in Megalopolis the population is on the average healthier, the consumption of goods higher, and the opportunity for advancement greater than in any other region of comparable extent.

Thus the type of urban growth experienced here generates many contracts, paradoxes, and apparently contradictory trends. It calls for debate and naturally excites passionate opinions for and against it. Are its results for the better or for the worse? It is not for our generation to moralize on the matter, but to strive to make the outcome be for the better, whatever obstacles may be in the way. Megalopolis stands indeed at the threshold of a new way of life, and upon solution of its problems will rest civilization's ability to survive. In the search for such solutions there will be found no easy keys to success, no "gimmicks" or "open-sesames." Solutions must be thought out, ironed out, and constantly revised in the light of all the knowledge that can be acquired by all concerned. It is the author's hope that this report, a systematic and sometimes critical analysis of the past and present of Megalopolis, will contribute to the gathering of such knowledge and to its distribution. At the same time, it will tell the story of an extraordinary region as its people have made it.

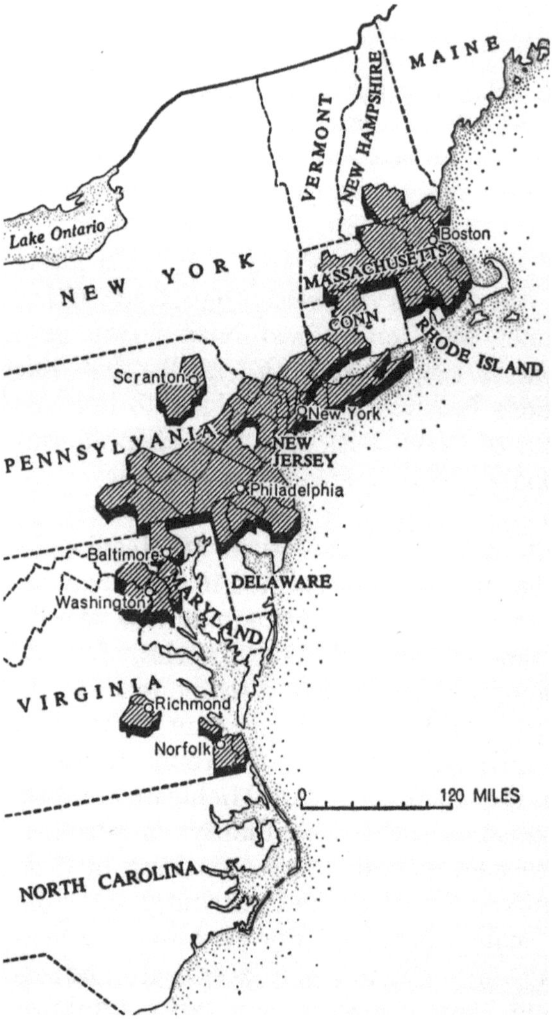

Fig. 13.1 "Map of the East Coast of the US showing the Megalopolis from Boston to Washington". Source: Gottmann, J. (1964), «Megalopolis, or the Urbanization of the Northeastern Seaboard», in: Mayer, H. H., Kohn, C. F. (1964—first edition in 1959), *Readings in Urban Geography*, Chicago: The University of Chicago Press, p. 47. © Clark University, reprinted by permission of Informa UK Limited, trading as Taylor & Francis Group, www.tandfonline.com on behalf of Clark University

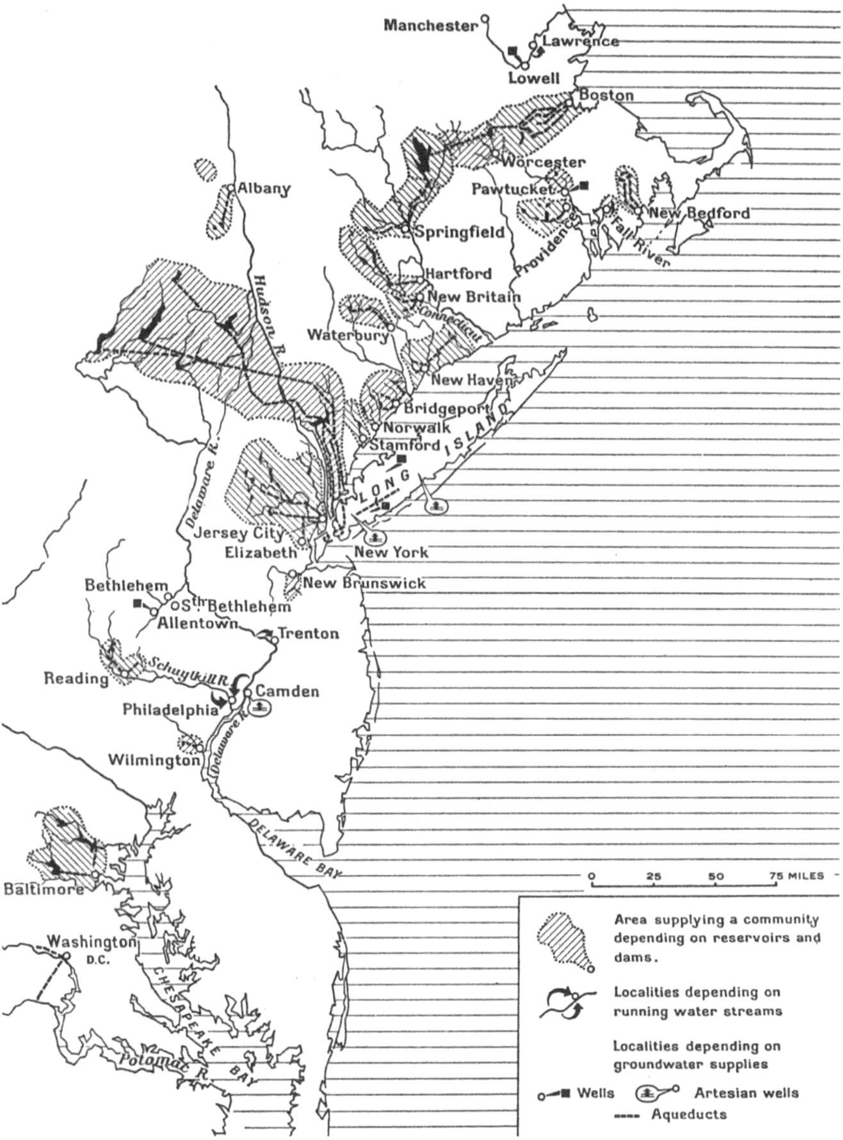

Fig. 13.2 "Water-supply systems of the main cities in Megalopolis. The map, prepared with the help of Prof. Gordon Wolman of the Johns Hopkins University, does not attempt to show the details of the various systems; it only hints at their intricacies with a few symbols; the shaded areas are not watersheds but the general areas in which a city has established facilities for its water supply; note the entanglements of the Springfield-Worcester-Boston area in Massachusetts." Source: Gottmann, J. (1966). *Megalopolis: The urbanized northeastern seaboard of the United States* (p. 734). MIT Press. Reprinted courtesy of The MIT Press

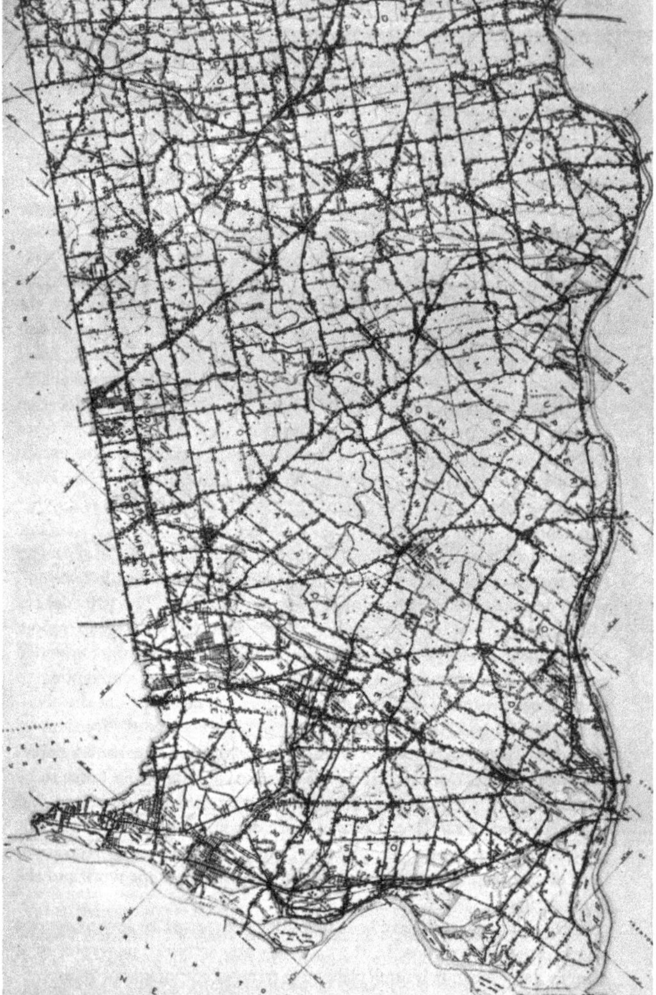

Fig. 13.3 "Dispersal of buildings in the southern part of Bucks County. Reduced from county maps of the State of Pennsylvania."
Source: ibid., p. 243. Reprinted courtesy of The MIT Press.
"Following the attraction of many people from rural areas to crowded urban centers, the modern evolution of American life has involved the recent resettling of a good deal of the previously emptied rural countryside with sparsely distributed residences. Because these have held much attraction for families who can afford them, many rural roads have been transformed into loosely and irregularly built-up avenues." (ibid. p. 239)

Fig. 13.4 "Land use generalized from a map by the Graduate School of Geography of Clark University, 1958."
Source: ibid., p. 240. Reprinted courtesy of The MIT Press.
"The affluence of society, the excellence of the highway network, the widespread ownership and use of automobiles, the rapid expansion of retail trade services in rural areas, the growing disadvantages of city middle-class residences – all these contributing factors have concurred to make this wide dispersal easy throughout a region where woods were increasing in area. Obviously it could not go on and on indefinitely without running the present fragile balance of rural and urban [...]." (ibid., p. 239)

Chapter 14
La Città Territorio (The City Territory) (1962)

Giorgio Piccinato, Vieri Quilici, and Manfredo Tafuri

Introduction[1]

Paola Viganò
EPFL, Lausanne, Switzerland
IUAV, Venice, Italy
paola.vigano@epfl.ch

In 1962, Piccinato, Quilici and Tafuri, representing the Roman studio AUA, introduce in Casabella Continuità *and in the Italian debate the image of the* Città Territorio *that indicates not only a new vantage point, but also a shift of scale in the urban articulation. The article is published the same year of the conference in Stresa on the new urban dimension (organized by Ilses in 1962), where other large interpretative images contain reflections on changing lifestyles and modes of use of the territory. In this occasion, among others, Giancarlo De Carlo picks up the image of the city-region (from Geddes and Mumford) and links it to increased affluence, the acceleration of social and territorial mobility and the resulting multiplication of the possibilities of choice.*

According to Piccinato, Quilici and Tafuri, the incapacity of urbanism and planning to interpret the new phenomena is seen as eminently ideological, with the latter

[1]The text is an extract, with slight changes of the paragraph 2.2 Images of the city territory in: Viganò, P. (2016). *Territories of urbanism—The project as knowledge producer*. EPFL Press Routledge.

G. Piccinato (✉)
Roma Tre University, Rome, Italy
e-mail: giorgiopiccinato@gmail.com

V. Quilici
Dipartimento di Studi Urbani, Università degli Studi Roma Tre, Rome, Italy

M. Barcelloni Corte and P. Viganò (eds.), *The Horizontal Metropolis*,
https://doi.org/10.1007/978-3-030-56398-1_14

engaged in elaborating "microcosms, 'neighborhoods' that are absolutely detached from the surrounding environment in all its aspects" (Tafuri, Piccinato, Quilici 1962:16). Advancing a strong critique of the intimist solutions of neo-empiricism, the authors point not to a problem of tools of intervention or investigation, but to one of identification of themes and their relationships, which would require the development of a new "structural pattern." The AUA[2] *group, like De Carlo, starts with the consideration of the raising of the level of life, but introduces a new focus on the theme of free time—the spread of loisirs, the rise of the second home, the role of television—and the gradual dispersion of production that "evaporates" in the territory and abandons its proximity to the urban center. The accent is on the acceleration of transformations, on territorial dynamics, the mutations of ways of living together. These positions insist on the risk of a lack of comprehension of the transformations in progress, on the need for new techniques of investigation, new readings and survey, and on the urgent need to reformulate the tools of the project, a question that is much more serious than that of the ageing of institutional and legislative measures.*

Around the image of the "City Territory" an idea of the project as process takes form, which no longer has the possibility of approaching the whole urban entity, and where the individual exercises freedom and autonomy: "The City Territory shifts its field of application from total urban planning to the identification of points on which to exert leverage [...] though this certainly does not imply rejection of urban planning on a territorial scale" (Tafuri, Piccinato, Quilici 1962: 17).

The dilation of the scale of the urban phenomenon does not imply more widespread planning, but the selection of places against the backdrop of an urban project on a large scale: at that period the construction of the Trentino plan on the part of Samonà (see chapter 15), the first territorial and development plan in Italy prepared by urbanists and not just economists, was still to come.

The territory that begins to be described in those years is simultaneously the result and the motor of important changes in society (Ardigò 1967).[3] *The problem, for Tafuri, Piccinato and Quilici, consists in giving a "democratic direction" to such potentialities to meet the demand of the new affluent society: "to rediscover, in every condition, the variety of contacts and choices the city offers". The idea of a movement in the direction of "more highly evolved forms of territorial organization" is shared by many, though it is not clear whether or not these are socially positive forms. Unlike the Modern Movement which had proposed and overlaid a model of progressive development, AUA underlines the impossibility of maintaining this ambition, evoking the crisis of a constructivist position and proposing a way of getting beyond a rationalist attitude, to "instead draw on a continuous process of rationalization." The possible emergence of a new form of associated life and the new territory should be kept in a state of observation and subjected to constant verification.*

---

[2]AUA (Studio Architetti Urbanisti Associati) was founded in 1961 and its manifesto was signed by M. Tafuri, L. Barbera, B. Rossi Doria, S. Ray, M. Teodori, and later by E. Fattinnanzi. The projects of the group were often done together with other professionals: including V. Quilici, S. Bracco, A. Calza Bini, M. La Perna, C. Maroni, G. Moneta and others.

[3]Ardigò, A. (1967). *La diffusione urbana*. Roma: Ave-an Veritas Ed.

14.1 The City Territory: Towards a New Dimension[4]

The term city territory represents an initial, timid attempt to respond to a question that seems more urgent than ever today: what should be the fundamental dimension of reference in our hypotheses of urban development? And what is the structure that frames our formal research?

The term city territory already indicates a change of scale in the investigation of structures, not just a different angle of viewing. Nevertheless, we encounter our first difficulties as soon as we try to further specify the meaning to be attributed to the generic formula of the "change of scale," and thus to try to define in what direction that change has been implemented and is being implemented.

The way things have reached their present condition is already singular. Not that there was any lack, ever since the heroic days of modern architecture, of intuitions regarding a dissociation of city-countryside, city-territory, production-services-housing, that somehow had to be surpassed. Nor was there a lack of continuous urging to connect urban planning to concrete economic planning, without which the "reality of things" would frustrate all efforts.

But these attempts—whose results should be carefully examined—seem to accompany an evolution of the situation and the structures that happens in a completely independent way, making the voiced complaints seem rather sterile.

This way of proceeding through successive approximations, through increasingly futile attempts to approach a reality that is changing too rapidly, points to the inadequacy of a type of culture that to get closer to the individual has wound up overlooking the society or, at least, its physical and historical "environment." Of the two terms, of what Quaroni calls the ambivalent attraction towards the big city and towards spaces commensurate with the individual, the latter has always prevailed.

The appeal, for example, of a communitarian culture organized around the primary school and the community center, in the years of major internal migrations, abandonment of entire territories and the violent growth of new urban centers, with all the socio-economic changes involved, takes on a pathetic and untimely hue to the point of becoming noncommittal, in the final analysis.

But we do not want to be misunderstood: the idea here is not to oppose a certain type of civil organization that could also be safeguarded, as we shall see, inside the structure of the city-territory; instead, the task is to clarify the fact that the defeats of (not just) Italian architecture and urbanism are not due simply to a lack of operative tools, from legislation to means of investigation.

If there has been a failure to impact the Italian situation, we believe this depends to a great extent on the failure to grasp the fact that what is inadequate is not the mode of intervention, but the very interpretation of the real phenomena to which the intervention is applied. In short, that famous "engagement" with the reality of the country so often invoked in the discussions of the 1950s has been definitively

---

[4]Source: Tafuri, M., Piccinato, G., & Quilici, V. (1962). La città territorio: Verso una nuova dimensione. *Casabella Continuità, 270*, pp. 16–25.

lacking. While the evolution became increasingly impossible to reverse, the cultural forces sought refuge in a cry for planning that was inadequate due to its ideological limitations or, more concretely, in the elaboration of microcosms, “neighborhoods,” absolutely detached from the surrounding environment in all its aspects.

Need for a cultural “leap.” The first attempts

It matters little if this is hindsight. Instead, it is necessary to transform the recognition of a defeat that involves everyone into a more authentic and grounded cultural perspective. Because what is at stake, it should be emphasized, is not a mere updating of the tools of intervention or investigation, but the identification of a whole series of new themes and, even more so, new relationships between them: which take on the characteristics of a true structural plot, constituting the skeleton of the new realities. In this way the process from past experiences to new ones does not seem continuous, and in fact a leap seems to be undeniable, in the very type of problematic issues now facing us. Somehow we have reached these new convictions, and to find the salient steps along that path has a value that goes beyond the mere chronicle.

We are convinced, in fact, that the change of scale to which we refer has somehow been proposed, when not imposed, from the outside. Once again, that is, reality has evolved in a much more rapid and, in the end, creative way than what could be foreseen by the cultural forces. The split that has happened between a professional practice strictly limited to the proposals of the market (in the absence of centers of research and experimentation) and a work of missionary preaching on more general themes (exhausted for the most part in the short time span of congresses and conferences) has prevented the rise of that exchange between the two planes of discussion that would be the only way to produce concrete prospects for action.

The obstacles to limited planning, in both horizontal and vertical directions, gradually became more evident; for example, as the compromises and unnerving negotiations to which municipal master plans were subjected became common knowledge; and as above all, with the passing years, even the plans considered to be the most outstanding failed to be implemented, got distorted or clearly proved to be inadequate, the faith that in the early stages had sustained our culture of urbanism began to dwindle.

At the same time, with the concrete formation of so-called “neighborhood policy” in the construction of a whole series of coordinated nuclei, an awareness grew of their impoverished, closed existence, which was of no use to the urban environment in which they were inserted. At the same time, though more from abroad than from our own country, the first dissenting voices were heard regarding urbanist theses of Anglo-Saxon and Scandinavian origin.

The plan of London—and Carlo Doglio was one of the first to grasp this – though for reasons very different from our own, had been substantially undermined in its final revision: the accent is placed on "replanning" in an even emphatic promotion of urban values. The masterpieces of Chandigarh proposed an expressive yardstick that was much more dilated, fertile and restless than the intimistic solutions of neo-empiricism, which thus revealed their evasive affectedness.

All these factors, combining in an admittedly confused way, have not remained without consequences. The symptomatic manifestations can be seen on various occasions; we can mention the book by Samonà, the Round Table at the 7th INU conference in Lecce, the CEP competition in Venice, certain indications of the Urban Planning Code, the formation of institutes like ILSES and IRES.

Setting aside the latter, for the moment, which deserve particular attention in a wider framework, it is worth reviewing the most significant reasoning put forth in the former.

Samonà, referring to recent urban planning developments in Paris, attempts an already precise hypothesis on the city-region of the future, though based on mostly visual impressions: "The results would be different if the radials of rapid penetration and the wide-ranging ring roads, which ought to form the main grid of the system of circulation, were able to insert themselves so closely in the fabric of the future city—as envisioned by the new plan—as to amplify relations, rather than growing apart from them; were they able, that is, to permeate in their varied structure the multiple activities with a distribution descriptive of its characters and its needs of local life and movement". Doing all this while maintaining the circulation flows in a linear pattern that makes fast movements easy between even distant zones, for the great masses engaged every day in movement from one point to another for reasons of work, socially interested—in a metropolis like Paris—in getting free of the tyranny of the constraint between place of work and place of habitation.

> The creation of this system could constitute the main motor of knowledge of a new large urban dimension, to be implemented through close relationships between the large circulation grid and the areas included inside it, if in the structural saturation of the various facilities, which constitute the slow putting into focus of the environment, the generalizing action of the plan and the individualizing action of private subjects (indispensable here to characterize the complexity of associated life) are able to find a balance, in harmony with the dimension given to the elements of the city by the road pattern and the new destinations, for which the new plan with its intentional flexibility leaves the definitive solutions open. (1)

A few months after the publication of Samonà's book, the national competition for the residential development of Venice-Mestre at the Barene di San Giuliano provided an opportunity to specify—though not to go deeper, given the short available time span—certain by now widespread convictions regarding the inadequacy of coordinated and total interventions.

The best solutions, in fact, rather than grouping the constructions of a public character in an island, opted to scatter them in the urban context, emphasizing the formal qualities to make them, in urban and visual terms, into the fundamental skeleton of urbanization. In more than one project the streets, junctions and waterways have a strong impact on the visage of the new city; which, in turn, is envisioned as a service center for a "hinterland" that extends well beyond the planned adjacent construction.

In the round table of the conference in Lecce, also in 1959, with Giancarlo De Carlo, Piero Moroni, Eduardo Vittoria and Ludovico Quaroni, the following statements were made, among others:

> The scale has changed in the urban setting, just as in life, in works, in physical, real space, and in the conception of these things… The result, or what could result, is the fusion of cities in the city-region, a practical plurality of functions and aspects, the physical and not only physical dimensions of the open city and large spaces, the wealth and continuous integration of the new life environment, complex and complete, a very efficient if not perfect tool for the resolution of some of the most serious social problems. (2)

We are still at the level of intuitions, and rather cloudy ones at that; nevertheless those reasons that lead to a new integration of urbanism and architecture, basing the expressive research on sociological and technological considerations, have already been indicated.

At the same time, but with a more official tone, the Urban Planning Code was being developed, which soon appeared as the conclusive act of the whole political-cultural battle waged in these years by the Istituto Nazionale di Urbanistica. An attempt is made with the Code, and with the latest ministerial drafts of urban planning legislation—though with much more extensive compromises—to insert the various sectorial plans in a general programming frame.

So in this field we are at a turning point of fundamental importance. But this does not mean we can avoid asking ourselves: will the new legislation be sufficiently ductile to favor the formation of the city-territory we can begin to glimpse today? Or will we run the risk of soon finding ourselves, once again, with an obsolete instrument?

At the present state of our knowledge, in any case, it is hard to hypothesize a replacement solution; we can nevertheless recognize that the city-territory shifts its field of application from total urban planning to the identification of leverage points, in the sense of direct intervention on the development, for example, of infrastructures; though this certainly does not invalidate the idea of an urban plan on a territorial scale.

We believe, however, that the urgency of specifying the new dimension of the city-territory has been imposed from the outside, as we mentioned above. This is due to the clarity with which a sum of technical and socio-economic transformations has emerged: making it possible to observe in the most direct way, that is, the inadequacy of some of our development hypotheses; to observe, without the possibility of doubt, how territories for which our plans envisioned timid and long adjustment processes have suddenly become the most productive ones only because it has been possible to locate certain new activities there.

We are thinking, for example, about the very great interest and weight of the formulation of the theory of poles of development (derived from the thinking of

F. Perroux) and of marginal areas in the last two years of policy regarding southern Italy; the ANIC-Gela took form very rapidly in this context.

On the other hand, we are thinking about which and how many changes have happened in our associated life; the destruction of the landscape of the coasts is undoubtedly the result of unbridled speculation, but it is undeniable that the thrust has come from the spread of a certain type of mass recreation that implies the expectation, on the part of many people, to find the variety of contacts and choices offered by the city in every other condition. Television spreads and unifies the same models and myths across the country; human relations and job evaluation, on the one hand, automation and cybernetics on the other, mystifying or objectifying, tend in any case to raise the standard of socio-economic life of certain sectors, though making the contrast with underdeveloped areas more strident.

Need for more suitable administrative tools

Other even greater difficulties arise in this context, which local administrations encounter in their economic and urban planning policy. Nevertheless, it should be clearly understood—in the structure in formation we are examining now—that the municipal districts, for example, will be of less and less significance, since the task is to supply each inhabitant with an equal quantity of possible choices, independent of the location of centers of production and of service. Nor can we imagine regressing from the condition of an affluent society in which we find ourselves, at least in the most advanced zones, with all the connected psychological and sociological implications.

So we will be faced with the much more complex problem of the control and democratic direction of these potentialities, not their rejection or passive registration.

Awareness of a "new" expressive necessity

All this clearly leads to a different approach to the problem of form. Action painting has already quickly shifted from a sort of post-Expressionist witnessing to an objectivized state, suggesting that of a thing amongst things. In Rothko structural research is essential, while in the younger painters the impasse of the gesture even tends to be overcome by a "new figuration."

From this confused, tangled situation, we can nevertheless glean an indication of faith in the possibility of achieving an expressive ethics, beyond industrial alienation.

We can understand the interest triggered by the debate that has begun in *Menabò* no. 4 and no. 5 on the theme of "Industry and Literature" in not so different and even more explicit terms. Vittorini's invitation to men of letters to also imaginatively engage with "the change on an industrial level that the industrial 'thing' implies in every other sort of thing" can be shifted almost intact into Italian urbanism and architecture, which faced with such new realities seem to be quite distracted and, at best, descriptive.

"The industrial world, though it has replaced the 'natural' world through the work of man, is still a world that we do not possess, but that possesses us, exactly like the 'natural' world. The former has inherited from the latter the old power of determining us, all the way into our ability to gain advantage from it, and therefore it must undergo an ulterior transformation that will divest it of the power to influence our choices and to determine us.

> Now a literature fully up to coping with the situation in which man finds himself with respect to the industrial world would consider the need for this ulterior passage, not ideologically, but in things, for the very fact that representation always implies … at least a minimum project, such as for example the project of freedom, new each time, which every image of servitude that is not approximate nor abstract has always included in its lines, etc., etc. (3)

The developments of the contemporary city

If the historical development of the city as it has taken form from the first industrial phase to the present is not to be accepted as is, nor to be rejected as a whole, because it contains contradictory elements, today we can recognize that the neo-capitalist development of the contemporary metropolis, also at the cost of serious social iniquities, is moving towards more evolved forms of territorial organization that are not, in the end, such a far cry from certain forecasts and prophecies of modern urbanism. At the same time, however, we cannot know a priori if the city-territory will take on socially positive forms of evolution, or their opposite.

Our concern today is to provide, and not just for the cultural forces, certain indications capable of orienting and determining in a progressive sense the future development of the city. While the ideological content of the rationality proclaimed by the Modern Movement has been dissipated as society has absorbed it, nevertheless it is still undeniable that—in that particular historical moment—the Modern Movement knew how to identify the elements that could act in a progressive direction, and it matters little if those very elements ("functionality," for example), in a different society such as the present one might be, have played or play a completely different role.

The present evolution of the urban structure is also confirmed by the rapid change that has taken place—from the early phase of industrial development to the present—in the housing-production-services relationship. The city has gone from a compact configuration—which gathered the entire multiplicity of functions inside a single system, characterized by close and direct relations of ranking among classes and immediate contact of urban zones of production and consumption, residence and commerce—to a more open configuration, in which production zones spread into the territory, causing gaps and large "voids" inside its structure.

Furthermore, while in the traditional city one could distinguish between three zones, in a chained relationship of rank—the center, the periphery, the suburban territory—with a clear prevalence in the development of production activities as opposed to services, in the present city we can see an expansion of the center, with a marked increase of administrative and commercial activities that can be traced back to the service sector, while the buildings that contain them are concentrated in certain zones. At the same time, we see a decentralization of production activities, which tend to spread outward into the territory.

These rapid changes in the location of activities and settlements correspond to other, more complex phenomena in the way of life of the different classes inside the metropolitan area, as we have already mentioned.

While the limited development commercial construction could have in the traditional city was accompanied by the relative poverty of consumption, limited to the producing classes alone, its expansion in the contemporary city is instead determined by the acquired ease with which the working classes are able to make rapid movements inside the entire urban and semi-urban area, and to take part in a perceptible way in the consumption of industrial products.

Today we are seeing a true disruption of the landscape—both urban and territorial—and the appearance of a new scale in the types of settlements and their interrelations; above all, however, we are seeing a substantial change in the relations between the city and the countryside. The passive relationship of subjection of the latter to the former is being replaced, and not without very serious economic imbalances, by a widening of the fabric of the city that leads to a gradual, total urbanization of the land. There is a process of enlargement of the city, within increasingly wide grids that cover the territory.

New indications of method

This is the reality in which to intervene and to provide those indications of method and of scale discussed above. For us, faced with a phenomenon we can consider historic, of structural transformation in the type of settlements, it makes sense to speak of intervention, now more than ever before. To provide, that is, the data and the programs that can lead to one form of the city-territory as opposed to another.

If it is true that the new type of city is the "place of settlements of various specialization made homogeneous by the existence of social and economic interrelations," and that this place is "configured as a site of homogeneous situations in constant evolution, where each part is combined with the others according to a relationship that is modified with every phase of growth" (4), this should not lead us to consider the territory as a surface on which every point is perfectly equivalent to all others and ready to welcome the function or purpose that, in time and space, can be attributed to it. We do not, that is, want it to become something like an electrical panel—in keeping with an extremely significant image that has been used—into which a plug can be inserted at any time.

In our view, it is instead necessary to reach the point of prompting the necessary differentiations between zones within the territory, according to well-defined relationships, also of a formal nature. What is needed, then, is to specify the concept of homogeneity of the territory, which in our opinion should not lead to an undifferentiated form.

If there are not to be qualitative discrepancies between the various zones (in the sense that through the intense development of means and networks of communication, the complementary nature of the functions of the various points of the territory—and of the activities within its overall economy—it is possible for each part to be positioned at the same level of potential), this does not mean that there will not be differences between them. Each zone, that is, can be assigned a different "situation" (though inside an elastic, transformable system).

Furthermore, we think that the definition of the territory—being the place of production activities—and of the city—as the meeting point of administrative and service functions—leads to a redefinition of the two terms that implies a democratic solution to the city-countryside conflict, when we come to grips with the totality of the problems involved: "the idea of city-territory is not a simple codification of current phenomena that already, in part, reveal positive aspects, and of a movement that spontaneously leads towards new forms of productivity; it also implies the necessary achievement of a new social order and of those forms of dynamic equilibrium of the economy which on their own are a guarantee of demographic progress" (5).

The same potential of movement that can be obtained in every point and that will practically place every settlement in the same conditions of equality with respect to productive, residential, administrative/business and recreational zones, erasing the differences related to distance between them, should not deceive us into thinking that it is possible to automatically eliminate depressed areas. It is clear that if very precise choices are not made in the context of complementary relations among the individual parts—to supply each of them with a qualitative indication—there will be the re-formation on a vaster scale, this time, of the imbalances of social and economic depression in the current peripheral zones.

For example, small residential centers set in the territory and gravitating around the main urban nucleus may be absorbed by it, or find their own particular function if a choice has or has not been made in their regard within the vast territorial system.

This problem, in particular, can emerge more vividly in the zones of metropolitan growth of Central-Southern Italy, like Rome or Naples, where the situation reflects different characteristics than those of the so-called industrial triangle. Here, in fact, many aspects continue to exist of the traditional city, which in a relatively short period of time has undergone major physical expansion and economic growth: most of the production activities still take place up against and inside the city, while in large portions the territory displays serious symptoms of decline, and in others it conserves the rural characteristics of its origins. It is clear that in these cases in which all the conditions for development towards the structural forms of the city-territory do not yet exist, it is nevertheless possible to intervene in order to bring those conditions about. Here we can consider the impulse to make new ties with zones not in immediate contact (Valle del Sacco and Val di Chiana)—after the opening of the Autostrada del Sole – in a traditionally isolated city like Rome, or simply the industrial development happening in the Pontine region, clearly generated by the proximity of a large center of consumption.

We believe, in definitive terms, that through the restructuring of the urban and territorial environment of the metropolis, within an elastic system whose dimensions themselves are impossible to specify, the process should move towards maximum characterization and qualification of both urban and territorial values. That it is possible, in other words, to recover the social function of a place of exchange of cultural values and economic interests that has traditionally belonged to the city, and to supply—with the positioning of production activities in the territory—that vast range of choices that must be available to the working classes, relying on their freedom of movement that has now been achieved. This would imply not only recovering certain typically urban functions or increasing the degree of freedom of the masses; it would also mean reconfiguring the physical arrangement of the settlements themselves.

The city remains, for us, not only the central driver of social life and the main economic nucleus of the territory, but also a point of physical concentration of architectural and spatial values. And this is independent of its size, its structure and the growth envisioned for it: all the data pertinent to the problem of determining its optimal configuration, for which a single model obviously cannot exist, given the multiplicity of the cases. In this sense we think Quaroni's words can be of use: "The idea of choice itself, which forms the basis of any conscious act and particularly of the idea of form and planning, does not necessarily have to mean immobilization, the stopping of a process. Actually, it should mean liberation from the obstacles that impede that movement, that evolution: and the idea of choice does not necessarily have to lead to 'unicity,' because it could instead lead to multiplicity, or at least a certain multiplicity, a certain integration" (6).

Relationships between city and territory

It remains to be seen what can be the relationship of the weights and quantities between the urban zone and the territory. And here it is clear that a second series of cases of multiplicity will arise. We will have to distinguish between urban residential zones that do not necessarily have to coincide with the main center, true city zones as centers of service industry activities, and—finally—zones of administrative/business centers. Some people are opposed to the very concept of such centers, asserting that they represent a traditional type of thinking and imply the formation of a new periphery. But this can no longer be true, if we think of the administrative center—indeed a ganglion of the life of the city-territory—as the point of pure quantitative condensation of a qualitatively homogeneous structure, without it having to present itself, then, in a position of hegemony with respect to the remaining zones. The administrative/business center, seen in these terms—whether it is posited as part of the territorial structures, along continuous lines of growth, or as the localization, in nuclei, of places of condensation of traffic—will become the expressive, symbolic site of the multiple urban functions.

It is at this point that the discourse concludes on the predictive and "pedagogical" function of language, which began with awareness of the inadequacy of the present culture. How can we forget the utopian (if you will) but very fertile message of Kenzo Tange, and Louis Kahn, in the plans for Tokyo and Philadelphia, and of Chandigarh, when we imagine a metropolitan structure that necessarily has to grow by means of modules related to the territorial scale?

The field in which we can make fundamental choices today is that of the large structures, which are those on the strictly architectural plane of the "large containers," and on the plane of territorial organization are those of the routes of communication and the service centers; and it is the field of formal qualification of these elements, that are the true "constants" in the new reality.

Form of the city territory

An initial evident formal characterization of the city-territory (meaning, by "formal," the individual qualities of our hypothesis of urban organization), can be suggested by the different cultural approach from which the hypothesis of the city-territory itself takes form; i.e. the scale of the dimensions to which the territorial problem is related and the method behind the approach itself.

Having gotten beyond the systematic approach of rationalist origin concerned with subdividing any problematic complex into particular problems each of which can be resolved separately (unless their synthesis is recreated in an a posteriori organization), to draw directly on the complexity and contradictory nature of a reality in which to operate (to insert a given modifier entirely taking part in the dramatic situation of contemporary society), the first characteristic, by now traditional in modern urbanism—namely the character of the "operative model" assigned to each working hypothesis—is automatically invalidated.

European urbanism, due to its Enlightenment orientation, even prior to its rationalistic character, has not understood until today how to do without a methodology based on the direct relationship of every operation to a formal model in which the complex of basic hypotheses formulated previously was taken to its synthesis: the same can be said of the various 19th-century utopias, from Fourier to Owen, all the way to the Garden City of Howard, and of the hypotheses already more methodologically formulated by more recent urban planning culture, from the *Ville radieuse* to the nuclear city, with all the resulting derivations and banalizations.

To place the problem of territorial settlement in a wider and different dimension should therefore lead, as a consequence, to the introduction of a working hypothesis that does not translate into a model, but is organized and structured in a configuration that ensures flexibility not only on the scale of the single urban sector, but also and principally on the scale of the overall organism.

The methodological issue therefore takes on a new value: we have already made inappropriate use of the term "configuration" to indicate the new dynamic arrangement of the settlements when even at this point, doubt might be cast in some places as to the possibility of the city-territory assuming an individuating form of its own. The new urban dimension, in fact, though flexible and dynamic, variable in time in its mobile structures and marked by its capacity to adhere to the mutability of situations and relationships between the single sectors, will be able to conserve formal qualification at the level of interzonal relations and organization of its fundamental structures.

Ideological value of the city territory

But from what has already been said it should be clear that the city-territory will no longer be able to be a "conformation," at least not in the traditional sense of the term, and will instead take on the dimension of a method of growth.

So the particular value is justified, in our discussion, of those fixed elements that form the overall skeleton of the city-territory, meaning the major networks of communication in which the "time" dimension becomes concrete and takes form, and the administrative structures, placed as a hinge between the various qualities of settlements (residential, productive, commercial, etc.), which therefore constitute ganglia of functional and expressive concentration placed at crucial points of the urban organization.

Now let us examine the consequences, in ideological terms, of such an innovation of the concept of the urban form.

Undoubtedly the most authentic value of those schemes and those operative models, which we have rejected for our hypothesis of the city territory, lay in their most specifically political premises: the city, seen as the maximum social and cultural concentration, as a reinforcement of political and social energies, could not help but be viewed as the best tool for those who intended to act on its structures and on the institutions that take concrete form inside it; and the only way to attempt to control the evolutions and processes that take form in the city seemed to be that of action on its physical dimension, on its figurative aspect. This led to an identification of sociology and urbanism, establishing a parallel condition between phenomena caused by social relations and those belonging to spatial relations.

Such a relationship between forms and modes of life and behavior, after all, belongs to the entire cultural attitude that can generically be summed up by the term "constructivist," whose crisis coincides with the decline of the myths that have presided over the planning of the city to date.

So the city-territory, precisely because it is a method of dynamic development, precisely for its new form that even becomes concrete inside a variability of its own limits, in its total adherence to the continuous and variable processes of reality, is the negation of any utopian or idealizing abstraction.

It is, so to speak, the surpassing of a rationalistic attitude that claims to impose a priori solutions on structures—which are not analyzed case by case in their complexity – to arrive, instead, at a continuous process of rationalization that leans towards constant verification of its own tools of analysis and synthesis.

This is not in conflict with what was stated above regarding the need for a new utopia, given the fact that the application of utopia is at present a mere means of visualization of a reality that struggles to find concrete form only due to the underdevelopment of our political and economic tools of programming.

Tools of implementation

Hence the extreme importance assumed by the study of the means of implementation: it is clear that faced with the dimension of the city-territory, the entire political organization and socio-economic arrangement now in place become anachronistic and ineffective. The achievement of a new form of associated life, in fact, implies the study of suitable tools that correspond on the level of the political framework to the demands expressed in that new form, giving it a content and an effective reality.

In this sense it is evident, for example, that the present dimension of administrative decentralization and autonomous power of local entities no longer seems timely; the new urban dimension will have to find corresponding decentralized units of a different type and quality. In a certain sense they will be much more limited in terms of responsibilities and territory, in order to be able monitor, control and orientate an urban and administrative policy on the scale of the various qualified and specialized sectors of the city-territory. This is also to safeguard the flexibility and variability of its physical dimension subject to change in time, given that in this way the coordination of the various zones for the provision of shared frameworks or at least shared programming would be greatly facilitated. Such decentralization would naturally take the various differentiations of quality of the city-territory into account, not only in the general sense, but also in keeping with the more complex characteristics we have already examined in descriptive terms.

Alongside this small-scale decentralization, on the contrary an expansion of local powers should be extended to the whole territory identified, case by case, according to the variability of the situations in time: with all the difficulties such flexibility can imply, from an administrative standpoint.

In any case a new dimension, also in this sector, is clearly auspicable, and the solution – in spite of the necessarily generic character of the indications that can be made in this regard at present—can be identified in the creation of a series of planning authorities capable of programming, monitoring, controlling in time—and, if necessary, modifying—the programs themselves, in order to finally generate that continuous planning which by now has been historically demonstrated to be the only valid form of democratic planning. It should be warned that to ensure this the city-territory is simply one possible tool, the only one that seems timely today: but it is still a tool, which to be fully and authentically made into a progressive structure of liberation from the alienating myths and techniques of the classist society, requires a new political dimension, as well as a renewed administrative status of representation, where democracy is no longer a slogan but a true achievement.

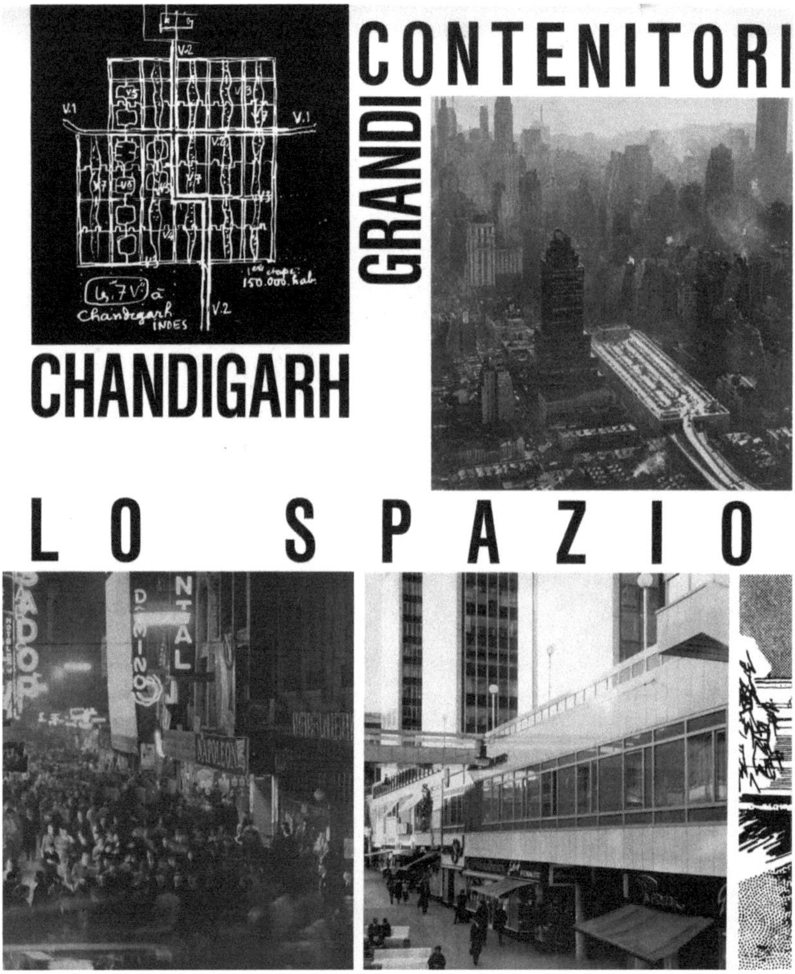

Fig. 14.1 *Casabella* article's original layout (p. 19). *Source* Tafuri, M., Piccinato, G., & Quilici, V. (1962). "La città territorio: Verso una nuova dimensione". In *Casabella Continuità*, 270, p. 19. Reprinted courtesy of *Casabella*

Fig. 14.2 *Casabella* article's original layout (p. 21). Text in the caption: "The *fosso dell'Aniene* in Rome. Renovation of Bethnal Green in London and the New Town". *Source* ibid., p. 21. Reprinted courtesy of *Casabella*

Fig. 14.3 *Casabella* article's original layout (p. 22). Text in the caption: «The Autostrada del Sole, petrol stations in Florence and in the Po Valley». *Source* ibid., p. 22. Reprinted courtesy of *Casabella*

Nella pagina di fronte: L'Autostrada del Sole, stazioni di servizio a Firenze e nella Val Padana. Qui sotto: Gli stabilimenti industriali dell'ANIC a Gela; Le Corbusier, « Le condizioni di natura sono ritrovate ».

I NUOVI CENTRI INDUSTRIALI

LE CONDIZIONI DI NATURA SONO RITROVATE

Fig. 14.4 Casabella article's original layout (p. 23). Text in the caption: «The ANIC industrial plants in Gela; Le Corbusier, "Nature conditions are found again"». *Source* ibid., p. 23. Reprinted courtesy of *Casabella*

Chapter 15
The Urbanized Countryside (1968)

Giuseppe Samonà

Introduction

Roberto Sega
EPFL, Lausanne, Switzerland
roberto.sega@epfl.ch

In the late 1960s, the Autonomous Province of Trento was the protagonist of an important page in the history of Italian urban planning: Bruno Kessler, the young president of the Trentino, gathered around himself architects, urban planners, economists, sociologists and demographers to conceive the first Italian attempt to manage urban growth on a regional scale. The coordination of the Plan was assigned to the architect Giuseppe Samonà.

The National Urban Planning Law of 1942 had prepared the ground for a particular kind of planning, capable of overcoming the opposition between city and countryside and significantly extending the focus from the city to the "City Territory", the city within its territory.

In the same years Samonà was invited to direct the Institute of Architecture of Venice (IUAV). His reforming will, breaking with the past, led to the recruitment of

M. Barcelloni Corte and P. Viganò (eds.), *The Horizontal Metropolis*,
https://doi.org/10.1007/978-3-030-56398-1_15

young Professors, creating an avant-garde academic milieu that contributed to re-discussing the very role of the urban discipline.[1]

The construction of the Trentino Plan, addressing the issue across a large territorial scale, has definitely benefited from this particular academic and legislative moment of transition.

The Plan pointed out the need for a renewed interaction between urban and agricultural areas. Samonà put forward the concept of Campagna Urbanizzata *(Urbanized Countryside), a "system of technologically and culturally advanced service locations, that makes living in rural areas similar to living in a town". This concept reflected the impossibility of reasoning in terms of contrast between city and country side and between city and diffused urbanization. The Urbanized Countryside concept was born from the perception of the crisis of the countryside, from the felt risk of a marginalisation of traditional dispersed settlements, and announced a new territorial project.*

Inside the Plan, Samonà refused to use simplifying and levelling terms such as "suburban system" or "City Region" and proposed instead the idea of a "Region City", an "isotropic territory equally open to urbanization". In his vision for the Trentino, he combined a polycentric model of a network of cities, towns and villages with the hierarchical system of main transportation routes. The unità insediative *(proto-unit settlements), shaped by the morphological and environmental characteristics of the territory, were key elements to imagine new scenarios for a "horizontal" urban development. The coarse grain isotropy imagined by Samonà aimed at balancing economic resources and also the built environment. In fact, what the author had in mind was not a mere spatial configuration but also a new "distribution of opportunities".*

Unfortunately, the spatio-morphological idea behind Samonà's decentralised thinking has been later translated into excessively bureaucratic terms, resulting in the establishment of mainly administrative perimeters that have taken the name of Comprensori *(districts). Today, the widespread construction of disparate, yet highly "urban" elements into a predominantly rural fabric, profoundly muddles traditional categories of town and country, making Samona's intuition even more relevant and visionary.*

[1]Giuseppe Samonà was rector of the University from 1943 to 1971. These years are remembered as the period of the "School of Venice". The new teachers who were named were: B. Zevi, L. Piccinato, G. Astengo, F. Albini, L. Belgioso, C. Scarpa, M. De Luigi, Romanelli, E. Trincanato, I. Gardella, G. De Carlo.

15.1 *Piano Urbanistico Provinciale*, General Objectives (from: *Piano Urbanistico del Trentino*)[2]

Remarks on the Evolution of the Concept of Urbanism with Reference to National Experiences

To understand the Piano Urbanistico Provinciale (Provincial Urban Plan) in legal-institutional terms, we should first examine certain aspects of the situation that have made it possible to arrive at independent provincial legislation on urban planning, enacted on the basis of several fundamental driving principles.

Urban planning, in its technical components and regulatory aspects, represents a relatively recent discipline, addressing problematic issues that in recent years have given rise to widespread debate on cultural and scientific as well as political levels.

Activities of "urbanism" in the wider sense of the term, i.e. activities related to a certain regulation of urbanized areas, can be seen ever since in the first inhabited zones and the first means of transport led to the formation of roads and paths; at their intersections the first villages began to spring up, in which with a certain design the centers of administration and trade of the social community took shape. Across the various epochs of history, and with the diversities due to cultural and political situations, the need for sufficiently rational organization of urban settlements can be seen, at least within certain limits, as something that has always existed.

Nevertheless, the first necessities of a modern rationalization of territorial development in relation to human settlements had a more proximate cause, in Europe, in episodes connected with the industrial revolution, which together with radical changes in more precisely economic factors led to the abnormal growth of the old walled cities, the formation of vast working class districts inside which the living environment was extremely precarious, and to the opening of new routes of communication, with major migrations of the population. The delayed arrival of industrialization in Italy meant that a perspective on urban planning in the modern sense of the term did not take form until 1940, apart from the problems caused by the growth of certain large cities due more than anything else to the administrative activities that were being inserted in them, in relation to the political structures of the time.

Prior to this, provisions of an urbanistic nature were found in the unified legislation of 1865, which in the wider sector of public works introduced the definition of sectorial master plans which set out to safeguard particular aspects of technical

[2]Excerpt from: Samonà, G. Andreatta, N. Giovanazzi, S. (1968), "Report of the Provincial Law 12 September 1967, N. 7 "Approval of the Provincial Urban Plan" presented by the President of the Provincial Council Bruno Kessler" ("Relazione alla Legge Provinciale 12 Settembre 1967, N.7 "Approvazione del Piano Urbanistico Provinciale" presentata dal Presidente della Giunta Provinciale Bruno Kessler"), in: Provincia Autonoma di Trento. Piano Urbanistico del Trentino, Padova: Marsilio, pp. 41–92. Reprinted courtesy of Provincia autonoma del Trentino.

development, mostly in order to govern matters of hygiene and circulation, while actually leading—for example—to the destruction of many historical centers.

These norms, in fact, as was only natural, soon revealed their inadequacy to cope with the urban planning problems that were beginning to become more complex, and this led to the enactment of special laws and the approval of master plans extending to the entire urban sphere, also with certain remarkable foreshadowings of what was to become the overall discipline of Italian urban planning.

In the regulations in effect until this time period, there was an implicit opposition between city and countryside, with a clear differentiation between problems related to urban and non-urban territories, focusing only on those of the urban centers, where the frictions of the initial process of industrialization were concentrated. Only later did a progressive opening occur, bringing an awareness to some extent of the mutual interpenetration of the two conditions—urban and territorial—and the resulting recognition of the fact that the city could not be seen as a phenomenon having an independent life, detached from the surrounding territory, but instead had to be understood in its immersion in the territory, with the resulting gradual unification and integration of the city with the outlying zones regarding planning criteria.

The national urban planning legislation of 1942, alongside shortcomings of a scientific order, due to the incomplete understanding of various doctrinal concepts, but also of a political order, due to the regime still in power, undoubtedly laid the groundwork for planning that at least in part responded to the need for an urbanism configured not only as the science of urban aggregations, but also as the science of human settlements, both centralized and dispersed. A process of development of the object of urban planning thus began, which gradually extended from considering only the "city" to considering a broader frame in which the city was seen in the territorial context of the municipality, and later to increasingly large spaces, to the point of coinciding with the territory of entire regions.

The possibility of a territorial coordination plan as a strategic tool to develop directives of a general order for urban planning activity—seen as planning of the organization and growth of large territories in relation to all the activities variously inserted in them—represented an undoubtedly positive factor, on a par with the system of general and particular master plans, the first to be extended to the entire municipal territory.

Nevertheless, the law totally overlooked the socio-economic conditions, while the municipal level of planning, due to its limited range, immediately revealed the practical impossibility of considering vaster territories in a unified way. The law of 1942 itself already contained a timid acknowledgment of the insufficiency of the municipal context as the spatial level of urban planning, and had thus introduced the factor of intermunicipal planning. But also in this area, the basic concept remained intact, namely that attention should be paid only to the dynamics of urban development, without assigning any importance—in practice—to those social, economic and territorial factors that could have permitted overall planning.

Towards the 1960s there was thus the beginning, or in any case the reinforcement, of that movement of ideas that continues into the present, which has led to many attempts at radical renewal of the existing legislation in keeping with a conception of urban planning that could more precisely be called territorial planning, having an impact on the territorial organization of the entire country, in keeping with a planning method—defined as open and continuous—carried out through a process from the general to the particular, constantly updated and in touch with the varying conditions of the time.

A kind of urban planning, in other words, that sets out to implement an overall vision of the problems of the territory, capable of covering not only the aspects of the constructed environment but also the more specifically economic aspects, based on the premise that expansion of industry, tourism, commerce, trade and administrative activities, as well as environmental values and characteristics, constitute urban "functions" to be taken into consideration on a par with demographic and real estate expansion.

General Objectives of the *Piano Urbanistico Provinciale*

The intense work of exploration, analysis and composition is based, then, on a general hypothesis that should be considered for an initial identification of the contents of the Plan.

The Piano Urbanistico Provinciale has in fact been approached as an instrument at the service of a population which, across vast zones and in its intellectual capacities, has always been exposed to an alternative between remaining in precarious conditions in its place of origin or emigrating.

We do not have evidence to be able to reliably judge whether—given this alternative—emigration can be desired and even urged for a certain number of individuals; we do not know, in other words, if the opportunities offered by large settlements are more attractive than those offered by the places of origin, if that place were effectively organized and raised to greater efficiency and civil dignity. The terms of comparison for such judgments are lacking. We only know—and this is already sufficient—that the choice between remaining and emigrating is not a "free choice" in practice.

In these conditions, also with the aim of specifying the real will of the citizens with respect to this alternative that has long existed, no other path seemed to be open for an hypothesis that would also politically characterize the Plan except that of proposing the pursuit of a general territorial "situation" capable of offering, in perspective, freedom of choices in keeping with multiple advantages and for all the citizens.

Given this need formulated in its problematic terms and with the intent to safeguard the requirements of freedom of individuals and groups, it nevertheless seemed necessary to immediately clear away any overly simplified evaluations of what could be based on the hypothesis of direct research to clarify the premises and conditions for the formation of free choices between remaining and emigrating.

Some have thought, or perhaps still think today, that the implicit premise on which the Plan is based is that the countryside is abandoned only because agricultural does not pay, and because industry or other activities are not present there.

This way of thinking may also be suggested by everyday observations and pressures through which it would seem possible to conclude that people expect only industry to consolidate the local economy and to obtain, through a higher quality of life, the compactness of the community and, as a result, less willingness to emigrate.

Still at a level of general indications, this extreme simplification of the problems has been seen to be unreliable; the observations we have been able to make regarding given behaviors of our population tended, in fact, to explain the abandonment of the countryside on the basis of psychological as well as economic criteria. It seemed possible to observe that the aspiration to live in the city was felt above all due to the range of choices the city itself offers to individuals, give the fact that only in cities, unlike the countryside, is it possible to find a variety of offerings capable of covering a larger range of desires and aspirations.

But the progress of the cities resulting from this tension, besides inexorably translating into downgrading, impoverishment and emptying of the rural environment, threatened to produce problems and risks of congestion and tension which an increasingly widespread doctrine, closely linked to experiences and direct, timely observations, considered burdensome to say the least for the social organization of the city.

To this theoretical awareness, only partially applicable in the situation in the Trentino due to the size of the cities, were added ulterior elements of assessment that contributed to invalidate the hypothesis of a big city into which to bring together the great majority of the population, in the light of the theoretical benefits the city itself seemed to offer.

There was first of all a physical-environmental obstacle: the available spaces for a global concentration of the population of the Trentino are not sufficient to envision rationality of settlement and above all prospects of growth.

Even were this objective difficulty to be overcome, no one was able to foresee the timing, costs and frictions that would result from a perspective of maximum urban concentration, so such an indication would have remained the pure and simple record of a theoretical hypothesis, open to an infinity of unforeseeable issues.

Lastly, no one could imagine that in the face of such concentration a desert should be created, given the fact that at least one sector—tourism—was already thriving, with good prospects for growth. In this situation a proposal of global concentration would also become harmful for an important economic sector, leading to direct negative impact on the provincial economic balance sheet as a whole.

Faced with these questions, due to the complexity of the motivations involved in decisions of emigration, and perceiving the impossibility even on the level of pure hypothesis of an urban concentration in practice on the axis of the Adige, it seemed there was no other path of action in pursuit of a "situation" of freedom than that of creating other "cities," transforming the countryside itself into constellations of "cities." The "urbanized countryside" was thus the most realistic response to an objective of balance seen in terms of maximum social mobility, envisioning a type of urbanization that reconstitutes, in the countryside, that urban type of settlement capable of offering a range of basic choices resembling the ones found today in the traditional phenomenon of the city.

To achieve this objective, the Piano Urbanistico Provinciale divided the territory into ten districts, which correspond in various degrees to sociological, economic and geographical conditions different from one another, constituting the basis for the configuration of the new urban development.

The districts, in essence, had to be progressively preordained for the creation of sufficiently vast environments, capable of stimulating attitudes of social aggregation and taking on the driving functions of progress.

The urbanization of the countryside relies on the will and ability of man; but this, in turn, is influenced to a large extent by the external environment. The widening of horizons, the integration between centers and sufficient demographic consistency are the premises that make it possible to foresee the intervention of public and private initiatives, capable of expressing symbols, models and assurances with the potential of promoting the cultural homogeneity within which the intellectual elites of labor, production and politics can converge to accentuate the process of transformation and to exercise that form of "social control" capable of discouraging inappropriate initiatives while selecting innovative projects.

The basic objectives of the Plan can therefore be summed up in the balance as a whole set of structural forecasts set out to humanize the territory, in the sense of making it more hospitable to the free spiritual and civil progress of man. The Plan, in its outlooks and interventions that, due to their direction, are developed and focused on man—not in a generic sense, but in a precise way along lines of complete development, not at all visionary but realistic and civil— sets thus its objective or intermediate condition for progress.

Indications and Strategies of Intervention

In the framework schematically described above, the first strategy of intervention that has to be implemented concerns the formation and the life of the districts.

The first action in this direction has to do with the study and constitution of the district consortia, foreseen by the provincial urban planning legislation, for the adoption and management of district plans in which they have to specify and clarify the principles the Plan indicates as the basis for territorial reorganization.

In this regard it should immediately be stated, also to prevent misunderstandings, that two of the districts (and perhaps several others) clearly do not have the minimum size that would seem to permit, in specific conditions of intervention, that internal dialectic that has to play an essential role in the development process.

In this situation, and with the present state of knowledge, nothing more effective can be envisioned than the proposal of a mechanism of stable integration, to be identified and characterized in Trento, with all that city represents, as has been expressed in the concept of the "guide city."

In the implementation strategy of the Plan this seems to be the first concept to develop, identifying the quality and type of services the Adige district is called upon to provide, in its sociological identification as a center of integration.

Trento and the Adige axis appear to be too limited in size today for the role they are expected to play.

A requalification of Trento, however, should not be seen in this perspective as a pure and simple urban expansion that comes passively in the wake of a demographic increase derived to a large extent from immigration.

Trento should be reorganized mainly in qualitative terms, and an assessment of efficiency in its regard can be positively formulated only if inside it the conditions are created for the development of stimuli, indications and interventions capable of moving into the entire periphery, with a particular accent on those zones and districts in which, for various reasons, the local communities are slow to take on driving functions of progress.

Districts and interdistrict integration, therefore, represent fundamental factors in the strategy of implementation of the Plan.

The success of this first objective which involves the entire arc of public interventions, nevertheless, calls for the development of a procedure of intervention that gives rise to the most rational relationships possible among the various players.

One priority in this regard seems to be the solution to the problem of coordination and evaluation of all public interventions in terms of costs and benefits, to seek not so much in the more obvious need to avoid duplication of efforts or interferences, as in the assumption of a rational method that makes it possible to impose, at all the levels of public operation, homogeneity of technical evaluations on a wide temporal horizon, to distribute available resources based on criteria of general usefulness.

The Provincial Urban Planning Commission, in its vote on 8 May 1967, made express reference to this concept both for its implications of a technical character, which underlie the concept itself of elasticity of the Plan, and for its function of stimulating the pursuit and evaluation of better solutions.

In this context, and returning to the concept, it seems worthwhile to widen its meaning in relation to the basic objectives of the Plan.

The cost-benefit ratio should in fact be applied by all the public operators with reference to the creation of a new balance across the whole territory of the province, ratifying and supporting the proposals that are more generally and completely pre-arranged in this sense.

For a rational urbanization of the territory, based on its subdivision into districts, the presence of a driving urban center inside every district is indispensable.

The need for a reference point on which to gradually focus the attention of the district community depends on the very concept of urbanized countryside, which implies an internal specialization of the territory in keeping with models similar to those applied for cities.

The function of the urban center is that of an administrative seat; therefore a concentration of public and private services is postulated, with a sturdy infrastructural context, efficient organization of secondary and tertiary economic activities, to permit the gradual formation of a managerial group capable of developing ideas and proposals that go beyond the narrow municipal context and stimulate a sense of "belonging" to a community in the populations, on a wider scale that what is manifested today only in relation to townships.

The progressive establishment of a dialogue between district centers and the outside world will also prevent the transformation of the district into a "hortus conclusus" that elaborates and consumes its material and intellectual products entirely inside itself, in favor of stimulation of its participation in a perspective of specialization and differentiation with respect to the social context that surrounds it.

Specialization and differentiation are in fact functions that the districts, on the basis of the indications of the Plan grasped and given value by the driving forces mentioned above, have to specify and define in order to act as vectors of progress for the society of Trentino as a whole.

The location and organization of a center that responds to the needs and performs the functions described herein is the task of the districts, based on evaluation of reputation and renown of the various centers in the district.

It can only be stated that given the particular arrangement of the Trentino territory and the growth of the use of motor vehicles, the location of the center can also be organized across multiple urban nuclei, or distributed along axes with coordinated, interacting points of specialization.

Whatever the urban organization of the center may be, it nevertheless seems that at least two instruments should be evaluated in particular for the above-described purposes, due to their primary role: the public administration and education.

Where the public administration is concerned, considering the objectives and assumptions of the Plan in terms of equilibrium, it first of all appears that the district has to present itself as the minimum area of planning feasibility for interventions of higher-level public agencies, the exclusive context of their decentralized territorial functions, the seat of operative coordination in both the study/design phase and in the dimensioning of interventions, and for the timing of the expenditures in relation to the objectives to be pursued.

The progressive identification in the districts of the minimum and exclusive levels of territorial layout for all the various sectors of peripheral intervention, and the concentration of offices and management in a single center, will also have to take place in the expectation that there will be some friction or incompatibility regarding the single sectors, with the compensation in terms of overall rationality and efficiency that can be the result of the advantages brought about by the achievement of the objectives illustrated above. Such a unitary arrangement of administrative organization, given these premises, will furthermore have to involve not only the existing agencies and offices, but also the agencies that are to be additionally formed with the aim of aggregating traditional services or functions, or those based on a new conception to be applied inside a given territory, as well as on a wider level when particular phenomena extend to more than one district at the same time.

This is also the context in which the Piano Urbanistico Provinciale advances a proposal on the urban aggregation of multiple constituent townships, in the context of the district, into a homogeneous area to respond to the demand for unified community services.

The proposal, defined as a "settlement unit," is not regulated on a formal level, but simply indicated in its essential characteristics, and offered as a technical stimulus to be specified and clarified in its practical content and for the objectives schematically outlined in the process of district planning.

The second instrument, which also seems to be effectively available for the purposes of the formation of driving urban centers in the provincial territory, is that of education.

The role of education is of great importance, for multiple levels of reasoning, in the process of balanced growth of the society in Trentino.

The level of education in Trentino, after compulsory schooling, is lower today than the level estimated as indispensable for the constitution of truly stimulating and effective centers of cultural propulsion. Without these shortcomings—which have been repeatedly raised as issues—it would be possible for the Trentino Province to have a higher level of economic dynamism and cultural advantage.

On the other hand, it should also be observed that the population of the area produces a quantity of intellectuals greater than the number it is effectively able to employ or utilize. Therefore we are clearly faced by a classic "closed circuit" of interaction between culture and growth, which can be addressed by schooling only if it is completely inserted in the social context, taking part in its growth as a cause and driving principle, while at the same time growing as an effect or consequence of the growth of the community itself.

Due to its function as a multiplier of social effects, education is therefore a strategic tool of extreme importance and delicacy in the urban development of the territory, and the stimulation of driving forces for the functions of social progress that form the basis of evolutionary transformations depends to a great extent on this factor.

A distribution of schooling in the centers that are to represent the motor of local progress is thus imperative, not only through the foundation of new school facilities, which has already begun, but also with the formulation of an entire educational, cultural and sporting organization collateral to the scholastic centers, able to enhance with its breadth the formative values of schooling and its role as a foundation for civil development as a whole.

Fig. 15.1 "Val di Non: Nanno and Castel Nanno in the background".
Source: Provincia Autonoma di Trento. (1968), Piano Urbanistico del Trentino, Padova: Marsilio, p. 85. Reprinted courtesy of Provincia autonoma del Trentino.
«These are about a hundred small nuclei that have a very interesting traditional building configuration from which often emerge, older, the church and the bell tower and detach isolated castles with their architectural form fractioned and aggressive.» (Source: Provincia Autonoma di Trento. (1968), *Piano Urbanistico del Trentino*, Padova: Marsilio, p. 186)

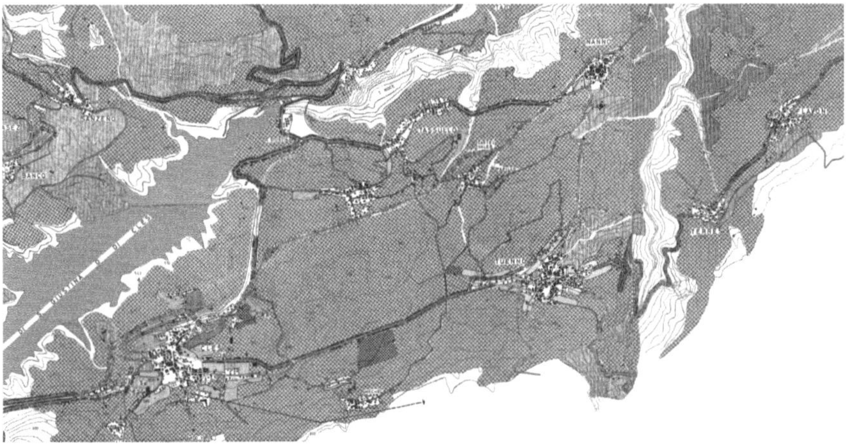

Fig. 15.2 Provincial Urban Plan of Trentino: Map C. 4.5 (Val di Non, excerpt)
Source: ibid., p. 241
Reprinted courtesy of Provincia autonoma del Trentino

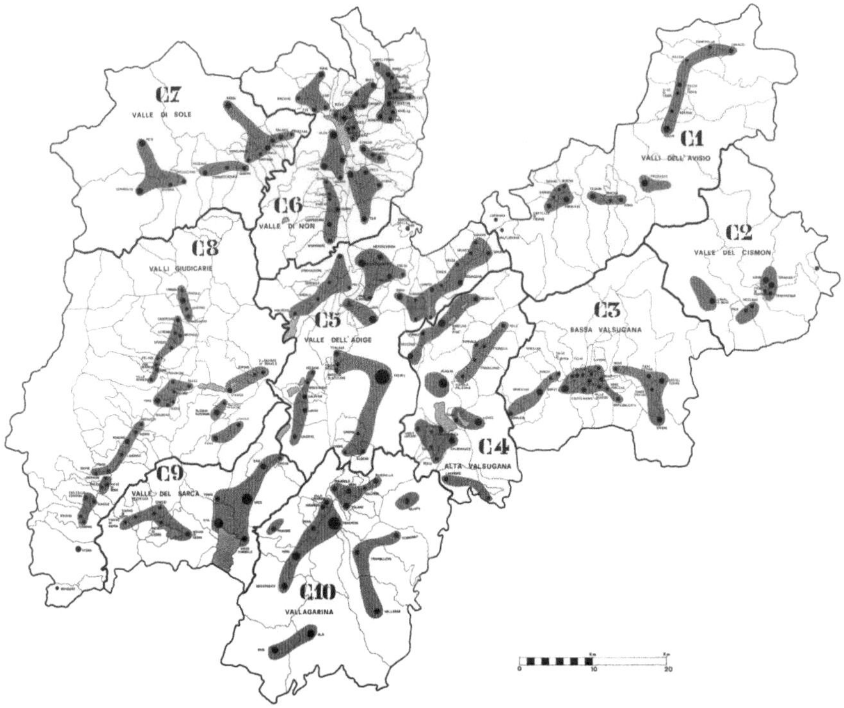

Fig. 15.3 "Trento Province: Map of the districts and settlement units" ("Carta comprensori e unità insediative").
Source: Ibid. pp. 148–149.
Reprinted courtesy of Provincia autonoma del Trentino

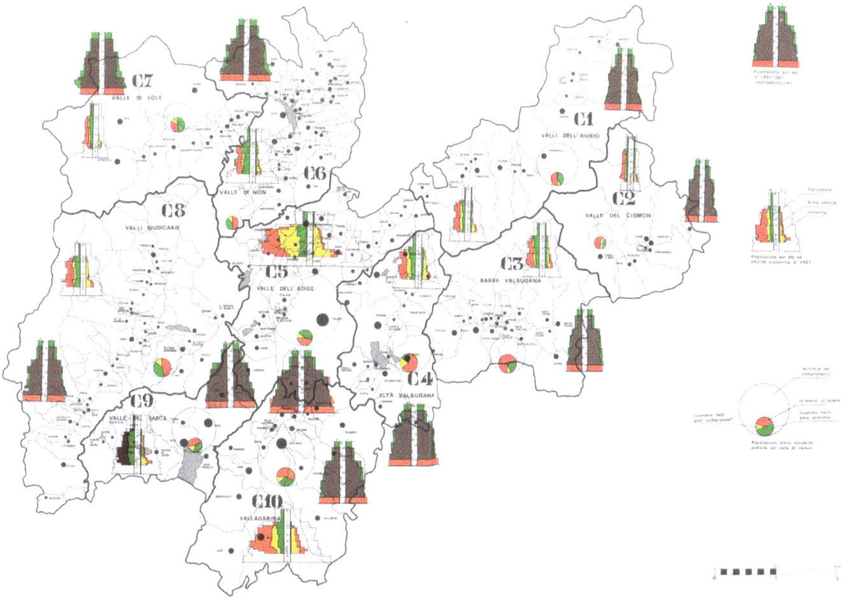

Fig. 15.4 "Trento province: Some demographic characteristics of the districts" ("Alcune caratteristiche demographiche dei comprensori della provincia di Trento").
Source: ibid., pp. 308–309.
Reprinted courtesy of Provincia autonoma del Trentino

Chapter 16
Rurbanisation (1976)

Gérard Bauer and Jean Michel Roux

Introduction

Marine Durand
EPFL, Lausanne, Switzerland
marine.durand@epfl.ch

La Rurbanisation, ou la ville éparpillée *is the first study published in France on the phenomenon of "imbrication between rural and urbanized spaces". Financed by the Ministry of Agriculture that wanted to understand what was happening in the French countryside where "new inhabitants coming from the city were building houses in the villages", the study was preceded (ten years earlier) by* L'habitat pavillonnaire[1] *which focused on the peri-urban phenomenon characterised by the emergence of single-family houses ("*pavillonnaire*") in continuity with city-cores. With the "*Rurbanisation*" instead, the two authors—the economist and agronomist J.-M. Roux and the architect and urban designer G. Bauer—identified a new phenomenon, a different form of urbanization that had appeared during the '60s.*

This phenomenon concerned new systems of individual housing built in rural villages "next to" but "in discontinuity with" large urban cores. By identifying this phenomenon, the authors went against the common idea that during the '60s the first allotments expanded from main urban cores in suburban ring/belts.

---

[1]Raymond, H., Haumont, N., Raymond, M.-G., & Haumont, A. (1966). *L'habitat pavillonnaire*. Centre de Recherche d'Urbanisme et Institut de Sociologie urbaine; Haumont, N. (1966). *Les pavillonnaires. Etude psycho-sociologique d'un mode d'habitat*. Centre de Recherche d'Urbanisme et Institut de Sociologie urbaine; Raymond, M.-G. (1966). *La politique pavillonnaire*. Centre de Recherche d'Urbanisme et Institut de Sociologie urbaine.

J. M. Roux (✉)
Grenoble, France
e-mail: roux.transversal@wanadoo.fr

M. Barcelloni Corte and P. Viganò (eds.), *The Horizontal Metropolis*,
https://doi.org/10.1007/978-3-030-56398-1_16

The study would be followed three years later by a report, the "Demain l'espace. L'habitat individuel péri-urbain"[2] *that will integrate the category of* rurban *as complementary to the one of* city-park, *forming together the* peri-urban. *Both studies garnered mixed reviews since the authors claimed for the necessity to deal with this on-going phenomenon via an interpretation of the same, its scale and its undeniable existence, while the institutions depreciated the idea of considering a phenomenon that was against the idea of concentration and centralization.*

Thanks to the empirical study of precise but limited geographical areas, the authors describe the emergence of a new form of urban growth, that appeared in France between 1960 and 1970. The future will teach us that the authors could have chosen almost any part of France matching the criteria of their investigations.

Firstly, the rural villages described by the authors and characterised by the rurban phenomenon were situated between 15 and 20 km away from the "big city".[3] *The phenomenon took part in a "flat territory, cultivated and, for this same reason, inhabited by regular and scattered agricultural communities".*[4] *Given the distances between human settlements, in the wide French countryside we will never witness the formation of "conurbations" either in the '60s nor, as the authors predicted, in the present day.*

Furthermore, the new inhabitants are defined as "urban". In reaction to the "Grands Ensembles" and to top-down policies, many people left the city in the '60s looking for a better quality of life and for a reconnection with "nature", but still dependent on the city for providing jobs or culture. This urban exodus is directly related to the massive use of the car that makes a new life possible.

The specificity of French "Rurbanisation", and what makes it deeply different from the "Rural development", is the fact of being based on individual initiatives and of concerning housing only.[5] *Differently to other studies as the ones on the "Ville-Territoire" (Corboz), the "Zwischenstadt" (Sieverts) or the "Città-Diffusa" (Indovina), the study on "Rurbanisation" focuses on a very specific and limited form of urban growth, even if it recognises it as being part of an "Agglomeration with no conurbation".*

[2]Report of the study mission chaired by Jacques Mayoux (*Rapport Mayoux*).

[3]A second phase of "*Rurbanisation*" would follow in France, touching villages up to 40 km from main cities.

[4]We recognize here the model proposed by Elisée Reclus (1895) and inspired by the French plains to describe the regular distribution of human settlements.

[5]In its early phases it is never related to the idea of diffused production, and as a result, economic activities and public services won't follow the "rurban" development.

16.1 *La Rurbanisation* (from: *La Rurbanisation ou la ville éparpillée*)[6]

Preface

Our cities do not grow solely by progressively engulfing the countryside immediately around them. They foster, across a broad area, scattered extensions where a growing percentage of new construction takes place. We will use a neologism to describe this phenomenon of interlinking rural spaces and urban zones: *rurbanization*. Detached homes, subdivisions, and "new villages" constitute the most visible manifestation of this phenomenon, but not the only one.

Current events confirm the relevance of our study. The most characteristic (if not the most desirable) form of the contemporary city, rurbanization remains unrecognized in France in spite of its vigor. It hardly interests researchers, who are more focused on the infrastructural foundations of urban growth than on its spatial modalities. In urban design and planning documents, it clashes with the all-powerful notions of concentration and centralization, which mobilize energy and capital to densify urban centers, build upward, and fill in the rare gaps in the urban fabric.

However, the *rurban* wave seems inextricably bound to urban growth, and nearly unstoppable if that growth continues. We analyze this form of development through examples and surveys. You will find no graphs of new urbanization, nor much critical analysis of its success at satisfying real needs. We will focus instead on the reality of the phenomenon, its extent, and its future prospects: before any policy can be suggested regarding it, it must be introduced into urban and regional planning discussions.

We have followed an empirical approach in producing this overview, which is the first one we know of in France. It is the fruit of novel observations of the local area undertaken during our studies of regional planning. It touches on a theoretical problem frequently discussed in France: the relation between economy, society, and urban space.

You will not find references here to more in-depth studies, which remain to be undertaken. The resources at our disposal have only allowed us to formulate a structured body of hypotheses.

Our ambitions have been constrained by the small number of quantitative documents on the subject. We have thus had to rely excessively on our own previous studies, which are far from providing an exhaustive overview of the spread of rurbanization: they only regard the areas around Nimes, Rouen, Orléans, Paris,

[6]Excerpt from: Bauer, G., Roux, J.-M. (1976). *La Rurbanisation ou la ville éparpillée*. Paris: Espacements (pp. 1–25). © Éditions du Seuil.

and Lyon. We do not yet provide a succinct theory of the bases and meaning of rurbanization. Our priority is to describe events: without such descriptions, a premature theory would accomplish little more than a reiteration of its hypotheses. We discard only a few presuppositions that seem simplistic to us.

We do not present it as a simple cultural phenomenon. On the other hand, we were in no way able to establish a strict and univocal dependence of rurban forms on an authoritarian economic logic. At this spatial and temporal scale, manifold, complex interactions occur between historical traditions, the various social groups (difficult to identify in broad analyses), and the economy. In a similar perspective, the organization of the territory is the cause, as much as the effect, of the structures and socioeconomic relations of production and reproduction.

Like all large industrialized countries (capitalist or socialist), albeit with its own distinct characteristics, France is subject to two opposing movements. An implosion draws the rural population toward urbanized zones while an explosion expands these zones outward. The latter dynamic is the focus of this book. We only measure the beginning of these displacements, which are accelerating.

We do not prejudge the relationship between a fact confined to a limited realm of time and space, rurbanization, and the logical right of the monopolistic capitalist system. If it exists, it is not necessarily (or solely) situated at the level of the principal markets and associated business sectors: construction, public works, financial exchange, etc. Spatial planning models are imposed on the decision makers, and one hardly knows, at this level, if the economic rationale creates the model or if it is the other way around. All causes of the phenomenon, cultural or economic, must be reviewed, their hierarchy currently appearing contradictory and confused. To use terms that are currently in vogue, we are navigating by sight in the second-to-last analysis.

Post Script This book was written prior to the general census in France in 1975; the bulk of the demographic statistics used herein date to 1968. At that time, the rurban trend had barely begun in our country. At the most, an observer in the field might have noted the burgeoning housing developments, and recognized that France would not be spared the spread of urban areas, something that was already quite advanced in northern Europe and in the United States.

At the time that the book was going to press, we learned of the first provisory and partial results from the 1975 census. While referring to the population as a whole, they are nevertheless quite significant. In brief, it seems that in seven years, the population in large urban agglomerations has stopped growing, that in remote rural areas has shrunk significantly, while the population in the rurban zones had increased quite a lot, well beyond our initial projections. These data only strengthen the theses of this book.

Definition

Interpenetration of Urban and Rural Space

"Rurbanization" is the result of the spread and scattering of cities in space.[7] Hence, according to an initial, approximative and provisory definition, a rural zone is "rurban" if:

1. It is close to an urban center and absorbs a new residential population, mainly coming from the city, and exhibits:

   - a positive migratory balance (permanent migrations);
   - a shrinking proportion of farmers and rural artisans in the population;
   - significant transformation of the real estate market;
   - significant construction of single-family residential (detached) housing in rural communities located within a half-hour radius of the nearby urban hub;

2. It is strongly characterized by the continuing prevalence of non-urban (i.e., rural) space. This is the principal distinction between the spatial organization of a rurban zone and that of a typical suburb. In effect, suburbs have taken form up to the present in full contiguity with the mother city, of which they are a spatial continuation; the inhabitants of these new neighborhoods are, in a certain sense, "sub-city-dwellers", awaiting the space around them to fill in with new urban development, slowly creating around them the truly urban fabric that characterizes older neighborhoods. However, when the usual peripheral extensions—the large housing developments and fabric composed of detached houses—are increasingly complemented (and contrasted) by new constructions located in and among the towns situated within a radius of fifteen to twenty kilometers from the urban center, then we have a specifically *rurban* situation. The key here is that, even accounting for the most extreme demographic growth, the filling of the interstitial space is out of the question, except perhaps if we take a timescale on the order of a century. And more importantly, "*[t]he city appears, more and more each day, to be the central district of a discontinuous agglomeration.*"[8]

The interpenetration of rural agricultural space and urban space thus becomes, on the level of the developer, a *permanent* aspect of the living environment, given that this type of growth cannot develop the urban continuity of typical suburbs due to purely quantitative reasons. Rurbanization thus creates a particular sort of settlement of the land whose consequences cannot be confused with those of the traditional suburb spreading out like an oil slick.

---

[7]Spread: to cause to open out or extend over a larger area. Scatter: to place (as buildings) here and there, distribute at irregular and widely separate intervals. *Merriam-Webster Unabridged Dictionary*.

[8]Gravier, J.-F. (1972). *Paris et le Désert français en 1972*. Flammarion.

The Most Recent Form of Urban Growth

Is rurbanization truly a new form of urban growth? A historical overview of urban growth in Western Europe (the conditions are different in relatively non-industrialized or relatively newly industrialized nations and in North American) leads us to identify *five* forms or phases of city extensions, both consecutive and cumulative, but distinct. Rurbanization is the most recent. They are not intended in any way to explain why a given city has a given form, their sole purpose is to illustrate the fact that a series of verifiable *morphological* transformations are associated with the introduction of new means of transportation, and that we are currently living through one of these transformations.

The model takes as its starting point a flat, cultivated territory dotted at fairly regular intervals with small faming communities. The development is as follows:

Phase 1 (to the mid-nineteenth century): in a village located at the crossroads of the major communication routes, commercial activities and crafts develop more than they do elsewhere. The settlement grows by simple *agglutination*. The size of this agglomeration is subject to a limit, which cannot be determined with any precision, deriving from the maximum amount of time it takes to walk from one side of the village to the other. As the village approaches this limit, the tendency to densify by clustering buildings more closely and constructing upward begins to prevail. The agglomeration is relatively non-polarized in terms of activities because no establishment, regardless of the intensity of relations it maintains with others, has a particularly marked interest in being located in precisely in the center. As a result, the coefficient of land occupancy is fairly even across the entire village, without a significant different between the center and the fringe. The network of streets tends to be laid out in a grid rather than as a spider web. We may still find well preserved examples of this archaic type of city, one of them being Bruges.[9]

Phase 2 (starting in 1850): the growth process described above continues, less by agglutination than by densification. However, now the railroad is introduced, drawing most new construction away from the center and clustering it around the stations in neighboring towns. This process may be called *swarming*. From that moment on, zoning and centralization, key aspects of modern urban planning, are facts of life. In effect, from that moment on, the difference in quality of services between the mother agglomeration and the satellites becomes such that an individual or organization that necessarily has numerous and frequent relations with others is ready to pay a high premium to be able to set up shop in the center, at the communications node. Thus, we have the debut of the suburbs.

Phase 3 (starting in 1900): a third type of expansion is added to the previous two. Motor-driven public bus service is established on routes converging on the city. This, in turn, drives continuing urbanization in the form of "glove fingers",

[9]This layout is described by Lewis Mumford, *The City in History* (Harcourt, 1961) and J.-C. Gutkind, *Le Crépuscule des villes* (Paris, Stock, 1966) on the urban level, and by B. Malisz, *La Formation des systèmes d'habitat* (Paris, Dunod, 1972) and W. Christaller, *Die Zentralen Orte in Süddeutschland* (Iena, Verlag Iena, 1933) on the regional level.

rather than the discontinuous "cluster" formation, the bus stations being more closely spaced than the train stations. "*Being served by a railroad line, with station stops from three to five miles apart, there was a natural limit to the spread of any particular community. Houses had to be sited 'within easy walking distance of the railroad station', as the advertising prospectus would point out; and only those wealthy enough to afford a horse and carriage dared to penetrate farther into the open country.*"[10] As regards zoning and centralization, this form of urbanization (glove fingers) had the same consequences as the previous one (swarming), compounding its effects.

Phase 4 (starting in 1950 in France): the spread of *individual* fast-transit vehicles (i.e., automobiles) gave rise to the third major transformation in the process of urban growth. In fact, automobiles and two-wheelers had two simultaneous but distinct effects:

- on the one hand, they pushed back the limits placed on agglomerations stretching along public transportation routes, reinitiating an earlier but nearly extinguished process of *agglutination*. The nucleus expands, the fingers of the glove lengthen and broaden;
- on the other—and this is what is truly new—they introduce all the outlying villages, which had remained out of reach up to this point because they were not served by public transportation, into the realm of convenient proximity relations. *All* the settlements in the area grew. And hence, for the first time, the territory within which city-dwellers could easily arrange their daily affairs was considerably broader than the area occupied by the city. That is to say—let us repeat—the interpenetration of agricultural space and urban space becomes a permanent aspect of the living environment.

Within a radius of 25 km, which is no longer enough to encompass France's largest metropolitan areas,[11] there is enough space to house 3.5 million people at a density of 5 detached homes per hectare serviced by a road network, parks, and fundamental amenities. Since this area already comprises a center and relatively dense adjacent suburbs, it is conceivable that the residential land use in more peripheral areas is highly discontinuous. In the most ambitious hypotheses of development, the urbanized spaces (houses + private gardens + amenities + local labor pool) in these peripheral areas will make up less than 10–20% of the total area

[10]Lewis Mumford, op. cit., p. 504.

[11]Average radius of a number of ZPIUs (Zone de Peuplement Industriel et Urbain – Industrial and Urban Population Zone, a disused term developed by INSEE, France's national statistics bureau, now replaced by *aire urbaine* [urban area], similar to a Functional Urban Area [FUA]): Lyon 25 km; Marseille-Étang de Berre 40 km; Bordeaux 25 km; Strasbourg 30 km. The ZPIU was developed to define geographical units that are larger than the city itself, reflecting a more extensive conception of urbanization accounting for three additional criteria: low proportion of resident population living off farming, high rate of daily commuting between home and work, demographic growth. The 1968 census provided the basis for identifying 812 ZPIUs: 84 were entirely composed of rural communities, of which 29 consisted of a single municipality; 249 were entirely composed of urban communities, of which 182 consisted of a single municipality.

from here to the end of the century [from 1975 to 2001—Translator's note]. This is a higher land occupation rate than in the classic countryside, but it does not support the conclusion that the interstitial areas are destined to be filled in.

In the conclusions to his recent book,[12] Jacques Jung, citing a German report, describes the agglomeration that he considers minimal in terms of employment and services: 20,000 inhabitants, a center accessible in less than 30 min by all residents. If we translate this definition into spatial terms, equating a radius of 15 km with 30 min of travel by moped, we derive an average population density of less than 30 inhabitants per square kilometer, equivalent to that of rural France. Motor vehicles, even the slowest among them, have thus fundamentally changed the ration of inhabitants to their geographical space, without developers and planners yet being aware.

This historical evolution is simply an observed fact. It implies no value judgment and we do not read into it any progress in the deterioration of our cities. Furthermore, no phase excludes another in time or in space. The four successive phases have only affected the metropolises, which were the only urban units able to develop the shared transportation networks underpinning the second and third phases. But the rurban explosion affects almost all cities, including small and medium-sized cities.[13] Conversely, the modern city is far from being confused with its rurban extensions. In France, especially in the last few decades, official urban planning has sought to extend the concentrated, dense city, following a policy of housing complexes. Rurbanization is visible, but it competes with other forms of dwelling space, and is countered by practice and most administrative regulations.[14] These principles do more than just slow rurbanization, which some may consider desirable, they also engender persistent ignorance of the process among both specialists and the public.

Limits to the Concept of Rurbanization

It is important to grasp clearly the specificity of the process we have outlined above, to mark the differences with respect to other models of urbanization that appear to be closely related. Thus, rurbanization cannot be likened—and we will explain this in greater detail below—to the phenomenon of *single-family housing developments*.

It contains this, but it adds its own morphological character, i.e., its being scattered across the countryside.

It is also distinct from *rural development*, of which we may see less in France than in neighboring countries. We are thus outlining a situation where, in spite of the strong reduction in numbers of people employed in agriculture, most of the people in a region are able to remain where they are because they can find other

[12]Jung, J. (1971). *L'Aménagement rural, une illusion économique*. Calmann-Lévy.

[13]CIEU (directed by B. Kayser), *Recherche sur les villes moyennes*, Université de Toulouse-Le Mirail, 1973.

[14]Less stringently in recent years.

employment without being forced to relocate. An increase in the percentage of workers in industry certainly induces changes in behaviors, which tend to become more and more similar to those of city dwellers. However, rural development implies only moderate growth (at times deliberately limited) of large agglomerations.

It is clear that rurbanization and rural development may overlap, depending on the chosen geographical scale of observation. Thus, if we situate the agglomeration community of Cholet—a rare example of a dense, industrialized rural area—in the West or in France as a whole, we note significant growth of Cholet itself, engendering rurbanization.[15]

Rurbanization is not to be confused:

- with city-countryside integration as conceived by Mao Zedong, namely "the deconcentration of urban development across the entire [Chinese] territory, the tendency to completely develop it as medium-sized industrial or agricultural units, adapted to peasant laborers".[16] Rurbanization in no way fosters an attenuation of socioeconomic differences between cities and rural areas. Of purely city-dweller origin and character, it is simply a consumption of rural space;
- with the *decentralization* of employment and employees in most European nations, and in France under the impetus of DATAR (Interministerial Delegation of Land Planning and Regional Attractiveness). Rurbanization must not be viewed on the national level, or on the regional planning level, but on a lower level, that of the "urban region", a notion that approaches (as we have seen) the concept of the ZPIU instituted by INSEE. It is situated *within* the process of concentration and polarization (creation of hubs) of the population in these regions. Besides, it essentially regards living space and not so much businesses, at least so far;
- or with the *socio-cultural urbanization of contemporary society*, as Rambaud,[17] for example, understood it. That is, the global application of models, attitudes, and aspirations copied from or inspired by those of city dwellers to the agricultural realm, all the way to the remotest rural zones: social division of work, distinction between work time and leisure time, demand for vacations, conspicuous consumption.

Finally, let us remember that rurbanization is not the only way for city dwellers to live a portion of their time in the countryside. It competes with the system of second homes, with a weekly alternation between the city and the country.

---

[15]CERU, *Études des besoins en logements en milieu rural. Le Chotelais, 1973–1978*, for the French Ministry of Agriculture (ACEAR), 1973.

[16]Luccioni, M. (1972). Processus révolutionnaire et organisation de l'espace en Chine. Vers la fin des séparations entre villes et campagnes. *Espaces et Sociétés*, 5.

[17]P. Rambaud, *Société rurale et urbanization*, Paris, Éd. Du Seuil, 1968.

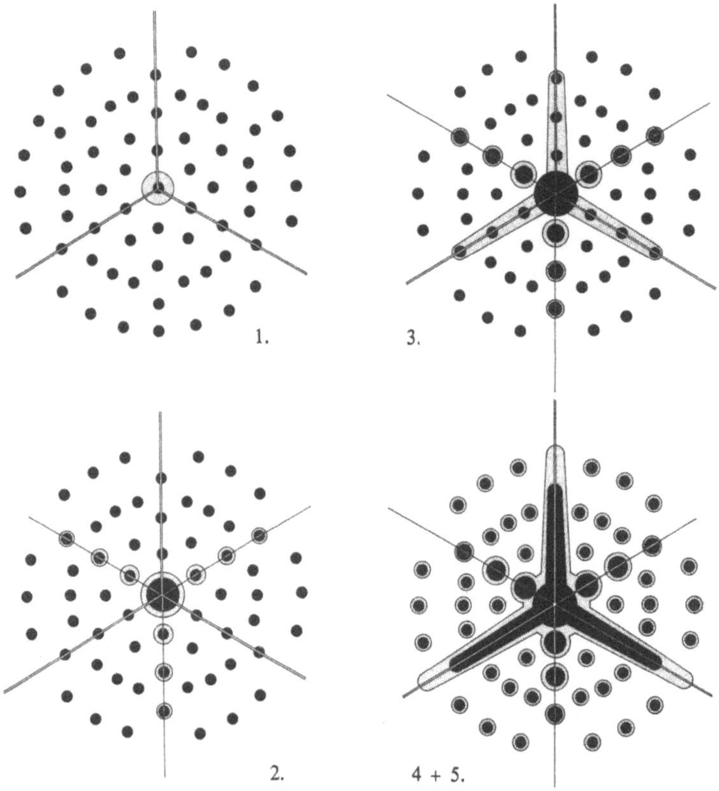

Fig. 16.1 Forms of growth
"1. Agglutination and densification—*Agglutination et densification*. 2. Star swarming—*Essaimage en étoile*. 3. Fingers—*Doigts de gant*. 4 + 5: Agglutination + *Rurbanisation*—*Agglutination + Rurbanisation*» Source: Bauer and Roux (1976), *La Rurbanisation ou la ville éparpillée*. pp. 18–19

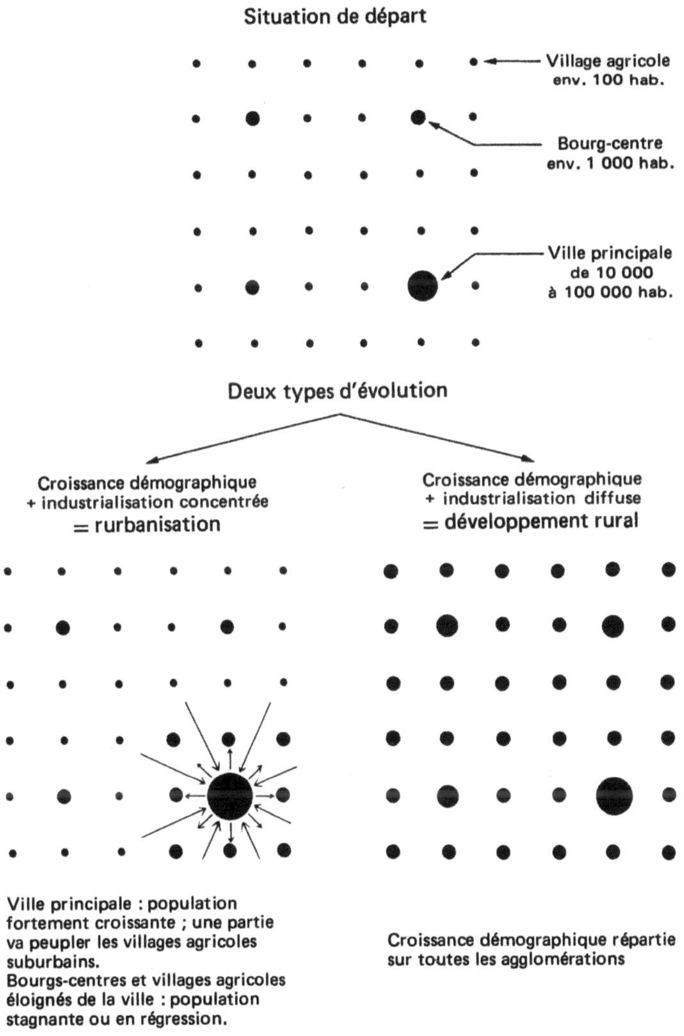

Fig. 16.2 Types of evolution
"Starting situation. Two types of evolution: Population growth + concentrated industrialization = r urbanisation / Population growth + diffuse industrialization = rural development"
Source: ibid., p. 22

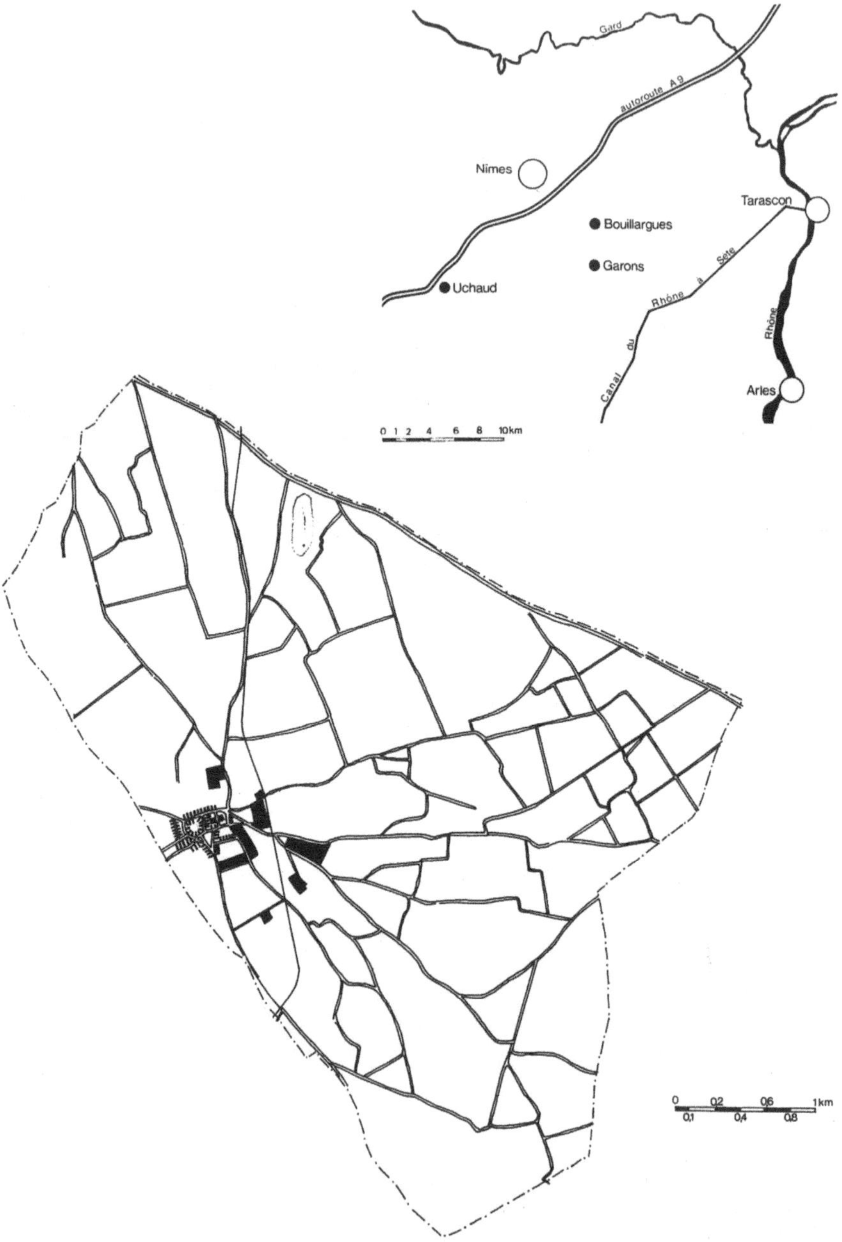

Fig. 16.3 "The outskirts of Nîmes: plots built between 1960 and 1970 in three communes taken as examples: Garons"
Source: ibid., p. 44

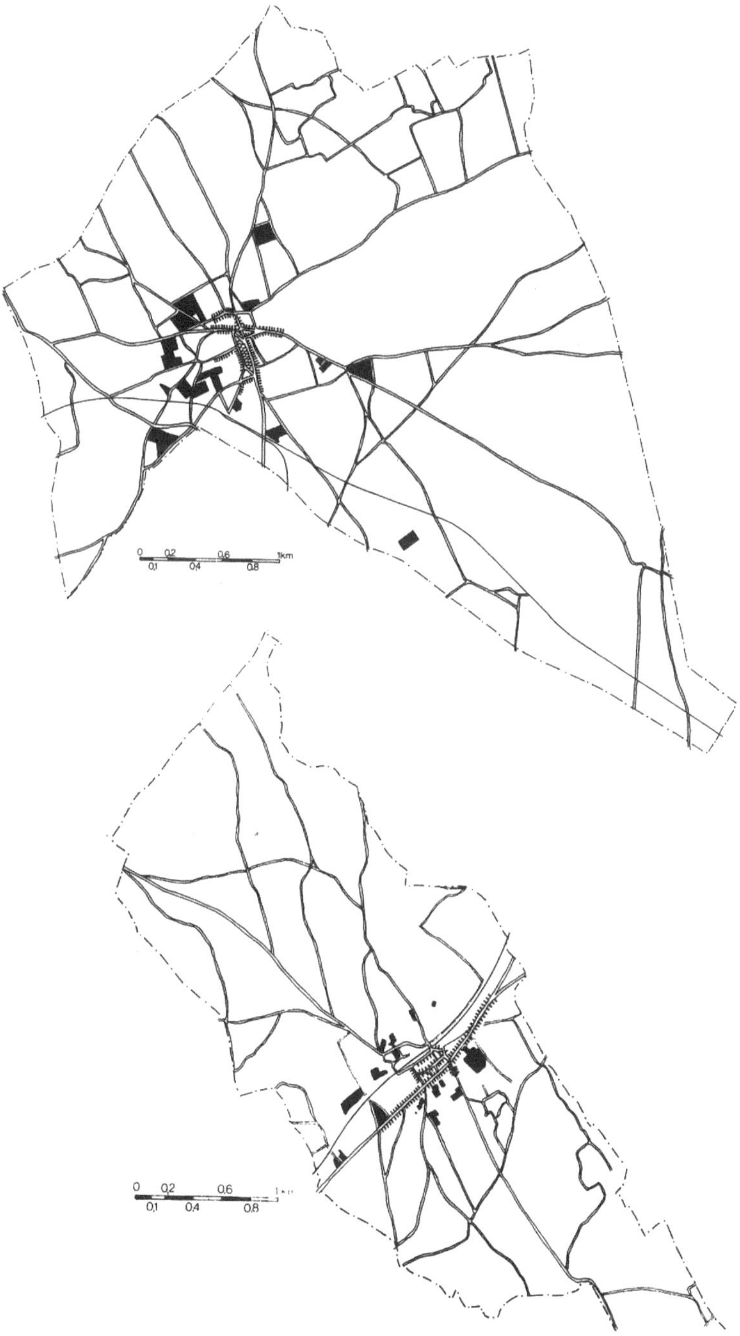

Fig. 16.4 "The outskirts of Nîmes: plots built between 1960 and 1970 in three communes taken as examples: Bouillargues (top) and Uchaud (bottom)"
Source: ibid., p. 45

Chapter 17
Agropolitan Development (1978)

John Friedmann

Introduction

Qinyi Zhang
EPFL, Lausanne, Switzerland
qinyi.zhang@epfl.ch

In the '60s and '70s, an accelerated industrialization process started to take place in many developing Asian countries including India, Indonesia, West Malaysia, the Philippines, Republic of Korea and Thailand.

In the following article, Friedmann describes and criticizes the "industrialization paradigm", both for its socio-economic (rise of import prices, decline of export markets, deterioration of environment and living conditions) and spatial consequences (an organisation privileging growth poles and metropolitan centres against the vast and densely populated rural areas, concentration rather than decentralization).

Friedmann's "Agropolitan Development", proposed as an alternative to the typical industrialization paradigm, focuses on an accelerated rural development and a new kind of social development where "the city-in-the-fields" is created by embedding key urban elements in dense rural areas of limited size.[1] *A new agrarian economy is imagined where rural development is conceived as decentralized, participatory, and deeply immersed in the local settings. Limited and specific human needs would replace unlimited, generalized wants; agriculture would be regarded as a leading sector of the economy; the attainment of self-sufficiency in domestic food production would be considered a high priority objective; existing inequalities in income and living conditions between social classes and between*

[1]An "Agropolitan District" has a 5–10 km radius and a population of 50,000–150,000.

M. Barcelloni Corte and P. Viganò (eds.), *The Horizontal Metropolis*,
https://doi.org/10.1007/978-3-030-56398-1_17

urban and rural areas would be reduced and facilitative measures to increase production of wage goods for domestic consumption given high priority.

The Agropolitan Development model was subsequently revised and developed further by Friedmann during his long career. Many of his works bear witness including Basic Needs, Agropolitan Development, and Planning from Below[2], The Active Community: Toward a Political-Territorial Framework for Rural Development in Asia,[3] and Modular Cities: beyond the rural–urban divide. *Through a long trajectory, the Agropolitan Development model evolves in time extending its scope from the rural areas to high density rural or peri-urban areas, and continues to challenge the urban–rural divide, imagining a "city-in-the-countryside".*[4]

This city-in-the-countryside or city-in-the-fields is more than a policy instrument. It refers not only spiritually but also materially to the utopian thinkers and reformers such as Kropotkin, Howard, Mumford, Wright, and Mao, to imagine a precise space: a decentralized territory with a mixture of rural and urban such as Friedmann could observe in Switzerland and Germany in the '60s/'70s and which can be observed today in many other parts of the world, such as Flanders, the Veneto Region in Italy or the Yangtze River Delta in China.

17.1 Agropolitan Development: Towards a New Strategy for Regional Planning in Asia[5]

The last few years have seen a reaching out for new paradigms of planning for development. The old paradigm is no longer compelling in its logic. The underlying idea was that human happiness and sense of well-being would flow, more or less spontaneously, from the relentless pursuit of economic growth; that economic growth would most readily result from rapid industrialization; that rapid industrialization would be accommodated most efficiently in a few metropolitan centers from where its benefits would "spread" to the rest of the national territory; that planning for growth, industrialization, and urbanization could be accomplished only through coordinated central planning; and that coordinated central planning was a technical task that would be solved through complex econometric modeling joined to sophisticated administrative technique.

---

[2]Friedmann, J. (1979). Basic needs. agropolitan development, and planning from below. *World Development, 7*(6), 607–613.

[3]Friedmann, J. (1981). The active community: Toward a political-territorial framework for rural development in Asia. *Economic Development and Cultural Change, 29*(2), 235–261.

[4]Friedmann, J. (1996). Modular cities: Beyond the rural-urban divide. *Environment and Urbanization, 8*, 129.

[5]Source: Friedmann, J. and Douglass, M. (1978). Agropolitan Development: Towards a New Strategy for Regional Planning in Asia. In Lo, F.-C. and Salih, K., (Eds.), *Growth pole strategy and regional development policy*. Oxford: Pergamon Press.

Spatial planners tried to find a place for themselves within this scheme. Their principal contribution to developmental planning was the notion of growth poles and growth centers (Kuklinski, 1972; Kuklinski & Petrella, 1972). They argued that investments had to be considered in a spatial dimension; that although the efficiency of investments might be maximized by concentrating them initially in large urban places, equity considerations called for a policy of concentrated decentralization, and that, in order to incorporate a spatial dimension into planning for national economic growth, planning would have to encompass at least three levels of spatial aggregation: urban, regional, and national. In this way, they claimed, the spatial integration of the national economy would be achieved (Lefeber & Datta-Chaudhuri, 1971; Misra et al., 1974).

In both its nonspatial and spatial dimensions, however, this paradigm has come under fire. On the one hand, it is argued that the paradigm has failed to bring about the results expected from it (Faber & Seers, 1972). On the other hand, the last few years seem to be signaling the transition of the international economy to a new state of affairs which suggests that the unmodified pursuit of the old paradigm would be, to put it bluntly, suicidal (Mesarovic & Pestel, 1974).

As a consequence of this double criticism, the contours of a possible new paradigm are beginning to emerge. Its primary objective is no longer economic growth but social development, with focus on specific human needs (Seers, 1970; Ul Haq, 1972). According to this paradigm, development must be fitted to ecological constraints; priority attention (in agrarian economies) must be given to rural development; and planning for rural development must be decentralized, participatory, and deeply immersed in the particulars of local settings. Planning, therefore, will have to be based on qualitative judgments as much as on quantitative techniques, and its style will have to be transacive (Friedmann, 1973a).

Our intention here is to propose a spatial policy for the emerging paradigm. The discussion is divided into two main parts. In the first, we assess the results of the strategy of accelerated industrialization in six Asian countries: India, Indonesia, West Malaysia, the Philippines, Republic of Korea, and Thailand. Two major conclusions are reached. The traditional paradigm, we argue, has generated contradictions of crisis proportions in the structure of national development. These contradictions of "dualistic dependency" are intensified by the present transition of the world capitalist economy from an era of high sustained growth to one of structural imbalance and stagnation. A strategy of accelerated rural development is seen as the most appropriate response to the emerging conditions.

After stating some of the structural characteristics of a policies framework which aims at fusing urban and rural conditions of living, the relevant experience in the six countries is reviewed. We conclude that current efforts are inadequate to meet the needs of the new strategy. The urgency of the present situation calls for heightened political commitment to an inward-looking and rurally based strategy of national development. In this way the contradictions of dualistic dependency may be overcome and national economies protected against the shock waves engendered by the structural changes inherent in the current world transition.

17.1.1 A Critique of Development Strategy Under Conditions of Dualistic Dependency

The Failures of the Industrialization Paradigm

Since gaining their independence, the six countries which are the subject of this study have chosen to follow a strategy of accelerated industrialization as the high road to economic growth and national prosperity. For simplicity's sake, we shall refer to this as Strategy I. In accordance with this strategy, the rapid expansion of a large-scale, corporate manufacturing sector was expected to bring bountiful benefits to the nation: a basis for further growth and rapid capital accumulation, full employment, higher levels of living for the entire population, increased national autonomy, a breakthrough into "modernity".

The strategy relied on four principal sources of financing. Three of them tied the country's development path to the world capitalist system: primary materials exports, private foreign investments, and international aid; the fourth relied on the country's internal capacity for savings primarily from within the agricultural sector. Public investments were to be used chiefly for building up the social infrastructure in support of private activity in the industrial sector. Where necessary, government might join with private capital in new industrial ventures; on the whole, however, these ventures were to remain in private hands.

Initially, Strategy I was oriented towards the substitution of manufactured imports, but the installation of only a few industries quickly saturated domestic markets. As a result of market limitations (and the difficulties experienced in achieving a vertical integration of industries), many countries began to shift to a policy of encouraging the export of manufactured products overseas.

Whereas the earlier policy of import-substitution had resulted in high-cost and relatively inefficient production, the change to export orientation made it imperative to cut costs to competitive international levels. In both cases, however, technology had to be imported from abroad. The chief competitive advantage of Asian countries was, therefore, their abundant supply of low-cost labor. By applying low-cost labor to a capital-intensive technology and by maintaining high production standards, Asian countries could attempt to supply Western markets at a competitive advantage. In addition, countries such as India began to promote the export of labor-intensive products such as textiles, electronics, and traditional handicrafts.

Unfortunately, the results of Strategy I were not at all what had been hoped for. By the mid-1970s it had become clear that, far from being the high road to national development, the strategy of accelerated industrialization had generated, and was continuing to generate and reinforce, a homologous set of economic, social, and spatial structures that confounded the rhetoric of national objectives. We shall call them the structures of dualistic dependency (Paauw & Fei, 1973).

In the "vertical" dimension, dualistic dependency displays a set of opposing structures. The corporate sector exercises dominance over the traditional household sector of the economy. An urban-based elite dominates the growing proletariat and subproletariat in cities no less effectively than it does the masses of peasants and landless laborers in the countryside. A capital-intensive imported technology successfully competes against a labor-intensive indigenous technology.

In their "horizontal", or spatial dimension, dualistic dependency structures find expression in one or, at best, a small number of highly urbanized "core regions" which, in turn, control the development path of regional economies in their periphery.

Dualistic dependency, however, is more than a set of dichotomous relationships in contradiction. It is, as well, a particular mode of integrating a national economy into the system of world capitalism (Wallerstein, 1974). The vertical components of corporate production, urban elites, and capital-intensive technology are, for the most part, concentrated in the core areas of their respective countries. These areas fulfill a double role, however. On the one hand, they function as nodal points in the periphery of the world capitalist system; on the other hand, they dominate (and exploit) their own peripheries in much the same way as they, in turn, are dominated (and exploited) by the world core regions oil Western Europe, the northeastern part of the United States, and central Japan.

The emergence of dualistic dependency structures has had profound consequences for national development. Space does not allow for a careful tracing through of causal effects, but the major trends may be noted.

Hyperurbanisation and Rural Densities

The concept of hyper-urbanization refers to the concentration of populations in dense urban settlements at a rate that, except on a basis of inequality and exploitation, renders their integration into the emerging social order impossible. The resulting "crisis of inclusion" forces governments to adopt policies that either undermine the basic premises of Strategy I by shifting resources to pacify social demands or lead to political repression.

In an earlier study, Friedmann suggested a doubling period for urban population growth of less than 20 years as a more or less arbitrary standard for identifying the existence of hyperurbanization (Friedmann, 1973b). Translated into rates of growth, this is equivalent to a decennial increase in urban population of at least 41%.

Several conclusions may be drawn from these data. First, all countries had doubling rates equal to or in excess of the 20-year threshold of hyper-urbanization. Second, urban growth in every country was greater than the growth of total population, reflecting large-scale rural—urban migration (Intermet, 1973). Third, the populations in the six countries were still predominantly rural at the end of the decade.[6] The high level of urbanization in Korea may be more apparent than real; labor statistics show that 55% of the work force in 1970 was agricultural (United Nations, 1974).

A corollary of these conclusions may be stated at this juncture: between 1960 and 1970 *the bulk of the population increase had to be absorbed in rural areas.* This was especially true for India, Indonesia, and Thailand where around two-thirds of the demographic increase was so accommodated. In India, at least, this densification of rural population led to an eight percent decline in physical productivity per male worker (Hayami & Ruttan, 1971). Unless a major policy reversal occurs, rural densities will continue to rise throughout the region with results for labor productivity that are at best uncertain (United Nations, 1973c).

Spatial Concentration of Population and "Modern" Activity

Planners typically reside in cities. They therefore tend to see a country's development through the distorting prism of city life. But the fact is that what passes for "modern" development is highly concentrated in only a few urban centers, while the rest of the country is relatively isolated from the social and economic changes that so preoccupy the planners' attention.

To illustrate: the top four cities in five out of the six countries accounted for between one-third and two-thirds of total urban population in 1970. Moreover, there was a pronounced tendency for larger cities to grow more rapidly than smaller ones, and for the degree of urban primacy to increase over the preceding decade. By 1970, the largest city in four out of the six countries was between 1.2 and 9.0 times the size of the next three largest cities.

Data pertaining to the spatial distribution of manufacturing are not readily available. Korean experience, however, may be illustrative of the more general case. As one proceeds from all manufacturing to the elite industries of the corporate sector, the concentration in the primate city increases in steady progression. In 1968, 81% of corporate employment was located in Seoul.

Unemployment and Underemployment

Policies under Strategy I aim at accelerating investment and production in the corporate sector of cities—particularly in large-scale manufacturing and supporting infrastructure and services—as well as at expanding commercial agriculture. What consequences has this strategy had for employment?

In the preceding sections we have shown how Strategy I promoted the rapid transfer of rural people to cities; nevertheless, a large proportion of the population had to be "absorbed" in agriculture, contributing to a sharp rise in rural densities.

---

[6]The definition of "urban" finds little consistency from nation to nation. The Korean definition exaggerates urban population size by using township boundaries which are much broader than actual built-up areas. Over 50% of the population in the smaller "cities" is in agricultural employment (United Nations, 1973a). A definition, such as the one adopted in India, which requires both a minimum density and specified non-agricultural employment levels, seems most appropriate under current conditions of urbanization in Asia. For a full discussion of the problem of urban definitions in Asia, see Bose (1971).

The urban economy, however, was by no means ready to provide full, productive employment to everyone. A significant part of the labor force remained unemployed. Oshima (1971), for example, estimates the general level of open unemployment in Asian cities at 10%, and the ratio of unemployed has been projected to rise to within 15% by 1980 (United Nations, 1973a). Individual cities have already reached or exceeded that level.

These are orders of magnitude only; precise measurements of open unemployment do not exist and, in any event, are greatly dependent on the formal definitions used. In Korea, for example, a person is counted as employed even if he or she worked only one hour per week.

The concept of underemployment is widely used in conjunction with open unemployment to provide a more complete picture of how labor is actually being used in the project of national construction. But the reliability of figures here is no greater than in the case of unemployment. Table 17.1 is an attempt to overcome some of these difficulties and to suggest the consequences of Strategy I for the employment structure of five Asian countries. Studying this table, we note:

(1) The dominant role which agriculture still plays as a labor-absorbing sector of the economy and the correspondingly small contribution of manufacturing to total employment.
(2) The large proportion of wage employment in agriculture, reflecting the high ratio of landless laborers in all countries, excepting Thailand and Korea.[7] Yet rural wages are typically only a fraction of what comparable wage levels are in cities.
(3) The large share of manufacturing employment accounted for by the non-wage sector, most of which represents employment in individual and small-scale family enterprises.
(4) The size of the trades and services sector which, in some cases, is beginning to rival agriculture as a source of employment (Malaysia, Korea).

In general, non-wage employment displays a level of productivity (reflected in remuneration) which is very much lower than that in the wage economy as a whole. This may be illustrated with data from the Philippines where small manufacturing enterprises (much of whose employment was unpaid family labor) not only declined under the "involutionary" pressures of growth in the labor force (McGee, 1974), but was only three percent of the productivity in the corporate sector of manufacturing. This would seem to be the rule rather than the exception, and Friedmann and Sullivan (1974) have argued that consumption levels in the informal sector of the urban economy (including the unemployed) tend, as a result of poverty-sharing among kin and fictive kin, to be reduced to mere subsistence. This appears inevitable under conditions where the rise in output from manufacturing is

[7]In Indonesia and India, landless laborers account for 30% of the agricultural worker force (IBRD, 1970).

from two to four times the rate of increase in manufacturing employment (de Haan, 1974).

Government responses to the employment problem are themselves caught up in the contradictions of Strategy I. The recent shift from import-substitution to a greater emphasis on manufactured exports has led to a number of serious problems that had not been anticipated.

(a) Employment in many of the new industries responds violently to even small changes in external demand. In Malaysia, for example, saw mills began working at half shift in 1974, and loggers in Sabah and the East Coast of the Peninsula had to cut their labor force by 19,000 workers over 1973 (Malaysia, 1974). The same effect was felt by labor-intensive industries, such as textiles, throughout the Asian region.
Estate agriculture, which is a primary source of wage employment in rural areas, suffers a similar fate. During 1974 depressed auto sales in the United States translated into a steep decline in the demand for rubber, one of the major export commodities of Thailand, Malaysia, and Indonesia.
(b) Productivity increases under Strategy I are being achieved at the expense of employment. As an extreme example, rubber production in Malaysia rose by nearly 50% during the 1960s, *but employment dropped by 10%*. Rather than upgrading skill levels on a national scale, governments imported modern machinery and new technologies in agriculture that displaced both current and future members of the labor force toward low-income occupations in the informal sector (United Nations, 1973a).
(c) National plans tend rather simplistically to tie employment creation to GNP growth rates despite lack of consistent empirical support for such association (United Nations, 1973a; Ariff, 1973; India, 1972; Malaysia, 1973). The fallacy becomes evident when it is seen that the only sector growing significantly more than the labor force is manufacturing which employs only a small proportion of worker and displays a strong tendency for productivity to outpace employment.[8]

To summarize: Strategy I was reasonably effective in raising GNP; it failed in expanding economic opportunities at a rate sufficient for a rapidly growing population. Rural areas suffered the most from policy neglect, from densification, from expansion of the corporate sector, and from massive income transfers to urban areas. The latter performed little better in providing means of livelihood. The majority of workers came to be absorbed into low-level trades and services. At the same time, open unemployment was gaining, and underemployment came to affect as many as 40% of the existing work force. As a result of what McGee (1971, 1974)

[8]Thus Malaysia uses an annual manufacturing growth rate projection of 12.3% over the next 20 years. This, in turn, predicts a manufacturing employment growth rate of 7.5%, but the low base level of manufacturing elasticity, implying no increases in labor productivity, renders the projected relationship exceedingly doubtful.

calls involution, productivity in the small-scale, informal sector of the urban economy declined.

Income Inequality and Poverty
It is symptomatic of Strategy I that inequalities in levels of living and actual poverty conditions are measured in monetary income. If we accept this standard, rather than more appropriate physical measures, it comes as no surprise when we find that existing incomes in the six countries are highly unequally distributed.

> Social increases in income disparity are generally displayed in terms of a decrease in the percentage of income received by the rural poor who make up the vast majority of those in poverty (Adelman, 1975). In both the Philippines and Thailand the income position of this group deteriorated during the 1960s, although for poverty groups in urban areas the trend seems to be reversed.[9]

Income statistics do not support the hypothesis that the rich are getting richer. But such a conclusion must be treated with caution. The incomes of the wealthy are suspect of being grossly underreported and, in any event, do not reflect accumulations of wealth, In India, for example, 70% of personal wealth in cities is held in the form of residential property, and 80% of urban property may be held by as few as 5% of the people (United Nations, 1974).

With these qualifications in mind, it is revealing that close to half the incomes in countries for which data are available flows into the hands of the richest 20% of the population and that, at least in Malaysia and the Philippines, *the top 10% of the population receive as much as 40% of all incomes!* (The concentration of incomes for the same decile in Thailand, on the other hand, is reported to be "only" 29% for urban and 35% for rural populations. See Kerdpibule, 1975.)

Spatial comparisons of income distribution show more marked disparities. In every documented case, *rural incomes declined relative to urban incomes.* Furthermore, except for the Republic of Korea, rural incomes are only a fraction of urban income levels. In Malaysia, for example, rural people receive only one-third of the income of urban residents, and 87% of the poverty households in the country are reported to be in rural areas—a ratio that is significantly higher than the proportion of the rural population in the country as a whole (Malaysia, 1974). In India, rural poverty increased dramatically during the sixties. In Korea, *per capita* income for Seoul was twice the national average in 1968 (Republic of Korea, 1971). And Indonesian data similarly reveal enormous rural–urban income disparities (Esmara, 1975).

Returning briefly to the question of employment, the dualism suggested previously relates directly and most importantly to the distribution of income. The generation of low-productivity work for an increasing number of people, the denial of access to "modern" employment and its commensurate higher wages, and the

---

[9] In India the official poverty level of 40.6 rupees per month (in 1972 prices) yields a forecast that about 30% of the population will be living in poverty at the end of the current planning period (1986). But this projection does not take inflation into account (India, 1972).

low returns to agriculture outside the corporate sector, all of which are associated with Strategy I, display themselves in terms of marked income inequalities. The highest incidence of poverty is among the self-employed and landless laborers. In Malaysia, this group constituted 55% of those living in poverty (Malaysia, 1974).

Persistent Food Shortages

The almost singular interest of Strategy I in promoting development of the corporate sector has resulted in a substantial neglect of domestic food production. Moreover, the emphasis on exports has led farmers to allocate their labor and land away from food to the production of so-called cash-crops. Although all six nations have repeatedly aimed at self-sufficiency, Thailand is the only country in Asia which is still able to meet its own requirements for rice (between 1968 and 1970, Thailand's ratio of production to food imports was 119% compared to a low of 68% for West Malaysia). During the 1960s *per capita* food production declined in both India and Indonesia, remained stable in the Philippines, and increased in the remaining countries. Four out of the six countries are obliged to spend large sums for the importation of basic foods: for the period 1969–71, India and Malaysia led with 21% of their total import bill, followed by Korea (16%) and Indonesia (13%). In more recent years, these percentages have risen still further.

Statistical measures of food shortages obscure the widespread prevalence of malnutrition and hunger. Although each nation claims close to a sufficiency in grain supplies, inclusive of imports, undernourishment is reported in many regions (most strikingly in India) and, according to an estimate for 1966, at least one-fifth of the population in poor countries was calorie-deficient, while about one-third suffered from a lack of protein (United Nations, 1973b).[10] Of critical concern, then, are both supply and distribution.

Dealing first with supply, Asia will have to increase cereal production by 40% during the present decade simply to maintain current (and mostly insufficient) consumption levels (United Nations, 1973b). Given that arable land is relatively fixed, and that Strategy I is reluctant to convert land in cash crops to cereal production, much faith has been put into the new technologies of the Green Revolution.[11]

The Green Revolution brings the issue of supply and distribution together and illustrates the contradictions inherent in current structural relationships. Three points summarize the effects of adopting the new technologies without changing the patterns of dualistic dependency. First, the Green Revolution has given a temporary

---

[10]Using a *per capita* nutritional adequacy standard of 2250 cal, 40% of rural and 50% of urban populations were receiving less than satisfactory nutrition in India in 1971 (Dandekar & Rath, 1971). In Indonesia over one-half of the rural families are short on calorie and protein intake (*Far Eastern Economic Review*, Sept. 5, 1975).

[11]The question of arable land is a complex one, involving such issues as the availability of water for irrigation, resettlement, and long-distance migration. For Korea and Java the physical limit of crop land appears to have been reached already. Elsewhere, there exists at least a theoretical possibility for expanding agricultural land.

respite from, but has not ended, endemic food shortages (United Nations, 1973b). Second, increased output can result in loss of income to those farmers who are institutionally denied access to the new techniques for increasing production on their land (Dourer, 1972). Third, the net effect is that, over the long run, the Green Revolution may fail to keep pace with increasing demand and, over the short run, will serve to benefit only the relatively few farmers who have sufficiently large land holdings and resources to take advantage of the new technologies.[12] The FAO has concluded darkly that rural poverty for small farmers, sharecroppers, and landless laborers, i.e., for the majority of people in Asia, excluding China, is likely to increase. The portent is one of increased malnutrition and hunger in both relative and absolute terms (United Nations, 1973b).

Deteriorating Material Conditions of Farm Population

The decline of rural (relative to urban) incomes has already been noted. But this is not the only source of impoverization which the faithful pursuit of Strategy I has imposed on rural people. Another, and far more serious consequence of dualistic dependency, is the adverse terms of trade that peasant farmers have to accept in their dealings with metropolitan economies. Data supporting this proposition are unavailable for the six countries, but theoretical grounds for accepting it are strong and deserve to be explored.

The argument can be divided into two parts. According to the first, inflationary pressures are generated in core regions and are then "exported" to the rural periphery. As a result, the prices of commodities bought by farmers, such as farm implements, irrigation pumps, fertilizers, insecticides, textiles, shoes, bicycles, etc., are tending to rise faster than the prices they receive for their own produce. The second line of reasoning suggests that the price of agricultural staples (primarily rice) is frequently "stabilized" by governments to the advantage of the growing masses of urban workers whose interests are more "visible" to politicians than those of a peasantry that is not only physically remote but is also less readily mobilized for political ends (hence, their characterization as "peripheral and dependent"). With steadily rising prices for manufactured products and the relatively stable prices received by farmers, the "terms of trade" of at least the food-producing sector will obviously be adverse and declining. This is especially true at a time when world commodity prices (and the prices of essential raw materials that have to be imported, such as oil) are advancing at annual rates greater than 10%.

More generally, it can be claimed that the material conditions of the poor are nowhere in the world greatly improving. Under conditions of dualistic dependency, they may, in fact, deteriorate.

As we have already noted, poverty in Asia is most pervasive in rural areas. To come to grips with this phenomenon, the term "peasant farmer" must be subjected to a series of refinements. For example, cash-crop farming must be separated from

[12]It is a matter of dispute whether the Green Revolution has actually pushed small and poor farmers off their land and into the cities. In any event, the Green Revolution introduces capitalist practices into a traditional rural economy and thus encourages the formation of rural wage labor.

food production; estate farming (which, in Malaysia, is 60% foreign-controlled) must be separated from non-corporate types of farming; tenants and owners must be compared; and farm sizes must be assigned to their relevant population size groups. In other words, terms of trade for a land-holding elite must be distinguished from those for the majority or peasant farmers. Some examples will help to illustrate this point.

Given present rural densities, average farm sizes in the six countries are not only small but getting smaller. On Java, for instance, farms are scarcely more than garden plots, or half a hectare in size (50 by 50 m); the situation is only slightly better in Thailand, where the typical farm measures 3.5 hectares. Other countries fall between these two extremes. As for tenancy relations, their incidence in Asian countries is extremely high and rising. Over a period of only six years, for example, tenancy rates in Thailand increased from 29% in 1963 to 40% in 1969 (Baldwin, 1974). And tenant farmers are typically obliged to pay from one-third to one-half their crop value to owners (Horii, 1972; Marzouk, 1972).

Neither the declining size of farm holdings nor the increases in tenure (and consequently absentee landlordism which underlies the process of income transfer from the periphery to the core) are conducive to rural savings and to their eventual reinvestment in land and farm technology. Average-size holdings in Thailand and Malaysia actually showed dissavings in recent years. For all these reasons, then, many farmers in Asia are too poor to join the Green Revolution.

External Dependency

Countries pursuing a strategy that succeeds only in saturating domestic markets, which pay for this "development" primarily through exports and which enlarge their corporate sector through the importation of foreign capital, become increasingly dependent upon their presumptive benefactor. They gravitate in the outer periphery of an economic system that is dominated by the advanced industrial nations of world capitalism.

Poor and dependent, they supply essential raw materials to these nations, provide an opportunity for profit-making investments from abroad, and present a valuable market for core—region exports. In all these ways they help sustain high growth rates in the world economy and, above all, a level of affluence in the industrialized countries that stands in stark contrast to their own poverty.

Dependency relations can take many forms. Among the most important is the dependency of peripheral economies on export markets. Its consequences may be illustrated with the example of Malaysia, 44% of whose GNP in 1970 was concentrated in export production (Malaysia, 1974). Between 1960 and 1970, Malaysia's rubber production increased 50%; yet a fall in the international price for rubber reduced this otherwise remarkable expansion of physical output to a *negative* growth of 1.5% in value added.

Elsewhere the story is similar.[13] Although the international system is able to absorb price fluctuations of single commodities without much notice, such changes can have dramatic effects on subregional economies that specialize in the production of the commodity in question.

The "solution" offered is typically one of diversifying the country's (or the region's) exports. Thus Malaysia adds palm oil to its list of exports to ease the shock of a drop in the price of rubber, and the Philippines emphasizes its "top ten exports" to do the same. But the results of this solution are, at best, ephemeral. At worst, they leave the economy more vulnerable than before.

A policy of "export substitution" may be tried for the same reason. According to present practice the products of light manufacturing—electronics, plastics, textiles, and the like—are aggressively pushed onto international markets. For the most part, however, their number remains small, competitive advantages favor foreign-controlled investments, and dependency on international markets becomes more entrenched.

Dependency mediated through trade relationships, of which export-dependency is a part, will thus have adverse consequences for the peripheral economy. In the most general case it will create and reinforce dualistic structures whose most immediate consequences have already been described:

(a) Hyper-urbanization, which tends to set the direction and emphasis for the nation's development process;
(b) the replication, on a *national* scale, of a worldwide core—periphery structure leading to a growing concentration of population and "modern" economic activities in a small number of metropolitan areas and their immediate surroundings;
(c) formation of national elites, which control a large and growing part of the nation's income and wealth and become coopted by their more powerful international counterparts whose life styles they emulate;
(d) rising urban unemployment and underemployment, which in a class perspective takes the form of an impoverished subproletariat that may encompass as much as 40% of the urban labor force but which, in an economic perspective, may be seen as a massive reservoir of cheap labor that helps to guarantee, by its mere presence, continuance of the incredibly high living standards that have become a way of life in world core regions (Emmanuel, 1972);
(e) progressive pauperization of the rural masses;
(f) growing food shortages, which deepen dependency relations to world core regions that, at least for the present, are still able to meet these deficits.

When understood in this light, the price that a nation is forced to pay for being externally dependent is, indeed, sleep. A vicious cycle is set in motion that, by

[13]Export stagnation in Thailand has been attributed to three external relationships: (1) international price fluctuations, (2) decline in US military spending, and (3) a decline in foreign investments (Thailand, 1972).

deepening dependency relations, renders the peripheral economy more vulnerable to exogenous change and reinforces its commitment to Strategy I which appears increasingly as the only path to national salvation. Foreign investments must be encouraged for reasons of becoming competitive in international markets but also because less and less can be squeezed out of a rural sector that is already close to subsistence. Foreign, capital-intensive technology must be imported for the same reason. Both policies tend significantly to raise the bill of imports, a trend which is intensified because of severe and growing shortages of food. The more specialized the economy becomes in the international division of labor, the more it is subject to periodic and violent fluctuations which are met, not with appropriate countervailing measures, but with policies that prescribe more of the same medicine. As inflation is imported, income differences become more pronounced, and the process of pauperization is speeded up.

Finally, political dependency, which follows in the footsteps of economic dependency, appears as almost a natural consequence, a way of life. The policies of peripheral development are being shaped by the interests of world core region elites that have little, if anything, to do with the true interests of the rural populations in whose name they are promulgated. Rising social unrest, which is the ineluctable consequence of contradictions inherent in the world periphery under conditions of dualistic dependency, is met by repression. The growing number of authoritarian regimes and military dictatorships throughout Asia attests to this trend with tragic regularity.

Dualistic Dependency and the World Economic Transition

The internal contradictions of Strategy I are often accepted with equanimity because, whatever else may follow, the strategy is believed to permit the achievement of high internal rates of growth in production. The empirical basis for this claim is not established unambiguously. But, except for India and Indonesia, our sample of six countries has, in fact, performed reasonably well over the recent period.

It would be committing a gross error, however, to assume that Strategy I was, by itself, chiefly responsible for the achievement of high growth rates. During the sixties and early seventies, peripheral countries benefited directly from the benign economic climate that followed World War II and the perverse opportunities created by wars in Korea and Vietnam. World core areas led the way with the sustained expansion of their economies. Demand was high, production costs were low, and West European economies actually experienced a prolonged period of overemployment. During the past several years, however, it has become clear that this era of economic euphoria has come to an end. World capitalism is in the midst of a major transition whose probable repercussions on its periphery cast doubt on the realism of the high growth rates that are still being projected by Asian planners.[14] If these "planned" production rates are not, in fact, achieved, the internal contradictions created by dualistic dependency structures will be so intensified as to raise serious questions about the continued viability of Strategy I itself.

What we refer to as the world economic transition is reflected in two major tendencies that directly affect the production possibilities and, hence, living conditions, in the world periphery: first, steep increases in world commodity prices, and, second, restricted expansion of export markets. A few pertinent observations are in order.

Commodity Price Inflation

This may be broken down into the major components of petroleum, other raw materials, food, and manufactures.

(a) *Petroleum.* Known world reserves of oil are expected to last for only another two to four decades (Mesarovic and Pestel, 1974). In the light of these projections, major national producers have begun to raise the price of petroleum. Because international demand for petroleum is relatively inelastic, these increases have come to be directly translated into the trade balances of countries whose economies are dependent on the continued supply of this vital energy source and industrial raw material.

This trend is not likely to be reversed. The IBRD expects the average price of crude oil to jump from $10.46 per barrel at the end of 1973 to twice this amount by the mid-eighties.

Present and prospective oil-producing countries, such as Indonesia, Malaysia, and Thailand, may derive substantial benefit from this situation. But this advantage may be more apparent than real, as higher costs are paid to buy back the same products processed and refined in industrialized nations. Non-producers—among them India, Korea, and the Philippines—have no advantage at all.

Petroleum is an input to general production, from power generation to food processing, fertilizers, insecticides, plastics, petrochemicals, and transportation services. Continuing price inflation of petroleum and its derivatives will thus be diffused throughout the entire domestic economy, raising prices across the board.

(b) *Other raw Materials.* Critical raw materials, other than petroleum, can also be expected to undergo dramatic price increases. World Bank estimates for 1985 put the world price index for minerals at nearly three times the level for 1973. A tin exporter, such as Thailand, is likely to see an increase of this magnitude as a gain. But tin-importing countries, e.g., Korea, may find it difficult to compensate for these increases with corresponding export earnings. In general, countries that control a significant share of world markets will be able to raise

---

[14]Reasons for doubting current planned GNP growth rates are reflected in the sober thinking of World Bank economists.

prices almost at will, though limits may be set by the possibility of synthetic substitutes, on the one hand, and the rate of resource exhaustion, on the other. In any event, resource-poor countries will be faced with import bills that are steadily gaining relative to the value of their exports.[15]

(c) *Food.* The story is similar with respect to staple foods. Recent events and projections reveal the stark magnitude of the crisis. The world is no longer merely a composite of independent units; a world system has come into being which makes American wheat sales to the Soviet Union impact profoundly upon people in areas of persistent food shortage in Asia and elsewhere (Mesarovic and Pestel, 1974). In sum, world food surpluses are being exhausted almost the instant they are produced, leaving nothing for nations suffering from crop failures and other disasters.[16]

The situation will become more severe during the present decade. By 1985 food prices are projected to rise 2.6 times above the level for 1973.

By the middle of the next decade world demand and surplus are expected to come momentarily into balance, after which demand will exceed supply. With growing inequities in distribution, however, and under conditions where demand begins to exceed the available supply on a regular basis, heavy food purchases will only accentuate the division between poor and rich.[17]

(d) *Manufactures.* Finally, there is the expected long-term price inflation for manufactured commodities. This may be attributed to rising fuel and raw material costs no less than to wage pressures generated by worldwide, secular inflation. What is likely to happen, however, is that export commodities from poor, peripheral economies will increase less in price than the products they must import, with the losses being absorbed by indigenous workers who are not only less well organized than their counterparts in world core regions but are also available in practically unlimited quantities. The terms of trade of developing countries have taken a sharp downward turn. During 1974 the prices of their imports rose by 40%, but their export prices increased only 27%. This year (1975), inflation is expected to add at least another 6% to the cost of their imports, but very little to the price of their exports (McNamara, 1975).[18]

---

[15]Increases in foreign aid from OPEC, DAC, and communist countries have failed to keep pace with the combination of inflation and import price increases. In the food sector alone, requirements outweigh aid (OECD Observer 74, March–April 1975).

[16]In cognizance of this problem, the United States recently proposed a comprehensive international cooperative approach to providing food security by establishing an international system of nationally held grain reserves. See speech by the US Secretary of State to the United Nations, September 1, 1975.

[17]One permutation of coupling contradictions was seen in 1972 when natural calamities in Indonesia and the Philippines increased demand for Thai rice. Prices in Thailand began to rise, leading the government to reimpose a rice premium which had previously been blamed for lowering production levels (United Nations, 1973c).

[18]A concrete example from one of our six countries may help to emphasize this point. During 1973 and 1974, Korea's terms of trade deteriorated drastically. The unit price of imports rose by 116%

Restricted Market Expansion

The world economic transition is changing as well the prospects for manufactured exports from the periphery. The sustained growth rates of the past two decades are no more. Growing unemployment, rising commodity prices, problems of environmental pollution, and impending zero population growth all contribute to a slowdown in the already industrialized countries. Pressure from organized labor in these countries, as well as a general trend towards self-sufficiency in large economic "blocks" such as the United States and western Europe, are beginning to lead to a reassessment of the time-honored free trade doctrine of economic liberalism In many cases, this will preclude the underselling of domestic products by equivalent imports from the periphery, whose major selling point is its ability to absorb tariff increases in the form of lower wages. The limit here, of course, is bare subsistence, or about US$1.00 per day.

This evolving situation has the further and vicious consequence of putting poor countries in direct conflict with each other as each tries desperately to outbid the others in essentially the same range of manufactured commodities, attempting to gain a tiny foothold in world export markets. Constant product innovation on the Japanese model would, in principle, be able to counteract this steady movement towards the restriction of export markets. The only other area of "opportunity" would be in heavily polluting industries that, foreclosed from further expansion in the industrial centers of the West and in Japan, may seek a more "receptive" climate in the poor countries of the world periphery.

Conclusion

In the light of these trends, continued adherence to Strategy I may become untenable. If this strategy created glaring social contradictions during the relatively "friendly" sixties, when growth rates in excess of 5% were not unusual, the contradictions will be intensified to the point of general crisis if a mere 5% growth rate should become no longer the rule but the exception.[19]

Rising import prices, declining export markets, and deteriorating terms of trade combine to make Strategy I inoperative as a guide to the future. We can look forward to a period when starvation will become general and endemic; when subsistence and below-subsistence survival in the countryside will no longer be overshadowed by relatively better living conditions in the city; when, even there in the metropolis, the economic environment starts to deteriorate.

---

while that of exports rose by only 57%. As a result, the terms of trade declined by 24% over 1972. The single largest reason for this increase in the price of imports may be attributed to the change in oil prices.

[19]According to rough estimates by de Haan (1974) for the present decade, sustained annual growth rates of 9 ± 1 in GDP would be required fully to absorb into non-agricultural occupations the net increase of labor force. Failure to achieve such growth rates would imply both increased rural densities (and growing poverty) and increased absorption into the individual and unemployed sectors of the economy of Asian cities.

For all these reasons, then, a case can be made for a thorough-going reassessment of national development strategies. The following policy elements would seem to be included on the agenda:

(1) *Limited and specific human needs should replace unlimited, generalized wants as the fundamental criterion of successful national development.* The old GNP standard, based on the idea that a sustained and rapid growth in income and production is the appropriate measure of development, would thus give way to new forms of social accounting that take as their starting point *guaranteed national minimum standards* across a wide spectrum of biological, familial, and communal needs. Although GNP accounts would be retained, they would require revision—to reflect the social costs of production; refinement—to reflect the social valuation of specific production sectors (one dollar's worth of food does not equal one dollar's worth of automobile); and expansion—to reflect the costs of production in terms of scarce energy units. Revised GNP accounts would be supplemented by social indicators showing progress made towards the attainment of specific national minima, such as protein intake and access to sanitary water supply.

(2) *Agriculture should be regarded as a leading or "propulsive" sector of the economy.* The social efficiency of investments in agriculture is thus assumed to be high (particularly with regard to the productivity of land) and capable of generating significant multipliers in agro-industries, commerce, transportation, and construction.

(3) *Attaining self-sufficiency in domestic food production should be regarded as a high priority objective.* An adequate supply of domestically produced food is a basic condition for meaningful participation in the world economy. Participation would be on a highly selective basis, however, and guided by national requirements.

(4) *Existing inequalities in income and living conditions between social classes and between urban and rural areas should be reduced.* The problem of distribution should not, however, be separated, either conceptually or in practice, from that of growth. Regional policy should be seen, not so much in terms of reducing disparities but of increasing generative growth through a full development of natural resources and human skills (Richardson, 1973, pp. 86–88). A concept of dynamic efficiency (Leibenstein, 1966), would replace the traditional criterion of allocative efficiency in capital allocation.

(5) *Facilitative measures to increase production of wage goods for domestic consumption should be given high priority.* This will become possible to the extent that agricultural and, more generally, rural incomes are improved. A national policy of encouraging the wage—goods sector, emphasizing small-scale production and labor-intensive methods, should help materially in the decentralization of manufacturing processes and in experimenting with new organizational settings.

(6) *A policy of planned industrial dualism should be adopted whereby small-scale production for the domestic market is protected against competition from*

large-scale capital intensive enterprise. The latter should be oriented primarily to export and/or to the production of basic materials and intermediate products. At the same time, the small-scale industry might be linked to corporate production through expanded systems of subcontracting.

We shall call a strategy which incorporates these elements a *strategy of accelerated rural development* or, in abbreviated form, Strategy II.

17.1.2 Towards a Spatial Policy of Agropolitan Development

Criteria for Spatial Development Policy

Just as growth center policies correspond to Strategy I (Hansen, 1975), so a strategy of accelerated rural development (Strategy II) requires an appropriate spatial framework for planning and development (Friedmann, 1975). This must be designed to be suitable under Asian conditions, including the prevailing low levels of urbanization (less than 20%), the high and steadily rising levels of rural density (in excess of 200 per km^2), a settlement pattern of clustered villages and towns, and conditions of extreme poverty and physical deprivation.

Commitment to Strategy II will confront policy planners with a series of practical problems that must be resolved if the strategy is to be made operational: how to engage in the development process the millions of rural households dispersed throughout the countryside, each a residential and production unit simultaneously; how to coordinate "sectoral" development in rural areas to achieve the broad purposes of the strategy; how to define, as a basis for accelerated rural development, spatial units larger than a single village; and how to organize on a territorial basis the new planning and development functions. A spatial policies framework for Strategy II will have to respond to these questions. In addition, it will have to be made consistent with a series of criteria that translate Strategy II into programs for implementation. Since none of the six countries under study have as yet committed themselves to the strategy of accelerated rural development, policy criteria will have to be invented in a vacuum of relevant experience.[20] Eventually, they will have to be refined in practice.

What would a strategy of accelerated rural development hope to achieve? This is simply another way of asking how we may determine its success or failure. The following statements are a first attempt to translate the idea of Strategy II into operational language.

[20]There is, of course, the overwhelming example of China's successful rural development effort (Gurley, 1975).

(1) To transform the countryside by introducing and adapting elements of urbanism to specific rural settings. This means: instead of encouraging the drift of rural people to cities by investing in cities, encouraging them to remain where they are by investing in rural districts, and so transmute existing settlements into a hybrid form we call *agropolis* or city-in-the-fields.[21] In agropolitan development the age-old conflict between town and countryside can be transcended (Stuckey, 1975).[22]
(2) To extend the network of social interaction in rural areas beyond the single village, creating a larger socioeconomic and political space or a*gropolitan district*. (Agropolitan districts can also be adapted to become the basic settlement units for large cities or metrocenters, especially along their growing edge.)
(3) To reduce social dislocation in the course of development, preserving the integrity of the family, strengthening psychological security, and providing for both individual and social fulfillment in the construction of a new order of community.
(4) To stabilize both rural and urban incomes and to reduce differences between them by diversifying opportunities for productive work and, more specifically, by joining agricultural to non-agricultural activities within the same territorial community.
(5) To use available labor more effectively by directing it towards a greatly intensified development of the natural resource base of each agropolitan district, including improvements in agricultural production, major conservation and water control projects, rural public works, expanded rural services, and agriculturally oriented industries.[23]
(6) To link agropolitan districts into regional networks by building up and improving physical channels of communication among agropolitan districts and to larger towns, and by regionalizing certain higher-order services and support activities that require a larger population base than is available within a single district.[24]

---

[21]An evocative image of agropolis is found in a Sung Dynasty painting reproduced in Mumford (1961), plate 64.

[22]This is not as crazy as it sounds. Under a capitalist framework, it has already been achieved in countries such as Switzerland and Germany, where the contrast between rural and urban has virtually disappeared. Rural, or rather "agropolitan" life there has become simply another "life style".

[23]Location economists have long argued the case for agglomeration economies. But not a single study exists which has succeeded in measuring them (Richardson, 1973). Actual locations, particularly in Germany, seem to belie the theoretical speculations of economists. Thousands of small- and medium-scale industries in Germany are scattered throughout rural areas even at a considerable distance from larger cities. But even if significant agglomeration economies existed for certain industries, they would be measured only in terms of their returns to private capital. A social cost-accounting might show a very different result.

[24]In line with an explicit policy of industrial dualism (see p. 182), Strategy II would be complemented by a policy of selective industrialization primarily for export and for basic materials supportive of Strategy II. The character of these industries, most of them of highly sophisticated and advanced design in line with international standards, is such that they frequently gain in

(7) To devise a system of governance and planning that is ecologically specific and gives substantial control over development priorities and program implementation to district populations. What is envisioned here is a system of governance in which effective power for decisions is devolved to agropolitan districts to enable them to take advantage of ecological opportunities where they exist (while being mindful of ecological constraints), to harness the richly personal, embodied learning of local inhabitants to the more formal, abstract knowledge of specialists and others experienced in agropolitan development, and to encourage a growing sense of identification of local people with the enlarged communal space of the agropolis.

(8) To provide adequate financing for agropolitan development by (a) ensuring the reinvestment of a large part of local savings in the district, (b) instituting a system of work-in-lieu-of-taxes for every adult member of the community,[25] (c) transferring development funds from metrocenters and specialized industrial areas to agropolitan development, and (d) reversing the adverse terms of trade between peasantry and city populations.[26]

---

productive efficiency by locating in close proximity to one another, forming industrial complexes or "growth poles". These "poles" would constitute regional enclave economies at major ports or in other suitable locations and would not necessarily be spatially integrated with agropolitan development. As a result of secondary urbanization effects, however, they would be expected to exert some influence on surrounding populated regions by providing both income and employment opportunities. For a detailed analysis of the regional effects of such a complex, see Kim (1975).

[25]Work-in-lieu-of-taxes may be regarded as a major source of capital formation under Strategy II as well as a form of building personal commitments to the tasks of national development. The basic concept is, in fact, quite simple, even though its execution will present considerable practical problems that will have to be worked out. Every ablebodied adult; from the Prime Minister on down to the lowliest street sweeper, would be liable for a fixed number of hours each year devoted to "service in national development". Time would be credited to public works projects or essential services, such as health or education. Skilled as well as unskilled labor would thus be equally valued.

[26]The Japanese economist Koichi Mera has raised the question whether, in fact, rural development and, in particular, small farmers, are not already being "subsidized" by the urban corporate sector (personal communication). This is a crucial issue that must ultimately be resolved on empirical grounds. In the absence of conclusive evidence, however, certain qualitative judgments may be ventured. (1) It would appear that in most Asian countries public investments are treated as positive externalities by private firms and are not reflected in the cost structure of private production. (2) Both transport rates and service charges for public utilities tend to favor large-scale corporate production and so contribute to the high degree of spatial concentration of "modern" activities. (3) Tax and other fiscal advantages abound in the corporate sector so that, in the typical case, complete amortization of the initial investment is accomplished in only a few years. (4) In many countries the private banking system serves primarily as a channel for mobilizing rural savings and for transferring them to the primate city of the country, where they become available for urban investments principally in the construction of offices and luxury apartments as well as for real estate speculation. (5) Regional terms of trade are generally against agriculture, particularly in the staple food-producing subsector, as governments tend to control the price of basic foodstuffs in the interest of a politically alert urban proletariat, while inflation in the manufactured goods sector is "exported" to rural populations. All told, the thesis that cities not only finance their own

The Agropolitan District

The concept of an agropolitan district may now be given greater substance. Like the traditional city, it corresponds to a certain size/density function, to the presence of services and conveniences that are appropriate at this scale and adjusted to prevailing levels of cultural and economic development, to the existence of substantial off-farm employment, and to the principle of self-governance.[27] In this sense, then, agropolitan districts may truly be called cities-in-the-fields whose population would place them in the middle range of a national size-distribution of cities.

In purely formal terms, agropolitan districts might be defined as rural areas that have an effective population density of at least 200 per km^2 (for prevailing densities).[28] A town of between 10,000 and 25,000 inhabitants would normally be found within the district, and district boundaries would be defined by a "commuting" radius of between 5 and 10 km (or approximately one hour's travel time by bicycle).[29] Such dimensions would yield an overall size of population ranging from 50,000 to 150,000 of whom a majority would be initially engaged in farming (Friedmann, 1974).

This is the abstract geometry of agropolitan units. Their actual definition in geographic space would, of course, be far more flexible than is suggested here, and would take account of existing patterns of social and economic relations, linguistic and religious boundaries, ecological complementarities, effective population densities, the distribution of traditional market towns, etc.

In addition, agropolitan districts may also be formed within the perimeter of large and growing cities (metrocenters). For just as agropolitan development attempts to bring the city to the countryside, so the countryside can be brought to the city ("fields in the city"). In other words, a modified agropolitan model can be introduced to help restructure the form of existing large cities by opening the built-up areas of the city to fields and urban farming. In this connection it may be noted that most Asian cities will more than double their populations in the next 20 years. The possibilities for restructuring their physical form as well as form of governance would thus appear to be quite considerable.

Although agropolitan districts may have both a town center and a dispersed village population, the key to successful agropolitan development is the treatment

economic expansion but actually transfer resources to help in rural development is, at least in the case of developing Asian economies, an exceedingly dubious one (Harvey, 1973).

[27]Throughout Asia, many so-called cities of up to 100,000 population may have a large proportion of their population classified as agricultural.

[28]According to common practice in the United States, an area is considered "urbanized" when it lies adjacent to a city of at least 50,000 people and has a minimum settlement density of 100 housing units per square mile. Using the 1970 average size of household 3.17, this translates into a population density of 120 persons per km^2. At the older ratio of 4.0 persons per household, the density would be 152 per km^2.

[29]For an equivalent definition of "urban fields" in the United States based on travel distance by automobile, see Friedmann and Miller (1965).

of each district as *a single, integrated, and self-governing unit*. Correlative with this idea of agropolitan development is the requirement that each unit has sufficient autonomy and economic resources to plan and carry out its own development.[30] In this scheme the tasks of central government would be to support locally initiated developments with financial, material, and technical resources to undertake projects of national significance, to ensure interregional equity in the allocation of development funds, and to maintain system-wide balances in the major macro-economic parameters. In addition the central government would monitor the achievement of the guaranteed national minima and assist with accelerating progress in areas of marked shortfalls from the established standards.

Problems of Implementation: An Evaluation of the Asian Experience

Assuming a political commitment to Strategy II, a thoroughgoing land reform appears as the first and quintessential step in a practical program of implementation. The relation of agropolitan development to land reform is summarized in the following imperatives:

(1) Local autonomy, resource management, and a sense of personal involvement must be engendered to sustain development. This cannot be accomplished unless landed and other wealth is returned to the control of members of each agropolitan district. The question of how much land should remain in individual ownership will have to be determined on a case-by-case basis. It may be

[30] To render the idea of agropolitan development a bit more clearly, a sample of specific developmental actions is listed below:

(a) Intensive development of land, water, forestry, and fishery resources for sustained yield; land reclamation schemes; conservation practices; development of small livestock.

(b) Agropolitan electrification; water supply system.

(c) Construction of all-weather roads and bicycle paths (capillary penetration). Inter-agropolitan transport systems as well as good transport connections to major trunk roads and to larger cities.
(d) Improvements in agropolitan communications: microwave telephone and radio; bus services; postal offices.
(e) Provision of basic agropolitan social services: primary, secondary, and technical schools, child care centers, training and research centers, libraries, cultural and recreational facilities, health services, and family planning centers.
(f) Construction of weather- and vermin-proof storage facilities for agricultural products (to reduce crop losses and stabilize supply) and of distribution centers for agricultural implements, pesticides, fertilizers, etc.
(g) Implementation of environmental sanitation programs.
(h) Formation of cooperative marketing services. Credit facilities.
(i) Introduction of light, labor-intensive industries oriented primarily towards the domestic market and producing wage goods, farm implements, processed agricultural materials, and subcontracted parts for final assembly in large-scale, export-oriented industry (Sigurdson, 1975).

noted, however, that unless a substantial share of the total farm land comes under effective communal control, the old inequalities will sooner or later reassert themselves, cancelling the initial effort at redistribution.

(2) Minimum and maximum limits of size holdings must be established in order to permit more rational and productive use of land resources. Fragmented land holdings must be reassembled for efficient production, preferably on a communal basis.

(3) Greater reinvestment of profits from local activity into agropolitan development must be made to raise productivity and to create environments conducive to sustained economic growth. To do this, the flow of resources to large cities, especially in the form of rental payments to absentee landlords, must be reversed.

The extent to which present patterns of land distribution inhibit accelerated rural development is suggested by Table 17.2. Except for Thailand, average farm sizes are near or below minimum levels for productive farming.[31]

Tenancy rates, which translate into a loss of from 30 to 50% of crop values for the farmer, range from 27 to 50% in the six nations. And the richest 5% of the rural population in countries for which data are available control from one-fifth to over one-third of total farm land.

Arguments for land reform have been widely accepted in theory and intention. Every non-communist nation in Asia has legislation which grants government the right to expropriate land (United Nations, 1973d), and all six nations in our sample have, at one time or another, expressed land redistribution as a goal.

Performance, however, has been poor. Land redistribution schemes are not proceeding either widely or effectively. The much-heralded programs carried out in the Republic of Korea in the early 1950s have not been continued; shrinking farm sizes are re-emerging as a problem in agricultural development (IBRD, 1974). Thailand and Indonesia have opted for reliance on new, Green Revolution technologies requiring large-scale mechanized cultivation. In lieu of land reform, such reliance will most likely increase, rather than decrease, rural inequalities by giving cumulative advantage to owners of large farms who represent only a small portion of the rural population.

Malaysia has been able to avoid land redistribution policies by opening up remaining areas of virgin land for the rural poor from already developed areas. India and the Philippines are the only two nations that currently stress land reform as a central planning issue. Indian progress, however, is mired in its bureaucracy, and rents and tenancy insecurity are increasing (IBRD, 1974). The Philippines promulgated a national policy of land reform in 1972. Some two million hectares are to be transferred to one million farmers. Minimum and maximum sizes have been set.

[31]Though such minimums vary according to local situations, in Malaysia the suggested level has been set by one author at 3–4 hectares (Kedhra, 1972); in Indonesia, 1 hectare has been considered as a minimum to allow all year-round rice consumption for a farm family (Tjondrenegoro, 1972); in the Philippines the minimum size has been set at 3 hectares (Philippines, 1972).

Although assessment of the Philippine policies of land reform cannot be made at this time, the approach is in the desired direction.

Despite the low priority assigned to land reform in five of the six nations, all countries are pursuing policies which, in one form or another, seem related to agropolitan development. They are briefly described below.

India. Under the Fourth Five-year Plan (1969–1974) a pilot program for rural growth centers with an accent on rural—urban integration was initiated. This beginning effort was to be buttressed by a "minimum needs program" focused on educational, water and health facilities, home sites for landless laborers, rural roads, and rural electrification (Shah, 1974).

Indonesia. In the current Plan (1974–79) the government has outlined a concept of Village Working Units, constituting groups of villages that cover an area from between 600 and 1000 hectares (or from one-fiftieth to one-tenth the proposed area of an agropolitan district). Within these units, agriculture and credit extension, as well as systems for processing and marketing, are to be formed. Depending on local capabilities, small-scale industry may also be developed.

Malaysia. With calls for the eradication of poverty and a restructuring of society to correct economic imbalances, a New Economic Policy was announced in 1971. Farmers were to be transferred from disadvantaged areas to regions with higher development potentials. Although the primary activity was to be farming, a series of rural cities were to be built to establish the new settlers in urban surroundings.

So far the Pahang Tenggara project is the most ambitious of the new settlement schemes: half a million people are expected in the area by 1990, many of them residing in towns. By 1974 the largest of these already had a population of 28,000 (Salih, 1974). In all, no less than 29 New Towns have been projected.

Philippines. Besides its heavy emphasis on land reform, the government is planning an acceleration of rural infrastructure development, including road networks based on farm size ratios and irrigation projects. Added to this is a push for development and reform of cooperatives with an ultimate goal of raising farm incomes and encouraging savings.

Thailand. The Second Five-year Plan (1967–71) declared that its objective of reducing geographical imbalances and income inequalities would be reached through direct investment in rural development. As with the Philippines and other nations, these programs include investment in infrastructure, the building up of cooperative movements, the provision of credit facilities, agricultural extension, and land development by government. These programs are being continued with a leaning toward growth center strategies for regional development.

How relevant are these policies to the theme of agropolitan development? How far do programs go in effectively reducing the contradictions created by the continued reliance on a strategy of accelerated industrialization? Though our intention is not to evaluate the programs of each nation, we may offer partial insights by noting common features:

(1) Rural development programs in Asia are inadequately funded and coordinated.
(2) Rural development is viewed as an auxiliary policy of Strategy I and not as its replacement.
(3) Rural development is conceived either on the scale of pilot experiments or as a process that occurs "spontaneously" in the hinterlands of cities that are vertically integrated into a hierarchy. The latter view is consistent with the growth center concept and is a tacit justification for the continued neglect of rural populations in policy and budgetary allocations.
(4) The command style of planning and decision making is still predominant, and there is little devolution of effective power to local communities. If any trend can be detected, it is towards an increasing concentration of decision making in central ministries.
(5) With the exception of the Philippines, land reform is receiving little attention. Yet without it, a strategy for accelerated rural development cannot succeed.

Our argument is straightforward. Without a comprehensive land reform program on a national scale, a strategy of accelerated rural development cannot succeed. But land reform cannot be separated from the simultaneous creation of agropolitan districts and the related policies of agropolitan development. Merely to redistribute land without, at the same time, undertaking the many complex tasks involved in raising agricultural productivity and in "urbanizing" the countryside, is to condemn the entire effort to failure. An agropolitan policy must thus be carried out across the board, in poor areas as well as rich, though always with due regard for local circumstances and conditions. As a result, new streams of rural to rural migration might be generated.

Where possible, they should be directed into unoccupied or sparsely populated areas to open up new land for settlement (e.g., Malaysia, Indonesia, the Philippines). The dangers of swamping the positive effects of rural development in specific localities through an influx of migrants from less developed, less productive areas must be avoided.

17.1.3 Concluding Observations

Our original intention was to review and assess the growth center experience of Asian countries. We might easily have done so, pointing to failures and successes. But in so restricting our horizon we should simply have added another chapter to the already voluminous growth center literature (Moseley, 1974). We should not have put the question that we finally came to ask, whether the development strategy to which growth center policies are meant to contribute is also a viable strategy for the future development of Asian countries.

As we delved into the record of national accomplishments, it seemed to us that the universal adoption of Strategy I had created serious contradictions in relation to the benefits it was expected to yield. Although the numerical coefficients were different for each country, the general pattern was unmistakable. Strategy I had brought into being dualistic dependency structures that, while they had helped in achieving respectable growth rates, had also created a set of related and potentially explosive problems, including hyper-urbanization, increasing rural densities, a spatial structure of dominant core and dependent periphery, widespread urban unemployment and underemployment, rising income inequalities, persistent and growing food shortages, and deteriorating material conditions in the countryside.

So long as overall economic growth could be sustained at high levels, these contradictions were perhaps not critical for the maintenance of political stability. At any rate, part of the generated surplus (or aid from outside sources) could always be used to ameliorate the worst conditions and so avoid more fundamental, structural changes in economic organization.

We then tried to place national economic growth into a world context. We concluded that the limited success of Strategy I had, in fact, been the result of a general economic buoyancy in the core regions of world capitalism that had been mediated to their periphery through international trade. By the mid-seventies, however, the world economic system was itself undergoing a major transition. As concerns the peripheral countries in Asia, this crisis was translated into higher

Table 17.1 Agricultural employment by wage and non-wage sectors

Agricultural employment by wage and non-wage sectors (percent)

	India 1971[b]	Indonesia 1971	West Malaysia 1967	Thailand 1970	Republic of Korea 1972
Wage	49	23	44	4	14
Non-wage[a]	51	77	56	96	86
Percent agriculture of total employment	71	62	50	75	49
A Manufacturing					
Wage	19	12	15	18	21
Non-wage	19	9	8	6	9
Manufacturing as percent of total employment	10	7	3	4	13
B Trades and services					
Wage	39	39	52	47	42
Non-wage	23	40	25	29	28
Trades and services as percent of total employment	16	26	31	13	30
Total non-agriculture (A+B)[c]					
Wage					
Non-wage	58	51	67	65	63
	42	49	33	35	37

Source: ILO, Labour Statistics Yearbook, 1974

[a] "Non Wage" includes self-employment and family workers.
[b] Data are subject to extreme bias in that 70 percent of the labor force was "status unknown" and could not be included in the calculations
[c] Mining and construction are not included

Table 17.2 Patterns of land distribution inhibit accelerated rural development

Farm size, tenency and concentration of ownership

	Average farm[a] size[a] (hectares)	Farms in tenancy [b] (percent)	Farmland in tenancy[b] (percent)	Percentage of land held by top 5% income group[c]
India	1.1 (1969)	27.3	n.a.	36 (1960)
Indonesia (Java)	0.5 (1969)	35.9	25.9	31 (1963)
West Malaysia	2.7 (1972)	31.2	15.7	
Philippines	2.7 (1970)	54.3	40.4	
Republic of Korea	2.1 (1961)	n.a.	n.a.	
Thailand	3.5 (1970)	40.0 (1969)	n.a.	20 (1969)

[a] United Nations, 1973, World Economic Survey
e IBRD (1974), Land Reform, World Bank Rural Development Series, Washington DC. Tables 5 and 10. Figures are from latest available dates in 1960s. Data refer to Indonesia and all malaysia.
[c] United Nations, 1973, Econonimic Survey of Asia and the Far East, ECAFE. p. 29

import costs, a worsening of the terms of trade, and restricted access to world markets. In combination, these trends suggested to us that the high rates of GNP projected by national governments were most probably in error. The long-term prospect is rather for a decline in the "normal" level of economic growth to only slightly more than increases in population.

Accepting this prognosis, we then suggested that under conditions of relative economic stagnation Strategy I would become untenable. Its place would have to be taken by an alternative strategy of accelerated rural development that was oriented to human needs, a more equal distribution of economic benefits, the direct involvement of local people in the process of development.

In Part II we considered the spatial correlate of such a strategy. We called it a policy framework for agropolitan development. This envisioned creating "cities-in-the-fields" by embedding some of the key elements of urbanism in dense rural areas of limited size.

The agropolitan district appeared to us as the appropriate unit for devising a policy of spatial development through decentralized planning and decision making.

Finally, we considered some of the concrete measures which our sample of six Asian countries had already implemented and which might be seen as related to the suggested policy of agropolitan development. First and foremost was land reform.

But we could find little evidence for substantial changes in patterns of land ownership. Second, a variety of recently adopted rural development programs was briefly surveyed. Here, too, we were led to conclude that none of them represented more than token efforts, and that the programs were viewed by governments as complementary to accelerated industrialization rather than as harbingers of new directions.

Our conclusions must be carefully weighed. If we are right, Asian economies will experience a deepening of the present crisis. The alternative would be to shift to a radically different strategy of accelerated rural development within the framework of agropolitan development. If we are wrong, the ultimate crisis may be avoided, even though the specific contradictions we have identified would continue. Growth center policies might then be re-examined as a means for "spreading" specific economic benefits to depressed peripheral regions within each of the national economies. This is the purpose for which they were originally devised, i.e., primarily as a means for ameliorating some of the worst side-effects of accelerated industrialization under dualistic dependency. Whether growth center policies would, in principle, be capable of achieving even this limited objective, remains an open question.

References

Adelman, I. (1975). Growth, income distribution and equity-oriented development strategies. *World Development, 3*(2), 67–76.

Ariff, K. A. M. (1973). Economic development of Malaysia: Pattern and perspective. *The Developing Economies, 11*(4), 371–391.

Baldwin, W. L. (1974). The Thai rice trade as a vertical market network: Structure, performance, and policy implications. *Economic Development and Cultural Change, 22*(2), 179–197.

Bautista, R. M. (1973). Anatomy of labour absorption in Philippine manufacturing, 1956–1966. *Economic Bulletin for Asia and the Far East, 24*(2/3), 12–22.

Bautista, R. M. (1974). Employment and labour productivity in smallscale manufacturing in the Philippines. *NEDA Journal of Development, 1*(1), 41–54.

Bose, A. (1971). The urbanization process in South and Southeast Asia. In Jakobson & Prakash (Eds.), pp. 81–110.

Cline, W. R. (1975). Distribution and development: A survey of literature. *Journal of Development Economics, 1*, 359–400.

Dandekar, V. M., & Rath, N. (1971). *Poverty in India, economic and political weekly* (pp. 25–48,106–146).

Davis, K. (1969). *World urbanization 1950–1970*, Vol. 1. University of California, Monograph Series No. 4, Berkeley, Cal.

de Haan, H. H. (1974). *The nature and magnitude of underutilization of labour in non-communist less developed countries*. Centre for Development Planning, Erasmus Universiteit Rotterdam, Rotterdam.

Emmanuel, A. (1972). *Unequal exchange: A study of the imperialism of trade*. Monthly Review Press.

Esmara, H. (1975). Regional income disparities. *Bulletin of Indonesian Economic Studies, 11*(1), 41–57.

Faber, M., & Seers, D. (Eds.). (1972). The crisis in planning, Vol. 2. Chatto Windus
Friedmann, J., & Miller, J. (1965). The urban field. *Journal of the American Institute of Planners, 31*(4), 312–319.
Friedmann, J. (1973a). *Retracking America: A theory of transactive planning*. New York: Anchor Press/Doubleday.
Friedmann, J. (1973b). *Urbanization, planning and national development.* Sage Publications, Chap. 5.
Friedmann, J. (1975). A spatial framework for rural development: Problems of organization and implementation. *Économie Appliquée, xxviii*(2–3), 519–544.
Friedmann, J., & Sullivan, F. (1974). The absorption of labor in the urban economy: The case of developing countries. *Economic Development and Cultural Change, 22*(3), 385–413.
Friedmann, J., & Douglass, M. (1978). Agropolitan development: Towards a new strategy for regional planning in Asia. In F.-C. Lo, & K. Salih (Eds.). *Growth pole strategy and regional development policy.* Pergamon Press. *Proceedings of the seminar on industrialization strategies and growth pole approach to regional planning and development: The Asian experience*, Nagoya, Japan, 4–13 November 1975.
Friedmann, J. (1979) Basic needs, Agropolitan development, and planning from below. *World Development, 7*(6), 607–613.
Friedmann, J. (1981). The active community: Toward a political-territorial framework for rural development in Asia. *Economic Development and Cultural Change, 29*(2), 235–261.
Friedmann, J. (1996). Modular cities: Beyond the rural-urban divide. *Environment and Urbanization, 8,* 129.
Gurley, J. G. (1975). Rural development in China 1949–72, and the lessons to be learned from it. *World Development, 3*(7/8), 455–471.
Hansen, N. M. (1975). *An evaluation of growth centre theory and practice.* United Nations Centre for Regional Development.
Harvey, D. (1973). *Social justice and the city.* The Johns Hopkins Press.
Hayami, Y., & Ruttan, V. (1971). Agricultural development: An international perspective. The Johns Hopkins Press.
Horii, K. (1972). The land tenure system of Malay padi farmers. *The Developing Economies, 10*(1), 49–73.
IBRD. (1974). *Land reform.* World Bank Rural Development Series.
ILO. (1971). *Concepts of labour force underutilisation.*
ILO, UNDP (1973). *Sharing in development: A programme of employment, equity and growth for the Philippines.*
India, Government of, Planning Commission (1968). *Fourth five-year plan, 1969–1974* .
India, Government of, Planning Commission (1972). *Draft of the fifth five-year plan, 1974–79.*
Indonesia, Government of (1969). *Indonesia's five-year development plan, 1969–74.*
Intermet. (1973). Town drift: Social and policy implications of rural—urban migration in eight developing countries. In *Final conference on rural—urban migrants and metropolitan development*, Nov. 24–Dec. 1, Istanbul.
Jakobson, L., & Prakash, V. (Eds.). (1971). *Urbanization and national development.* Sage Publications.
Kerdpibule, U. (1975). Distribution of income and wealth in Thailand. In P. Sondysuvan (Ed.), *Finance, trade and economic development in Thailand.* Sompong Press.
Khera, H. S. (1972). *Problems and prospects of the malaysian economy.* Central News Agency.
Kim, A.-J. (1975). *Industrialization and growth pole development in Korea: A case study of ulsan industrial complex in the context of regional development in Southern Coastal Areas.* United Nations Centre for Regional Development.
Kuklinski, A. (Ed.). (1972). *Growth poles and growth centres in regional planning.*
Kuklinski, A., & Petrella, R. (Eds.). (1972). *Growth poles and regional policies.*
Lefeber, L., & Datta-Chaudhuri, M. (1971). *Regional development experiences and prospects in South and Southeast Asia.*

Leibenstein, H. (1966). Allocative efficiency vs. X-efficiency. *American Economic Review*, 392–415.
Marzouk, G. A. (1972). *Economic development policies: Case study of Thailand*. Rotterdam University Press.
Malaysia, Government of (1973). *Midterm review of the second Malaysia plan 1971–75*.
Malaysia, Government of (1974). *Economic report 1974–75*.
McGee, T. G. (1971). Catalyst or cancers? The role of cities in Asian society. In Jakobson & Prakash (Eds.), pp. 157–182.
McGee, T. G. (1974). *The persistence of the proto-proletariat: Occupational structures and planning for the future of third world cities, comparative urban studies series, school of architecture and urban planning*. University of California.
McNamara, R. S. (1975). *Address to the board of governors*. IBRD.
Mesarovic, M., & Pestel, E. (1974). *Mankind at the turning point: The second report to the club of Rome*, E. P. Dutton.
Misra, R. P., Sundaram, K. V., & Prakasa Rao, V. L. S. (1974). *Regional development planning in India: A new strategy*. Vikas Publishing House.
Moseley, M. J. (1974). *Growth centres in spatial planning*. Pergamon Press.
Oshima, H. T. (1970). Income inequality and economic growth—the postwar experience of Asian countries. *Malaysian Economic Review*.
Oshima, H. T. (1971). Labor-force "explosion" and the labor-intensive sector in Asian growth. *Economic Development and Cultural Change, 19*(2), 161–183.
Paauw, D., & Fei, J. C. (1973). *The transition in open dualistic economies: Theory and Southeast Asian experience*. Yale University Press.
Republic of the Philippines, NEDA (1973). *Four-year development plan, Fy 1974–77*.
Republic of Korea, Government of (1971). *Third five-year economic development plan, 1972–1976*, Kyongje Kihoegwon.
Richardson, H. W. (1973). *Regional growth theory*. Macmillan.
Romm, J. (1972). *Urbanization in Thailand*. Ford Foundation.
Salih, K. (1974). On urban development strategies in Malaysia. In *UNCRD seminar on urban development strategies for attaining desirable population distribution*, Nagoya, Japan.
Seers, D. (1970). *The meaning of development*. Agricultural Development Council Reprint, September.
Shah, S. M. (1974). Growth centers as a strategy for rural development: India experience. *Economic Development and Cultural Change, 22*(2), 215–228.
Sigurdson, J. (1975). Rural development in China: Approaches and results. *World Development, 3* (7/8), 527–538.
Stuckey, B. (1975). *From tribe to multi-national corporation: An approach to the study of urbanization*, Ph.D. Dissertation, University of California, School of Architecture and Urban Planning, Planning Program, Los Angeles.
Sundrum, R. M. (1975). Manufacturing employment, 1961–1971. *Bulletin of Indonesian Economic Studies, 11*(1), 58–65.
Thailand, Government of (1972). *Summary of the third five-year plan (1972–1976)*. National Economic and Social Development Board, Office of the Prime Minister.
Tjondronegoro, S. M. P. (1972). *Land reform or land settlement: Shifts in Indonesia's land policy, 1960–1970*. University of Wisconsin, Land Tenure Center.
Turnham, D., & Jaeger, I. (1971). *The employment problem in less developed countries: A review of the evidence*. Development Centre, OECD.
Ul Haq, M. (1972). *Employment in the 1970s: A new perspective*. Agricultural Development Council, Rreprint, October.
United Nations. (1973a). Manpower growth and labour absorption in developing Asia. *Economic Bulletin for Asia and the Far East, 24*(1), 10–33.
United Nations. (1973b). Population and food supply in Asia. *Economic Bulletin for Asia and the Far East, 24*(1), 34–50.
United Nations. (1973c). *Economic survey of asia and the far east*. ECAFE.

United Nations. (1973d). *Urban land policies and land-use control measure: Volume II, Asia and the Far East*. New York.

United Nations. (1974). *World economic survey 1973*.

Wallerstein, I. (1974) The rise and future demise of the world capitalist system: Concepts for comparative analysis. *Comparative Studies in Society and History*, 387–415.

Chapter 18
Towards a New City (1983)

Giorgio Piccinato and Giuseppe De Luca

Introduction

Tommaso Pietropolli
EPFL, Lausanne, Switzerland
tommaso.pietropolli@epfl.ch

After around twenty years from the conceptualization of the Città Territorio,[1] *Giorgio Piccinato writes about the diffused urbanization of the central Veneto area, together with Giuseppe De Luca. The occasion is an article on the "processes of urban diffusion in the Veneto region", appeared in 1983 in the journal of economic and social analysis* Oltre il Ponte, *edited by the IRES association*[2] *in collaboration with the CGIL labour union. The contribution has a substantially analytical character, with an economically-based stance, conforming to the general frame of the publication; highlighting a more general interest in the study of marginal, non-urban areas, the contribution goes beyond being just an analysis, becoming a highly interpretative description of the condition of Veneto's urban dispersion.*

Two intuitions, in particular, determine the originality of this contribution within the vast literature of urban and regional studies developed in the same years around the phenomenon of the Veneto region's urban dispersion and the reading of its urban structure. On the one hand, the authors recognized that such a "city in the making" was not just a matter of dispersion of housing in the rural landscape, but

[1]Tafuri, M., Piccinato, G., & Quilici, V. (1962). La città territorio: verso una nuova dimensione. *Casabella Continuità* 270, pp. 16–25.

[2]The Institute of Economic and Social Research (IRES) was an institute founded by the Confederazione Generale Italiana del Lavoro (CGIL) in 1979; it was mainly concerned with studying the world of work and its changes.

G. Piccinato (✉) · G. De Luca
Roma Tre University, Rome, Italy

M. Barcelloni Corte and P. Viganò (eds.), *The Horizontal Metropolis*,
https://doi.org/10.1007/978-3-030-56398-1_18

also the result of complex dynamics intertwining infrastructures, services, job opportunities and income prospects, which enabled to concretely realize a very particular urban lifestyle, taking place in a traditionally marginal (mainly agricultural) territory, spatially characterized by low density. On the other hand, they highlighted public services, welfare spaces and infrastructures as decisive elements in the construction of this dispersed territory; even though implicitly, the study in fact underlines the importance of the public project and public investment in proceeding towards a new type of city. According to the authors, far from the logic of "development as concentration", a different social and territorial policy (akin to the industrial one of the motto "a chimney for each bell tower"[3]*) could build the conditions for living the territory in a different way. They imagined a new model resulting not only from infinite individual "rationalities' (the choices of single inhabitants) but also from specific collective wills, supported by a specific spatial condition resting on the region's particular territorial structure, traditionally organized according to water management and mobility systems.*[4]

The author's considerations cross the reflections brought forward within the research on the Horizontal Metropolis, in particular the ones concerning the role of the Project for a changing territory. Some territorial characters highlighted in this text in fact became central in the reflection on the Horizontal Metropolis and represent perhaps the main legacy of this contribution.[5]

First, the flexibility and a certain predisposition to recycling that characterize the territory and the single elements/materials that constitute the territory itself: the possibility of accepting different socio-economic models is, according to Piccinato, related to the capacity of these spaces to adapt, to have intrinsically less resistance and inertia than denser spatial configurations.

Secondly, the role of the centres and of the different centralities: the polycentric character of the Veneto region's urban system designates to the centres a role of "coagulating elements" of a wider, diffused system rather than that of "distinctive and autonomous elements", centralising resources, services and opportunities for inhabitants. In such a spatial configuration, the different centres do not impose their dominant character on the marginal and less central places. The authors' reading evokes–thus–different possible hierarchies of spaces, in which urban centres orient and structure the territory without prevailing.

[3]A motto of the Italian Christian Democratic party, which encouraged, also for electoral purposes, a diffuse industrial development based on the structure of small centers linked to churches in the Veneto area.

[4]See Viganò P., Secchi B., Fabian L. (Eds.). (2016). *Water and asphalt. The project of isotropy.* Zürich: Park Books.

[5]Especially with the Venetian case deepened in the Horizontal Metropolis research: Tommaso Pietropolli, "Urbs in Horto: recycling the diffuse city"; 2015, master thesis, Iuav University of Venice, prof. Paola Viganò; now in Cavalieri C., Viganò P. (Eds.). (2019). *The Horizontal Metropolis, A radical project.* Zürich: Park Books.

18.1 Towards a "New City"? Analysis of Processes of Urban Diffusion[6]

In the literature we find an increasing emphasis on the arrest of urban concentration —in its consolidated terms of housing, productive activities and population—and of an inversion of the trend, of the growth of non-urban areas, revitalization of marginal zones, urban decentralization. It would appear that the foremost element of socio-economic and also cultural expression of industrial development, the city, is dissolving and breaking up across the entire territory, no longer focusing on a give portion that can be physically perceived with the naked eye.

The fact that processes of transformation of the urban structure are in progress, and that they have a particular impact on large cities, appears to be indubitable. Just as there can be no doubt if we observe the trend of evolution of the Italian society, as well as others, towards a system of territorial diffusion of life. But care should be taken here: this is not the "urban nebula" of which Gambi speaks;[7] that too exists, it is true, but it is the dense intertwining of availability of income, presence of employment, official or otherwise, creation of social expectations, leveling and multiplication of information, that permits even marginal portions of territory—or in any case those considered such in an analytical perspective closely tied to "growth as concentration"—to go through moments of extreme socio-economic and therefore also territorial dynamism.

Since the phenomenon manifests in the cessation of growth of urban and metropolitan areas, many have attempted to connect the two directions in pursuit of possible "models of consequentiality" of the inversion—as in the period of concentration—overlooking not only the dynamic element of the phenomenon, but also its territorial connotations, confusing causes with effects and local elements with general and "diffused" processes.

Of course the problem arises of the *measure* of the phenomenon, its impact in various territorial realities, its "spread." But this is not easily resolved with traditional, official data sources: either because they are aggregated and anonymous, or because they cancel out the possible relations between problems, phenomena and individual behaviors, or because the narrow territorial boundaries, even when marginal, have no reason to exist given the high level of economic, social and also cultural integration. Moreover, in our view, we are faced with a form of organization of space that is not obligatory, nor predetermined on the basis of explicit objectives, as might be the case in the period of concentration, making it hard to grasp within factors of general and quantitative synthesis.

Aware of this, we have attempted to initiate a local analysis through direct investigation in the field in two restricted situations of the Veneto. The work is simply an attempt to identify more articulate and specific systems of detection of the phenomenon in its territorial scope. We cannot help but point quite frankly at the

[6]Source: Piccinato, G., & de Luca, G. (1983). Verso una nuova città? Analisi di processi di diffusione urbana. *Oltre il ponte*, *2*, 9–19.

[7]L. Gambi, "Da città ad area metropolitana," in *Storia d'Italia*, vol. 5, book 1.

limits of the work and its approach, strongly focused on only one social segment: the population that lived in a house built or completely refurbished after 1970. But these have been intentional, pursued choices; it is in fact these figures that determine the transformations of space, and probably represent the tip of an iceberg that is still submerged to a great extent.

1. The urban structure in the Veneto has traditionally been based on spreading settlements with small centers scattered across the territory. Even in the periods of highest urban concentration in Italy (1950s and 1960s) the phenomena of centralization of the population have not been sizeable, with the *exception* of certain individual situations. The process of industrialization itself has not represented a moment of "traumatic" rupture, but has been extremely "cadenced," almost settling over the existing urban structure, reinforcing it without breaking it up. So much so that it has become common to speak of a "Veneto model" and a "polycentric urban system" in reference both to industrialization and to the territory.
   The present phase of ulterior growth of minor urban systems, or those that have remained at the margins of concentrated industrialization, has therefore not taken on in this region the clarity that can be observed in other territories. It is undeniable, however, that the entire dynamic of population growth is "concentrated" in terms of both residential and industrial use only in these urban systems.[8]
   But how is the territory structured? We can make reference to the two observed areas and to the sole issue of housing, which is what interests us on this occasion. The observations cannot help but be partial, but we think they are sufficiently explanatory. In the two areas we have sampled, one of recent growth, the other already consolidated in its development,[9] new construction is spread across the territory at a level of 65.8%, with greater weight in the area of Padua and less weight in that of Vicenza. In the total spread of housing, 46% of

[8]For example, see A. De Angelini, *L'evoluzione del sistema insediativo italiano negli anni '70; il decentramento e la crescitadelle aree rurali*, in 1a Conferenza italiana di Scienze Regionali, Naples, October 1981; and idem. *Lo sviluppo territorialedel Veneto negli anni '70: decentramento urbano e industriale*, in 2a Conferenza italiana di Scienze Regionali, Naples, October 1981; also B. Anastasia and E. Rullini, "La nuova periferia industriale. Saggio sul modello Veneto," *Materiali Veneti*, no. 17–18, 1882; G. Rizzo and G. Robiglio Rizzo, "Marginalizzazione e rivalorizzazione nel Vento," in Agei, *L'Italia emergente*, vol. 2, F. Angeli, Milano, 1983, although the latter, considering only the demographic dimensions as aparameter of measure, is insufficient to explain and interpret the real territorial processes.

[9]The townships covered by the study are: Campo San Martino, Carmignano, Cittadella, Fontaniva (later excluded for lack ofdata), Galliera, Grantorto, San Giorgio delle Pertiche, San Giorgio in Bosco, Santa Giustina in Colle, Tombolo and Villa delConte in the province of Padua; Agugliaro, Albettone, Barbarano V., Campiglia dei Berici, Noventa V., Orgiano, Poiana,Sossano, Villaga in the province of Vicenza. The research has been conducted through a questionnaire regarding 5% ofthe licenses and permits for new constructions or complete renovations issued starting in 1970 for a total of 310 (192 in the province of Padua, 118 in that of Vicenza). The data have been processed with the collaboration of G. Secco of the Institute of Geography of the University of Padua.

Table 18.1 Territorial location of surveyed residences

	Absolute values	Relative values (%)
Urban center	106	34.2%
Rest of territory	201	65.8%
Of which:		
– In nuclei	109	35.2%
– Completely scattered	95	30.6%

Total residences surveyed
Source Survey, November 1981

the units are completely isolated in the territory, while the remaining 54% are gathered in small nuclei of 4–5 dwellings (Table 18.1). Most of the construction activity of the last decade has thus been located on the territory outside the main urban center. But the same data on housing located in inhabited centers is relatively significant, since we are dealing with towns of modest size, relatively distant from one another. In fact, the entire urban structure of the two areas, more strikingly in that of Padua, represents a single organism of a territorially distributed urban fabric, where the original inhabited center—with rare exceptions—functions more as an element of coagulation of the whole diffused residential system than as a distinctive and autonomous element. The administrative functions of the entire life of the community are not always located inside it, but are often distributed across the whole territory or prevalently in certain areas—at the intersection of two or three streets that connect a comparable number of municipalities or inhabited zones—which may also not coincide with the original urban center.

This gives rise to an initial distinction seldom addressed in the literature on this subject: speaking of urban diffusion, in our view, should not imply a completely scattered location of housing, but instead a *territorially diffused minor urban system* where policies of social assistance, the presence of services and infrastructures and the continuous integration of the market have made access to goods and services independent with respect to the residential position on the territory. Hence the historic difference between city and countryside, to sum it up in linguistic terms, tends to be erased or is no longer as important as one might think.

In these systems, then, to speak of residences located in the urban center is simply a distinctive characteristic of the terminology of urbanism, rather than a true element of distinction between different morphologies. And, in fact, if we examine the typology, the dwellings located in the center are of the single-family type (67.9%), while the apartment buildings (14.1%) or other typologies (17.9%) are less numerous. The single-family typology, furthermore, is also prevalent throughout the zone examined (Table 18.2). This datum goes together with that of property use: over 90% of the surveyed residences are owned by their inhabitants, singularly (86.4%) or in co-ownership (4.2%) (Table 18.3). The rental system is not widespread, which means that movements

Table 18.2 Type of surveyed residences and their territorial location

Relative Values			Of which		
		Urban center	Nucleus	Scattered	
Single-family	76.0	30.5	35.6	33.9	100.0
Apartment building	8.1	60.0	36.0	4.0	100.0
Other typologies	15.9	38.9	32.6	28.5	100.0
	100.0	34.2	35.2	30.6	100.0

Source Survey, November 1981

Table. 18.3 Structure of residential property and type—relative values

	Ownership	Co-ownership	Rental	Other	
Single-family	89.2	3.0	2.4	5.4	100.0
Apartment building	84.0	7.0	6.0	3.0	100.0
Other typologies	81.4	2.3	14.0	2.3	100.0
	86.4	4.2	5.2	4.2	100.0

Source Survey, November 1981

from the larger urban centers towards the area are not caused by a tight housing market, even if we assume that law no. 392 of 1978, by excluding municipalities with a population lower than 5000 inhabitants from rent regulation, steers the demand for rental housing into these areas. If anything, a shift takes place in the municipalities in close proximity to the large urban center and not in other areas such as those we have examined. Areas we could define as intermediate, between the townships adjacent to urban concentration, reflecting all the ills of the latter, and the remaining, marginal municipalities.

The growth dynamic thus has connotations that are totally endogenous, presumably due to a sizeable increase of income and job opportunities existing in the area. The movements are therefore inter-area (Table 18.4), connected with processes of relocation and/or qualitative improvement of the dwelling. Here the subdivision of land ownership that exists in the region becomes the key element to remove one of the main costs of residential relocation: that of the land (Table 18.5). There have been many cases of new residential constructions alongside old settlements of an agricultural character. But this would not have been possible had there not been a remarkable increase of availability of income for each individual family nucleus at the same time. And in fact, in the families we have surveyed, which do not represent a sort of "*nouveau riche*"[10],

[10]Out of the 62.3% of the population over 14 years of age, 74% declares employment, of which 34.4% are self-employed; hired workers represent 28.6% of the sample. See G. Piccinato, G. De Luca, G. Cretella, "I caratteri dell'urbanizzazione in un'area periferica: analisi di un caso veneto," in III Conferenza italiana di Scienze Regionali, Venice, November 1982, nowin Bollettino Du, no. 1, 1983.

Table 18.4 Location of previous residence—relative values

Same neighborhood	45.8
Same municipality	33.9
Other municipality in district	10.0
Other municipality in province	5.2
Other municipality in region	4.5
Outside region	0.6
	100.0

Source: Survey, November 1981

Table 18.5 Motivation for choice of location—relative values

Land ownership	45.5
Land purchase	34.2
House purchase	6.4
Other	13.9
	100.0

Source Survey, November 1981

unemployment is practically absent: out of an average household size of 3.9 persons/family, on the average two of those persons hold jobs. If we consider the fact that the school age members are, in average, 1.2 per family, we can see that there is a high deployment of the entire potential labor force in the official job market. This is a telling factor for the economy of the area we have surveyed, or at least its families. The work opportunities themselves have not been created in parasitic sectors, but in productive sectors (51.9%).

We should not forget that precisely inside the central area there has been a marked trend towards the growth of small businesses in the proper sense of the term, aided by a series of structural conditions existing in the region.[11] This is joined by the agricultural sector which tends to take on the role of a reservoir of second jobs and therefore of added income: almost all the residences we have surveyed are located on agricultural lands that are cultivated by the families, or by others called in for the purpose.[12] This latter form of employment (and income) eludes any quantification because it happens on the family's own property and is therefore seen as an obligation/duty with respect to ownership. The presence of income probably also functions as an element of attraction, if we consider the fact that 10.3% of the families previously lived in municipalities outside the district. This is undoubtedly a small quantity that cannot be significant enough to represent a trend signal. If these areas have "found" the elasticity needed to attract new population, plausibly also returning to the area

[11]Cf. R. Camagni, *La dinamica delle localizzazioni industriali nel Veneto negli anni '70*, report to the Convegno CGIL-Veneto *Politiche di localizzazione industriale*, Venice, April 1981; B. Anastasia and E. Rullani, op. cit.

[12]The percentage is 74.5%, of which 80.1% in the Vicenza area and 68.9% in that of Padua.

given the fact that these were zones impacted by emigration in the 1950s and 1960s, it means that they are no longer "peripheral" with respect to the overall model of development.

Therefore, if on the one hand the growth has endogenous connotations (residential improvement), on the other it closely depends on exogenous factors (location of employment opportunities), consequences of the intense economic growth recorded during the course of the 1970s in the region, particularly in the central areas.

Here we encounter a second point for clarification. In our view, the growth of the modern capitalist economy no longer seeks a specific place in which to *concentrate* the multipliers by accumulation; instead, it "*distributes*" them, "*spreads*" them, "*organizes*" them across the entire territory. It is the territory per se that becomes the preferred place of all the factors of production, but also those of reproduction. This is joined by a strong tendency towards the "personification" of the lifestyle—particularly perceptible in social realities such as that of the Veneto—towards the segmentation of the decision-making power of the State on both administrative and managerial fronts, towards the spread of "social wellbeing" over the entire territory in terms of assistance, income and culture, towards the presence—also in marginal portions of the territory—of businesses and of all the products available on the market, with a high level of diffusion of service industries in the territory made possible by the spread of data processing.

Hence the possibility of extending the urban effect—not just in physical terms—over increasingly vast portions of the territory; an effect that probably does not create territorial systems from scratch, but relies (at least in the situations of ramified urbanization like the Veneto) on existing systems, exploiting combined factors like the increased productivity of agriculture, the presence of infrastructures of reference, etc., renovating them and "adjusting" them with respect to the new needs of production in such a way as to get beyond the typical rigidities of concentration.

2. It is therefore the possibility of access also to diversified sources of income and of deploying almost the entire workforce that determines—in social realities like this one where the family is still the central fulcrum of convivial experience—an elasticity in the use of savings that has an immediate impact on the house-asset. And this also is reflected in the general living conditions of the family, both in terms of the needs satisfied and the services installed in the residence. This discourse opens the way to new observations that the theme of this contribution does not permit us to explore. We can simply mention them: from private locomotion (the family-automobile ratio is 1:1.9, with a preference for medium–high engine size), to color television sets, walk-in freezers, etc.; thus to the house of large size (over 150 m^2 of living space in 80% of the cases), typically of the "suburban villa" typology of very bad taste, with two or three bathrooms.

A reality that is a "carbon copy" of the most detrimental citified civilization, where autonomy is confused with dependence and "wellbeing" with social conservation.

3. Naturally this type of settlement, which breaks away from the historic structure in settlement nuclei (and therefore in albeit minimum forms of concentration), leads to serious difficulties in terms of consumption of agricultural land, pollution (only 7.5% of the houses we surveyed are connected to the public sewer system, while the rest have cesspools, 89.3%, or nothing at all, 3.2%), and excessive infrastructure. What makes the problem even more complex than in the past is the fact that this "mode" of settlement is not simply the result of a sum of purely speculative behaviors—and thus liable to being addressed by effective standards and controls, were the political will for this not lacking—but of the form assumed in space by economic and social structures that differ from those of the past.

To distinguish and separate the different uses of the land, in keeping with the traditional indications of urban planning, somehow implies the existence of social conditions of production in which the functions are clearly marked and to some extent autonomous: exactly the opposite of what happens in the model we have described, where the range of activities in a single social cell is such as to deny the possibility of a real distinction. The proof of this can be found by examining the regional legislation on construction development and protection of agricultural zones (LR no. 58, 1978).[13] If the intentions—praiseworthy but perhaps not timely —were to regulate the structure of agricultural space, the result is instead that of temporarily freezing new construction, while permitting extensive restructuring of what already exists, with the transformation and expansion of agricultural annexes. Therefore, in practice, the outcome has been what was supposed to be avoided, in terms of undermining of roles—and in what has probably been the worst or in any case least controlled way. It would be hard to imagine that a legal standard could suffice to combat the final results of a deeply rooted structural transformation, which cannot help but be reflected in a different organization of space.

Here we get to the crux of the issue. What has been called the "regulatory obsession" of Italian urbanism has actually been a way to try to resolve by edict contradictions and conflicts that are hard to interpret, and even harder to control. This is the symptom of a serious difficulty in coming to terms with real complexity, clinging to a guarantism that has no effective impact because it is detached from the

[13]The regional regulation based its philosophy on a constraint of what "can be done," inserting itself in the logic of the distinctionof the territory in functions and assigning each of them a "quantum" of maximum construction "under certain conditions." Inpractice, by granting the Provincial Inspectorate of Agriculture the right of waiver of regulatory constraints and obliging the municipality to "take note" the result is to give rise to an intense, vast process of transformation of use (towards residence) of agricultural annexes. A research project we are conducting demonstrates that the applications for building permits for new constructions have practically vanished, while there has been a large increase in those for the expansion of rural outbuildings or houses, all requiring waivers.

processes of change in progress, and failing to construct the means and tools with which to grasp that change in a timely way.

Urbanism (like politics?) has lagged behind, for many years now, and by now even their exponents themselves seem to be convinced of this fact: not by chance, the national congress of the National Institute of Urban Planning, in 1983, addresses precisely these themes.

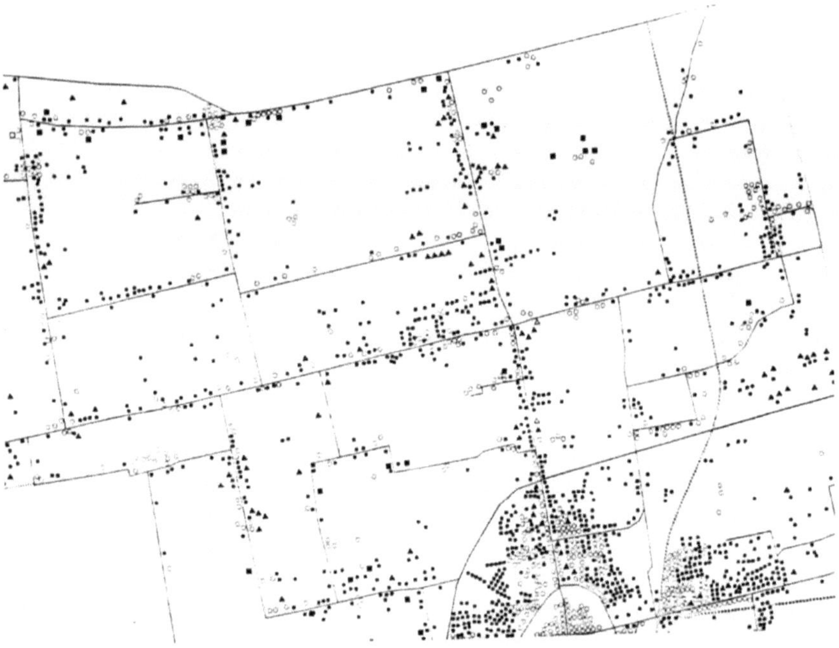

Fig. 18.1 Map of the structure and of settlement transformations.
(Triangle: craft/industrial/commercial buildings; Square: agricultural production buildings; Circle: residential buildings and other; Empty figure—existing in 1966; Filled figure—built between 1966 and 1981).
Source: Sartore, M, 1988, Forme e processi di urbanizzazione diffusa: un'analisi della morfologia degli insediamenti nelle aree rurali in un contesto territoriale del Veneto centrale (Master Thesis). Thesis supervisor: G. Piccinato, IUAV di Venezia. Reprinted courtesy of M. Sartore

Fig. 18.2 Aerial view of the Veneto Region, in the province of Padova, on the East of Camposampiero, over Baldassa, Rustega and Straelle, on the 16th of March 1983.
Source: Reven 1983 Venezia-Treviso, strisc. 13-foto n. 3564. © Area Infrastrutture e Lavori Pubblici; Direzione Pianificazione Territoriale; Pianificazione Territoriale Strategica e cartografia

Chapter 19
The Radiant Periphery (1986)

Marcel Smets

Introduction

Guillaume Vanneste
UCLouvain – LOCI, Louvain-la-Neuve, Belgium
guillaume.vanneste@uclouvain.be

In 1986, almost ten years after the publication of his doctoral thesis on the Advent of Garden-Cities in Belgium,[1] *Marcel Smets, then a young professor of urban design at the Catholic University of Leuven, published the small work* La Belgique ou la Banlieue Radieuse *(Belgium or the Radiant Periphery) in the catalogue of an exhibition held at the* Fondation pour l'Architecture *in Brussels.*[2] *Following a growing interest in city renewal and city project in Europe*[3] *during the eighties and*

[1]Marcel Smets defended his Ph.D. at the Catholic University of Leuven in 1976 and published one year later. See Smets, M. (1977). *De ontwikkeling van de tuinwijkgedachte in België. Een overzicht van de Belgische volkswoningbouw 1830–1930.* P. Mardaga, Collection Architecture + Documents, Brussels-Liège, 224 p as well as the separate French edition: Smets, M. 1977. *L'avènement de la Cité-jardin en Belgique. Histoire de l'habitat social en Belgique de 1830 à 1930*, by the same publisher.

[2]Smets, M. (1986). *La Belgique ou la banlieue radieuse*. In M. Culot, & C. Mierop (Eds.), *Paysage d'architectures*, Bruxelles: Archives d'Architecture Moderne, pp. 33–35.

[3]About the context of urbanism in Belgium in the '80s, the exhibition catalogue *Resurgam*, published in 1985 witnesses the focus on the history of reconstruction of city centre in the aftermath of World War II. See Smets, M. (Ed.). (1985). *Resurgam. La reconstruction de la Belgique après 1914,* (catalogue of the exhibition), Brussel, Crédit communal de Belgique.

Furthermore, the editorial work done by Marcel Smets in 1983 for the review *Storia Urbana*, an analytical and exhaustive panorama of present research in urban history in Belgium, is notable and

M. Smets (✉)
KULeuven, Leuven, Belgium
e-mail: marcel.smets@kuleuven.be

M. Barcelloni Corte and P. Viganò (eds.), *The Horizontal Metropolis*,
https://doi.org/10.1007/978-3-030-56398-1_19

at the same time as the operational law and tools of planning—les plans de secteur *—were being elaborated in Belgium, this very concise and narrative pamphlet framed the situation from a new angle. Marcel Smets pictured for the first time the conditions of dispersion that had not yet been addressed by the discipline of urbanism in Belgium in such a clear way.*

In his delicate and precise style, Smets retraces the path of desurbanisation[4] *in Belgium, between historical stakes, economic or political processes and urban design theory concepts. Looking at the Belgian territory, urbanization appears as a landscape of individual houses disseminated in the countryside. Coining the allegory of "la Banlieue Radieuse" borrowed from Le Corbusier to describe such a territorial condition, Smets points out how modernism in Belgium always intended but never succeeded in planning and organizing urban dispersion.*

Starting from Benjamin Seebohm Rowntree[5] *(Chap.* 7 in this book) and Emile Vandervelde[6] (Chap. 6 *in this book), two intellectuals who studied the organization of labour and land at the turn of the twentieth century in Belgium, Marcel Smets shows how industrialization, inheritance, tradition, mobility and work organization have gradually forged an implicit political vision of a* radiant *future both relying on and resulting in the urbanization of the countryside. A* banlieue, *a peripheral situation or condition, which, as suggested in the title '*la Belgique ou' *would apply to the whole country itself.*[7] *At the same time, the urbanisation he describes, this peripheral condition, by enabling a kind of liberalism, individuality and autonomy based on the private property, with the dense city in the background, had paradoxically become the landscape of emancipation of a whole nation. In the words of Smets:*

> That's what led to the great explosion. Those who remained in the village were building town houses. Those who had abandoned the city began erecting country houses. In both cases, it came down to the same thing: the isolated abode multiplied countless-fold, representing the ideal habitat dreamed by everyone. The problem of urbanization was definitively limited to the establishment of an infrastructure capable of making this ever-growing market of private pavilions work. The great 'Belgian Suburb' was born.

---

pursues the debate opened by la Banlieue Radieuse three year after its publication. See De Meyer, R., & Smets, M. (1983). "Recenti studi sulla storia urbana in Belgio per il XIX secolo e l'inizio del XX". In *Belgio: studi di storia urbana tra '800 e '900. Storia Urbana n°22/23* (dir. Smets, M.), Milano, Franco Angeli Editore, pp. 8–55.

[4]The term *desurbanisation* is used at the end of *La Belgique ou la Banlieue Radieuse* («*le caractère impromptu, éclectique et libertin de la désurbanisation belge*») and can refer both to the urbanization of countryside rather than the city centre during the twentieth century, but also to the individual initiative generating this specific Belgian landscape that Smets compares here to other European contexts.

[5]Rowntree, B.S. (1910). *Land and labour: Lessons from Belgium.* London, Macmillan and Co.

[6]Vandervelde, E. (1900). *La propriété foncière en Belgique.* Paris: Schleicher frères éditeurs, Bibliothèque internationale des sciences sociologiques.

[7]One can note the similitude with the text of Jean-Jacques Rousseau making the analogy between city and country for Switzerland, "La Suisse entière n'est pour ainsi dire qu'une grande ville", see Rousseau, J.-J. 1853. *Œuvres complètes*, Paris.

Today, Marcel Smets' essay remains a milestone in the description of the twentieth century urbanization of Belgium; both in the author's genealogy, emerging as an explicit and logical continuation to his first works on Garden Cities and as an introduction to the future and numerous writings he would publish on the production of Belgian urbanization and urbanism; but also in the corpus of national bibliographic references as one of the early concerns about city renewal, contemporary urbanization, production of peripheral condition[8] *and metropolisation processes that would later inspire a genealogy of academics, practitioners and thinkers in urbanism, inside and outside Belgium.*

In this sense, many further writings of Smets can be seen as responsive essays deepening themes announced in the work of 1986 such as the enlarging of scale and frame of urbanisation or the dual dynamic of national collective infrastructure production and private individual housing production. Among his academic researches and his practice of urban design, Smets is constantly in search of an understanding of the evolution of urbanism as a theoretical and disciplinary field, in particular, in "Grid, Casco, Clearing and Montage" in 2002,[9] *he gives an analytical glance at design operations in contemporary urban projects dealing with the "uncertainty" of today's urban conditions. Operations are classified through the four figures of the Grid, the Casco, the Clearing and the Montage, synthetizing the way they interact with and modify the new horizontal metropolitan condition. Beyond the categories proposed by Smets, he made clear for the reader that "the end of the old antagonism between country and city has blurred the distinction between landscape, urbanism and architecture". Consequently, the essay "la Belgique ou la Banlieue Radieuse" can then also be revisited as a starting point of Smets' continuous and growing interest for landscape as a medium of design and for peripheral situations such as places in search for representation and design.*[10] *By constantly questioning and experimenting the urban production and design,*[11] *Smets and la Banlieue Radieuse oriented the disciplinary debate towards the enlarged horizons of a nebulous town and suburban landscape, keeping the adjective "radiant'as a compass for experimentation in the urban field of Belgium.*[12]

---

[8]Heynen, H., Loeckx, A., & Smets, M. (1990). The periphery: An exploratory study. In *Green paper on the urban environment, Office des publications officielles des Communautés européennes* (pp. 19–29). Luxemburg [translated as separate edition in every language of the European Community].

[9]Smets, M. (2002). Raster, Hüsle, Lichtung und Montage/Grid, Casco, Clearing and Montage. *Topos, European landscape magazine, 38.*

[10]For further reading: Smets, M., & Shannon, K. (2016). *The landscape of contemporary infrastructure*. Rotterdam: Nai Publishers.

[11] Smets, M. (2016). *De Stad Ontwerpen. Een leven tussen theorie en praktijk van stedenbouw* [Designing the City. A life between theory and practice of urbanism] (95p). Leuven: Dutch original, Lipsius Publisher.

[12] Smets, M. (1986). La Belgique ou la banlieue radieuse. In M. Culot, & C. Mierop (Eds.), *Paysage d'architectures, Bruxelles: Archives d'Architecture Moderne* (pp. 33–35).

See also Dehaene, M. (2013). *Tuinieren in het Stedelijk Veld/Gardening in the urban field.* Gent, A&S Books.

19.1 Belgium or The Radiant Periphery[13]

In the voluminous report he wrote in 1910 on the relations between land and labor,[14] Seebohm Rowntree pointed out that Belgium, in his opinion, was the country that had succeeded in creating a true balance between the city and the countryside. Rowntree explained this phenomenon by the density of the railway and roadway network, which had made it possible to avoid the monstrous immigration into industrial centers that had afflicted Germany and Great Britain. Instead of being engulfed by the dehumanized black holes of the urban neighborhoods, the Belgians had been able—again, in his opinion—to savor the countryside while working in the city.

As Emile Vandervelde wrote in 1903, the introduction of commuter trains had "fostered the most profound revolution yet seen in the organization of labor in Belgium in a quarter century".[15] Although "the initiators […] had no other concern beyond mobilizing a certain number of country dwellers in order to […] provide the industrialists with docile manpower with few demands",[16] they also made it possible for the population to disperse across the national territory. This Belgian inclination, which made it possible to resolve urban issues by developing regional-level infrastructure, was in some way the logical response to a specific problem. It corresponded to the economic and geographical situation of Belgium and coincided with the well-known individualism and strong sense of identity of its citizens.

In this tiny country where the provincial centers are separated by no more than a day's walk, industrialization took place, in effect, in a much more dispersed manner than in neighboring countries.

Even in difficult economic times, the allure of the big city did not necessarily dazzle the entire countryside, and to dissuade those who were getting ready to leave, the government had instituted the "emergency residence" law, which required that new immigrants must reside in the city for six to eight years before being eligible for public welfare.

From an economic and financial point of view, we note that even the big promotional companies—which would realize the improvements and urban road networks in the second half of the nineteenth century—adhered to the tradition of private ownership. They always gained their profit from selling land, without

[13]Source: Smets, M. (1986). *La Belgique ou la banlieue radieuse*. In M. Culot, & C. Mierop (Eds.). *Paysage d'architectures* (Catalogue published for the exhibition "Paysages d'Architecture" organized by the Fondation pour L'Architecture and The Musée des archives d'architecture moderne June 21 to September 20, 1986 at La Fondation pour L'Architecture), Bruxelles: Archives d'Architecture Moderne, pp. 33–35. Reprinted courtesy of M Smets and Archives d'Architecture Moderne. Reprinted courtesy of M Smets and Archives d'Architecture Moderne.

[14]Rowntree, B. S. (1910). *Land and labour: Lessons from Belgium*. London, Macmillan and Co.

[15]Vandervelde, E. (1903). *L'exode rural et le retour aux champs* (p. 143). Paris: Felix Alcan.

[16]*Ibid*., p. 203.

worrying about creating homogeneous developments. This tradition of the small parcel of land went hand in hand with the extreme subdivision of land and the great number of small businesses and investors. Thus was the city fashioned, by accumulation of privately owned plots of land arranged so as to create roads.

The law of 1889, which granted workers the right to borrow money to build or buy a house, placed the seal on this predilection for individual ownership. The example of the middle class, who were already ensconced in the tranquility of the inner suburbs, was the one to follow. In spite of the "clean-ups", the public authorities had not completely succeeded in rehabilitating the bad reputation of the inner city. Even the allure of society life was no longer enough to hold them in the center: the newly organized transportation network allowed them to continue to attend the events that attracted them; and, as opposed to what was seen in large metropolises such as Paris, London, or Vienna, the inhabitants of the "*beaux quartiers*" (nice neighborhoods) also moved out to the greener periphery.

This was the beginning of a total upheaval. Those who remained in the village were building town houses. Those who had abandoned the city were building country houses. In both cases, it came to the same thing: detached houses came to represent the ideal abode, everyone's dream house, and their numbers multiplied. The problem of urbanization was definitively limited to the installation of infrastructure capable of ensuring the vitality of this continually growing market for detached houses. The great "Belgian periphery" was born.

It must be said that this de-urbanization satisfied the needs of a great many people. It gave them a quality-price ratio that would have been unthinkable in the city with the high price of land; it led to the de-concentration hoped for by hygienists and fostered social obedience via the obligation to make monthly payments. It also made it possible to contain wages because of the lower cost of life in the countryside and the economic benefits of the vegetable gardens that workers were expected to have. Lastly, it sanctioned a clear social distinction among the inhabitants based on the model of the detached house. It was simply a matter of adapting the size of the plots, choosing the location of the subdivisions, and attending to the layout of the outdoor space according to the future envisaged for the neighborhood. The suburb of detached houses thus allowed typological-morphological planning and introduced an instrument for managing land values.

All of these elements—as well as the other advantages of this dispersion for road and bridge builders, architects, and vendors of household appliances—are found in other countries. But they do not explain the tenacity by which Belgian lawmakers defended and subsidized this access to private ownership after the first law in 1889. They also do not account for the particularity of this typically Belgian setting. We scarcely find here the hodgepodge of standardized prototypes that characterizes the French suburbs is scarcely found here. We rarely find the planning stratagems which, as is seen in England or the Netherlands, cover the landscape with gloomy, predictable structures. We do not feel the same burning desire to conform to the norm and follow the latest fad that has seized the new German suburbs.

While these few stereotypical images do not truly reflect the suburban development in the countries I have mentioned, they at least help to better define the impromptu, eclectic, and libertine character of the Belgian de-urbanization process. In this curious mix, where the extremes feed off the same naïve and generous candor, the ascetic rules of an imposed architecture are overcome by one's own convictions. The singular interpretation of the resulting personalized work fosters an amalgam of pragmatic sense, rites, and meanings that constituted the wealth of the city of the past.

This apparent disorder troubled modern architecture, which attempted to counter the indistinguishable ambiguity between freedom and oppression in the countryside with the clarity of unquestionable schemas. It refused to allow the essential joys that were the prerogative of the entire community to be privatized.

All the efforts made in Belgium since the end of WWI aiming toward coherent urban planning follow these lines. From the heroic garden-cities to the Heysel model city, from Victor Bourgeois's "New Brussels" project to René Braem's linear city, the same concerns dominate. They all seek to contain the megalomaniacal spread of the suburbs. They attempt to stop the parceling out of land by opening up space, and to counter the systematic cutting up of the land with collective membership in a shared place. They transform space to suit a new generation where each individual identity will find expression while integrating into the greater society.

We know what this utopian reformism has led to. In the initial phase, it slowly freed itself from the restrictions of the real project only to fall into the trap of a self-legitimizing technocracy. In a second phase, the formal packaging of rationalizing schemas was gutted of its poetic content. Lacking this innovational enthusiasm, it was applied with ruthlessness. The complexes that came of it were imposed on those who could not afford to plan their own place. The dream of an egalitarian society became an oppression.

Against this current, the model of the "radiant periphery" was maintained. It had resisted all attempts at restructuring and continued to regenerate itself from within. For architecture in Belgium, it remained the privileged realm of experimentation.

Fig. 19.1 "View and plan of a cottage in 1908 as an ideal suburban habitat".
Source: *L'Habitation à bon marché*. Supplément n°4. 1908. Republished in: Smets, M. (1977). *L'avénement de la cité-jardin en Belgique, Histoire de l'habitat social en Belgique de 1830 à 1930*. Bruxelles and Liège: Pierre Mardaga, p. 53.
"Individual housing after the 1889 law [...][Housing] is a prisoner of the individualistic image of autonomous housing. It goes without saying that we return to the village house or, even better, to the isolated villa in the suburbs, as a formal ideal. The model house presented by CGER at the 1910 World Exhibition, as the ideal of cheap housing, shows just how much this was the case: it was a typical suburban pavilion, built in a wood, and therefore devoid of any urban reality. This same ideal of housing also emerges from the standard plan of a cottage included in the Journal du petit propriétaire founded in 1908 under the title *L'Habitation à Bon Marché*" (ibid., pp. 49–54)

Fig. 19.2 Photo by Philippi, F. M. (1964). *Lifestyle and housing during les Trente Glorieuses in Belgium*. © 2019, ProLitteris, Zurich.
"After the Second World War, with the massive introduction of the car in the 1950s, the dwelling ideal continued to inspire while being enlarged in scale and dimension since, at that time, all places became very easily accessible. This dynamic produced- to a certain extent-a loss of the local "togetherness", a territorial condition having more to do with what Melvin Webber called the non place urban realm, which guaranteed complete accessibility of people and activities through a perpetual connection to the network"

Chapter 20
Cidade Território (The City Territory) (1986)

Nuno Portas

Introduction

Chiara Cavalieri
Université Catholique de Louvain, Louvain-la-Neuve, Belgium
chiara.cavalieri@uclouvain.be

In 1969, the Portuguese architect Fernando Tavora declared the urgency of studying the human environment framed within its maximum complexity, that of the City Territory.[1] *This new perspective, continues Tavora, "implies the destruction of that city/countryside dichotomy in favour of the appearance of a unique entity".*[2] *With these words, Tavora wrote the introduction to the book* A Cidade como Arquitectura *[The City as Architecture] published in 1969 by Nuno Portas. With the publication of this book, Portas announced himself as the first urbanist that in Portugal started to look beyond that dichotomy, seeking for new tools able to inform the project and the transformation of this new kind of city. Portas was largely influenced by the contemporary Italian debate: among others, he knew* The Architecture of the City *by Aldo Rossi*[3] *(to which the title "The City as Architecture" certainly refers) as also the concept of* città-territorio *first defined in 1962,*[4] *around which he developed his researches. From that moment, during the following dec-*

[1]Tavora F. (1969). Prefacio. In N. Portas (Ed.). *A cidade como arquitectura. Apontamentos de metodo e critica* (p. 10). Porto: Livros Horizontes.

[2]Ibid. p. 11.

[3]Rossi, A. (1966). *L'architettura della città*. Venice: Marsilio.

[4]See infra, Tafuri, M., Piccinato, G. and Quilici, V. (1962). La città territorio: verso una nuova dimensione. *Casabella continuità, 270*, pp. 16–25.

N. Portas (✉)
Oporto University, Porto, Portugal

M. Barcelloni Corte and P. Viganò (eds.), *The Horizontal Metropolis*,
https://doi.org/10.1007/978-3-030-56398-1_20

ades, Nuno Portas kept analysing, exploring and ultimately describing the notion of urban diffusion in the Portuguese context. The City Territory described in his book is a specific city that roots in a highly dense mobility system, which connects two types of poles: the city-poles *and the* territory-poles. *According to Portas, the city-poles host mainly tertiary activities and medium high density residential tissues, while the* territory-poles *attract those productive activities that can't be placed in dense areas, such as big industrial platforms expelled from the traditional city, or more recently urbanised intensive agricultural fields. While, since the late 1960s, he reflected on the need for "new planning" as well as representational tools for the city territory, later, in 1986 he started focusing on the Medio-Ave region territorial model,*[5] *and patiently analysed the specific socio-economic and historical conditions that led that particular context to become a diffuse city.*[6] *The first results of this research are set out in the following pages, in the description of the Medio-Ave region as a continuous urbanization of more than 250 administrative units, differently aggregated by small and medium-sized urban clusters. Via the extensive use of statistical analysis, Portas explains the enlarged distribution of urbanity as the result of: (1) a strong and diffuse industrial support (the textile and shoe production chain); (2) a temporary (seasonal complement for family income) exploitation of a fine-grain and small-scale agrarian structure; (3) a low grade of tertiary activities. This original mix developed relatively quickly within a territory whose form reflected the geography of a valley: a higher degree of urban density downstream—along the main mobility axes, and a sort of urbanised rurality*[7] *upstream, shaping the territory in strips of homogeneous landscape. The result was a* rur-urbanized *region irrigated by a dense network of roads (more than 3,000 km in length).*

In Portas' further work, the attempt to define a spatial model of development for the Portuguese territory has been combined with the one of defining possible tools for governing it: during a conference in Turin in 1986,[8] *Nuno Portas and Manuel Fernandes De Sa presented a* Plano Director Municipal *as a means for experimenting inter-municipal policies.*[9] *The main hypothesis of the Plan was that of rooting the territorial project into the existing mobility network, introducing the idea of using a* lattice logic[10] *for providing equal-accessibility and thus contrasting*

[5]Three municipalities situated along Portugal's northeastern seaboard.

[6]Nuno Portas. (1986). Modelo Territorial e Intervençao no Médio Ave. In *Sociedade e Territorio*, November 5th: pp. 8–13.

[7]Rui Braz Afonso. (1990). Un percorso nel diffuso. In *Urbanistica* 101, December, pp. 52–55.

[8]Portas, N., & de Sá, M. F. (1986). Images produced for the Talk "A diffused redistribution, A case study in the «Media Ave» Region Portugal". Dipartimento Interateneo Territorio, Politecnico di Torino.

[9]Portas, N., & de Sá, M. F. (1986). ibid.

The images accompanying the text in the following pages are those presented in Turin (courtesy of the author).

[10]Referring to the idea of *semilattice*, as defined by Christopher Alexander in Alexander C., 1965, "The City is not a tree". In Architectural Forum, Vol 122, No 1, April 1965, pp. 58–62 (Part I), Vol 122, No 2, May 1965, pp. 58–62 (Part II).

polarization and hierarchisation of big axes and nodes. After having presented his researches in Italy, Portas translated and published the 1986 article in Italian in the issue number 101 of Urbanistica, *the last one directed by Bernardo Secchi.*[11] *Probably it is not by chance that different researches on the European city territory are enclosed in the same issue, where André Corboz had the role of sketching a first periodisation of the 20th Century Urbanism: according to him in the early '90 s Urbanism was entering into the period of the* nebulosa urbana.[12] *From this perspective, the contribution of Portas could be two-fold. On the one hand, the combination of morphological and statistical analysis—while using both a prescriptive and predictive approach—depicts a local economy that reinforces the idea of* site-specific *against that of generic city, and that ultimately displays the Portuguese City Territory as a unique combination of productive chain with a—temporary—agricultural landscape developed thanks to the presence of a historically dense mobility network. On the other hand, in the same act of seeking for the adequate tools of governance for such a new—both for size and density—territorial object, Portas discloses a sort of incremental dimension of this specific urbanisation process, whereas individual choices are suggested as a possible path for political choices.*

20.1 Territorial Model and Intervention in the Middle Ave Area[13]

1. In this short article, I will try to identify, albeit in a necessarily simplified form, the model of the territorial structure of a region not much studied as the Middle Ave despite its demographic and productive importance, as a cognitive condition to explain its dynamic so that we know, consequently, how we can act successfully upon it, from a given political orientation. In fact, more than in-depth knowledge, this region is often the subject of simplistic attitudes of acceptance or rejection.

   As known, a model is more than an orderly description of indicators, in the same manner as a structure, more than the sum of its parts, privileges its relations of interdependence. A territorial model, in this case, privileges the spatial relations between the activities and the way they are marked or leave a mark in the

[11]Portas N., De Sa M. F., Afonso R. B. (1990). "Urbanizzazione Diffusa nel Nord del Portogallo". In *Urbanistica* 101, December, pp. 38–51.

[12]Corboz A. (1990). "L'Urbanistica del XX secolo: un bilancio". In *Urbanistica* 101, December, pp. 7–12.

[13]Source: Portas, N. (1986). "Modelo Territorial e Intervenção no Médio Ave" ["Territorial Model and Intervention in the Middle Ave Area"]. In *Sociedade e Território* – 5, pp. 8–13. Porto: Edições Afrontamento. Reprinted courtesy of N Portas

territory—this last one understood as a resource, as a space–time, as an environment and also as a significant.

This model characterization task is even more indispensable the more the organization of the "region" in question distinguishes itself from the population settlement paradigms that usually serve as a referential to current theories and policies. Preconceived ideas, often misaligned with its original referent, due to a lack of confrontation with the dynamics of reality may become disastrous—namely, if the referent is not homologous to the one that served as a referential basis to those same preconceived ideas.

We begin to realize the singularity of the process and of the forms of settlement patterns of the littoral-north of the country throughout the work developed with Municipalities that are particularly representative of this "difference", and where the failure of the analysis criteria and planning and management options that were being adopted during the last two decades was quite evident, despite the will of local politicians and the quality of the technicians involved.

Given the evidence of that maladjustment, it was necessary to move away from the attitude that is dear to planning technicians, that it is the world that is wrong in face of the model of what is good, and that it is not the theoretical framework of technicians that does not adjust to reality. One sought to understand, then, whether this little "chaos" hides a consistent logic, and if and under what conditions the action to be taken will reinforce or inflect it.

It is true that the available information about the activities and its spaces—despite numerous studies related to plans of various nature—is not as significant as desirable in the case of the littoral-north and of the middle Ave in particular, comparing to what happens towards other areas of the country, where the descriptive and explanatory power of existing theories is better suited to the reality studied (or vice versa). Thus, we continue to speak, for example, about a Porto's "metropolitan area" when we should consider instead an urban region just relatively polarized and extensively rur-urbanised; it is common to address the littoral-northern region as a classical network of urban centres, missing out in that grid a true conurbation that includes, but not exclusively, its traditional centres (on the contrary, its major dynamics occur "between" those centres).

Besides the insufficiency of information—or, rather, its inadequacy to highlight the characteristic differences of this reality—there are the sufficiency and the prejudices of the experts that are called to explain and to deal with it. I could ironically say that planners' training is notoriously manichaeistic (the concentration of industry, of habitat… is good; dispersion is socially and economically bad… or vice versa), because it privileges certain variables, despising others. And if the planner is an architect, a mental reservation is added: the architecture of dispersion is not seen or is not drawn according to the terms that define the tradition of the discipline, which is based on the monument and on the compact city. Furthermore, to talk about the "city" is to talk about centre and concentration of buildings, of something that must have a defined perimeter, that one can be able to limit by green or black belts—especially when it is forgotten that the city par excellence, Athens, was not like this, but that it was quite well dispersed…

And if it comes to economists, agronomists, ecologists, sanitation or transport engineers, they are not exempt from such a priori: an urbanisation like this will be classified as an archaic structure unsuitable for modernization, a waste of natural resources, non-economical systems… in one word, dispersion is chaos—and the gods (as well as planners) abhor chaos.

Also, the proof by the evaluation of the outcomes that result from the application of conventional recipes to realities that reject them is not done: spatial planning is an activity practiced without an assumed "feed-back".

Despite this, some teams have been trying, from works developed with Municipalities of these regions or for research purposes, to characterize the territorial model and to define strategies that are different from those that were practiced in plans of the 60 s and 70 s. The present notes result from this work and assume a simplification that can only be excused if the statistical indicators and the cartographic schemes required by a deeper analysis are published elsewhere and by specialists.

2. The territorial model of the sub-region that has been called the "Ave's Valley" or "Middle Ave" (none of these designations can be taken in their literal sense) is formed as an urbanised continuum of over 250 civil parishes, of half a dozen municipalities, including small and medium sized urban centres and several thousands of extremely small size laid out along an extensive network of national, municipal and local roads that is probably over three thousand kilometres… With two urban centres generously estimated to be of between 40 and 80 thousand inhabitants, the rest of the borough capitals have less than 15,000 or even 10,000 people—which means that over half of the (almost) half a million inhabitants that populate the sub-region lives between the conventional urban centres.

   To be a little more concrete, take the example of the most paradigmatic municipality in what concerns dispersion: Guimarães. It is worth to verify that according to the last census one fifth of the households were classified as "isolated", while 40 in 100 lived in statistical places with less than 200 households, and 75 in 100 lived in settlements with less than 500… In fact, in a Municipality that reaches 150 000 inhabitants, only 17% of them live in settlements with more than 10,000 inhabitants (i.e., in the only city, properly saying)—and areas with more than 500 inhabitants and less than 10,000 are almost absent. This gives us a sufficient expressive picture of the dispersion. And if instead of the number of residents we took the workplaces of the active population, we would have, similarly, a widespread distribution all over the territory.

   Finally, if we overlay some indicators related to the dynamics of occupation of this territory over a static picture of it—note, for example, the construction of

detached housings, which over the last decade has got an annual growth of 3–4% (about 1000 dwellings/year), or the establishment of productive units about which we do not have reliable data—one reinforces the idea that the dynamic of the activities follows the same pattern of dispersion, not having reinforced the historical, more concentrated, urban centres.
The remaining municipalities do not present so extreme numbers, namely Braga (that owes its concentration mainly to the weight of services) or Santo Tirso, although Famalicão is close to Guimarães concerning these indicators.

3. Searching for the socioeconomic foundations that may explain, at least in part, such a peculiar spatial pattern, I will point out the following factors or components as mere hypothesis:

   (a) A strong industrial basis that occupies more than half of the active population, with an overwhelming predominance of one sector (textile-clothing-manufacture and footwear) [TCM + F] already traditionally established in a dispersed way due to the location near the watercourses for the textile and leather industry, and to the availing of the extensive road network in the case of the units associated to manufacture and footwear; the second most important sector, that of construction, does not also favour polarization.
   (b) An agrarian structure of variable size but in which the smallholdings predominate, often farmed on a part-time basis and as a supplement for households' resources, proceeding from the industry or other activity.
   (c) A significant rate of pluriactivity among the active population, which has been allowing the majority of families to reach reasonable levels of consumption, despite the low wages paid at the textile-clothing-manufacture and footwear industry and the also low agricultural productivity.
   (d) A low degree of concentrated tertiary activity if we consider the great importance of the productive sector, what lead us to the assumption that this activity is based outside the sub-region and, to a lesser extent, disseminated within the own facilities of larger companies.
   These four vectors would accentuate the already high dispersion of human settlements—characteristic of the pre-industrial settlement in the region—if we consider the dynamic that each one of them seems to show and, mainly, the interdependence relations between such dynamics:

      (a) Manufacture, a component of TCM—organized in small and medium-sized enterprises, formal and informal ones, resulting from the segmentation of supply chains and sustained by local networks of subcontracting—may prescind of watercourses, and to minimize the cost of the main roads' land, may also privilege the proximity towards residential areas, even smaller ones, where they can reach a familiar labour force. As this labour force is mostly feminine, it tends to look for shorter distances between factory-house-ateliers, not forgetting that the latter two may coincide.

(b) Part-time agriculture, as a complement or as a form of subsistence, demands contiguity between house and field; correlating this fact with the previous, if the trend to pluriactivity practices remain we will get the shrinking of the distances between the vertices of the triangle factory-house-field.
(c) The low tertiarization does not generate the preferential growth of historical urban cores in the region (with the exception of Braga) but rather of external administrative centers despite of the foreseeable progress of communications. Such stagnation, in turn, does not attract innovative activities, neither for production (in what concerns design, for example) nor for the cultural environment of the region.

4. It is not easy to quantify the intensity of those factors and their interactions, given the very nature of the dispersion, the presumable submersion of countless production units and the situation of tolerated illegality of a substantial part of the recent housing stock.
Nevertheless, it can be anticipated that the model has not only consolidated over the last two decades, as it is likely to be strengthened in the near future if we consider the combined effects of some of the predictable trends that strongly determine it, such as:

(a) The increasing segmentation of the production cycle especially in the TCM, and the transference of the labour force of conventional companies that are passing by a process of technological conversion to small units, formal or not.
(b) The subdivision of agricultural property due to inheritance sharing processes or its selling for construction or domestic farming, being likely that food sufficiency remains as a counterpart of the low wages' standard paid in industry, indirectly acting as one of its competitive factors when facing the New Industrialized Countries.
(c) The spread out of part of the tertiary sector driven by the development and implementation of telematics support near industrial facilities in conversion, to take advantage of the proximity to workers' residences.
(d) The propensity for construction by individual initiative with or without public support, and by means of various forms of self-construction or phased construction, in spots chosen by the owners themselves for sentimental reasons or by the contiguity with fields or other working places.
We can argue, without falling into fiction, that technological innovation favours the deconcentration of activities, changing to a lesser or greater degree the location factors which were dominant during the phase of industrialization.

5. Thus, the territorial model resulting from the tendencies pointed out above is characterized not only by a strong dispersion of the urbanisation, but also by the mixture of activities on the most capillary level, probably consolidating long rows or strips that are somewhat advanced in their formation, cutting through the

territory and skirting the meadows and hills of an irregular orography. It is difficult to foresee whether this movement neutralizes or not the growth polarized from the urban traditional centres, once that in a region with a strong demographic dynamics it will depend above all on the development of the administrative and commercial tertiary sector, and also on the capacity of consolidating the academic activity and research and design activities applied to production.
In any case, it is a model that results from the spontaneous formation and accumulation of acts of very individualized (and individualistic) agents with a powerful logic that, beyond the mentioned economic foundations, shows a very rooted cultural component that informs the behaviours of the population and of entrepreneurs concerning their current options, such as choosing a place to build up a house or to open a workshop, sending their children to school, setting up a shop, or, being a local elected representative, to open one more road.
The scale or space of decision of the agents is inevitably "micro" (as it is proper to this kind of model), dragging the very own action of the planning services and local instances of public investment decision. A micro scale that manifests itself not only on the extreme atomization of urbanistic agents, but also on the limited space whereon they try to solve their problems.
Economic basis and cultural patterns reinforce each other and they have been resisting to the fanciful attempts of introducing in the region, by conviction or by authority, different forms of organization of life and of production. At least until inhabitants and entrepreneurs realize that their own interests will be adversely affected in the medium-term, if behaviours that tend to reinforce the most "negative" or counterproductive traces of the consolidated model persist, by law or by transgression.
A consolidation which is clearly visible in the territory when one analyzes it by looking at the adequate geographical scales. In effect, what only seems disorder or chaos presents a logic of urbanisation or rur-urbanisation that must be understood before any evaluation, and must be evaluated with a minimum of common sense if we wish to find the means to correct the trends that, on behalf of majority interests, should be rectified. That logic has no correspondence in a formal reading of the landscape, it is a fact, especially when it is seen from the outside, at the distance.

6. The territorial logic we have alluded to consists of a partial linear-like organization of the dispersion and the mixture of activities that could be called urban. The potentially urban strips or rows—or where industries, facilities and houses are more likely to appear—twist throughout the territory in search of the perimeter of maximum contact with the fields, being supported on an increasingly tight network of roads, just as they had previously been supported on the watercourses from which they are now becoming able to release. When speaking of tendencies, it is important to note that these strips of urbanisation do not systematically cut through the meadows that are most productive in terms of agriculture, but rather prefer to choose the slopes, thus surrounding the best land.

They therefore use a principle that could be stated as that of the gradual use of the perimeter, forming something like irregular "blocks" that are irregularly filled according to the orography and the division of the property. Principles that after all are homologous to that which presided over the post-industrial expansions (clear in the radial growth of the city of Porto, to not use a far off example) and by which means in the peripheral situations the rural roads are occupied as "streets" with single family houses or premises for other purposes, allowing a penetration, among them, of kitchen-gardens or allotments up to the most central areas of the city.

The exploration of this principle of the urbanised perimeter appears as the logical solution to solve the conflict in the uses of the land which, at first sight, emerges from the tendencies that we pointed out at the beginning. From the agricultural point of view it would thus be an error to set the natural model against that of polarized and hierarchy-based concentration, which, as is known, also does not avoid competition of the urban and rural land rates around the urban outskirts, with loss for agriculture.

It happens, however, that this spontaneous model is not only contradicted by respectable technicians, defenders of natural resources, even by doing it from an inappropriate counter-model—it is also thwarted by spurious forms of real estate development or simply by land ownership interests of the local agents, that not only have been distorting its general logic, as have been turning into the major ecologic predators in the region. We refer, for example, to the private urban developments of appreciable dimension, with minimum sized urban lots for single-family detached houses, implying the opening of new streets supported on the main roads to invade the best fields or to climb up the hills once forested. As a result from the appearance of new sorts of demand—that of the emigrant who only sought to build a house near his birthplace or that of the qualified blue-collar worker of a big company—the business of urban lots in rural areas has spread, the cheaper the tract of thicket land ["bouça", in Portuguese] and minimal the requirements established by municipal authorities. Curiously, this kind of product faced lesser obstacles raised by official authorities (as long as the soil was not under any legal agricultural restriction) than the random construction of houses, even with larger plots of land, on the verge of existent roadsides.

Criteria or prejudices…

Besides from the fact that the residents of these private urban developments stay irremediably apart from the fields and become more distant from the companies, there was an increase in costs to municipalities towards intensive urbanisation forms, which suppose sanitation networks and conventional transport systems that cannot be ensured so soon, and that a linear dispersion could, under certain conditions, spare.

Another perversion of the model refers to the irrationality of the recent location of industrial units of appreciable size, not so much because they disperse but

rather because how they disperse, often sacrificing accessibility for a short-term gain with hilltop or hillside land plots, situated away from the main roads.
In a region with an already dense road network in poor conditions of design and conservation, whose improvement imposes on the Municipalities as a priority even over sanitation and public facilities given that this road network supports the intense economic activity, the trends reported deepen the needs, making it difficult to find the adequate answers to counteract them (such answers cannot be reduced to the refusal of licensing; nevertheless, they clash with the difficulties of supplying well located soil on cost terms that are able to compete with the "spontaneous" supply).

7. Being out of the question the proposal of a "counter-model" within the region's municipalities urbanistic policies, as the comparative advantages for much of the population and economic agents contribute towards strengthening the logic of dispersion/mixing of activities, it is nevertheless clear that those benefits suppose sacrifices, at least temporary, in relation to certain "standards" or services which may not have uniform coverage throughout the extent of urbanisation. Not least since we don't seem to be able to assess the financial or operational viability of implementing an alternative model, that would assure more satisfactory results.

The challenge is to find the institutional or technical forms adjusted to the existing system:

- Concerning transportation, sanitation, soil supply, location of public facilities, urban art, etc.—instead of envision solutions that may seem more rational from certain points of view, but that are unfeasible or undesirable in practical terms, when seen together with the other vectors.
  This is, however, the most common attitude amongst technicians:
- To isolate agriculture to postulate reparcelling as a condition for production profitability (but forgetting the part-time occupational regime, the pluriactivity or the subsistence farming);
- To isolate services to postulate the hierarchization of urban centres according to aprioristic levels of equipments or central functions (but forgetting that the majority of population and employment are dispersed over a few hundred civil parishes);
- To isolate the road network to postulate its speed and disengagement from marginal buildings, reducing it to the function of connecting the main urban centres (but forgetting that production activity and the demand for transport are mostly found between and not in the conventional centres, increasingly lacking more of an urban kind of connections at a capillary level—"street-connections", although located in the middle of the countryside);
- To isolate basic sanitation to postulate minimum thresholds of agglomerations' concentration in order to take advantage of economies of scale (but forgetting that the urbanisation, dispersed, does not move to where it should concentrate

and that, on the contrary, it is possible to rationalize multiple solutions of self-supplied sanitation);

- To isolate housing to postulate its arrangement in relatively compact neighbourhoods in the main agglomerations (but forgetting that such option may suit a demand occupied in tertiary services, of weak growth, but does not respond to the concrete needs of the majority, who demands proximity to workplaces and furthermore a kitchen-garden, not to mention the easing of self-construction or incremental construction…)
- To isolate industry to postulate its concentration in private urban developments or industrial parks (but forgetting that only part of the production units has interest in that solution, while another part, the most atomized and dependent of a chosen workforce, can disseminate and mingle with residential areas, avoiding transportation);

and so on…

In fact, all these examples of "rationality" could be relevant if they could be seen per se and if, above all, a physical, economical and social structure didn't already exist in the territory, structured in such a way that it rejects such therapies, which contribute to one other model, from the outset. Because the system of its sectoral logics is different, and also because the vast installed fixed capital shall not be moved, buried or replaced by following a political decision or due to the virtues of a plan, and much less in an acute phase of international competition and fiscal crisis. Thus, the efforts shall be oriented towards another direction: aiming to reinforce the installed model, by persistently correcting, in a participatory way, its deviations or perversions.

Nevertheless, this challenge finds the technicians particularly ill-prepared: with preconceived "solutions" but having evident difficulties at understanding the specificities and the differences of the reality they have before them. And one of the classic mistakes in these situations has to do with the voluntarism of "spatial planners" who arrogantly or naively tend to counter propose alternative schemes of socioeconomic and cultural organization based on the physical settlement of activities—in any case, not being able to affect more than the marginal additions of activities which can occur during the short life of a plan.

But it also finds the institutional organization non adapted to identify and manage the viable solutions. In the case of this region, not only one verifies the incapacity of the central government to find the appropriate measures for the productive sectors—despite its strategic importance in country's economy—as, once the local governments suffer from an evident handicap concerning the local finance distribution, they have been finding great difficulties in cooperating on joint projects, and each Municipality suffers from chronic shortcomings concerning their own technical capability to define projects and to materialize them in multiple fronts, as the ones imposed by the territorial model.

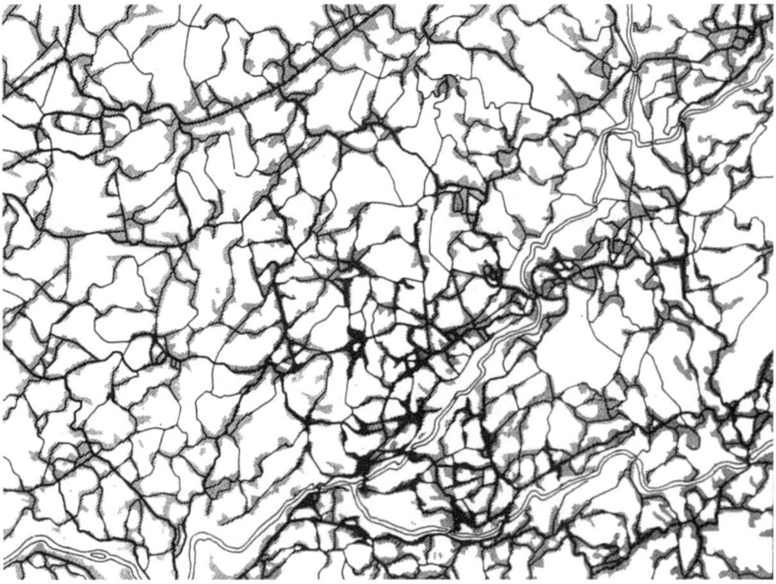

Fig. 20.1 "The intersection between the agricultural land and the diffused urbanization in an aerial view."
Source: Portas, N. de Sa, M. F. Afonso, R. B. (1990), "Modello territoriale e intervento urbanistico nella regione del Medio Ave". In *Urbanistica*, 101, p. 41. Reprinted courtesy of INU - INU Edizioni, Roma

Fig. 20.2 "The road network and the scattered buildings to the north-east of Santo Tirso, between the villages of Barrio, Riba de Ave and Selho".
Source: ibid. Reprinted courtesy of INU - INU Edizioni, Roma

Fig. 20.3 "Land use scheme (fragment)"
Source: Portas, N. de Sá, M. F. (1986). A diffused redistribution, A case study in the « Media Ave » Region Portugal. Images for the Talk given at the Dipartimento Interateneo Territorio, Politecnico di Torino Reprinted courtesy of N. Portas

"[…]we consider zoning as a delimitation of surfaces for continuous management without defining or dimensioning from the beginning the activities or levels and range of services. We consider three classes of areas […]:

(a) Urban areas, already consolidated by building density and mixture of activities, when we try to reinforce urban development.

(b) Agriculture and forest conservation areas where building and subdivision of property is forbidden.

(c) The remaining municipal territory not included in (a) or (b) is controlled through a set of negotiable regulations defined according to building and economic agents and to the land offer observed in (a)." (ibid., pp. 9–11)

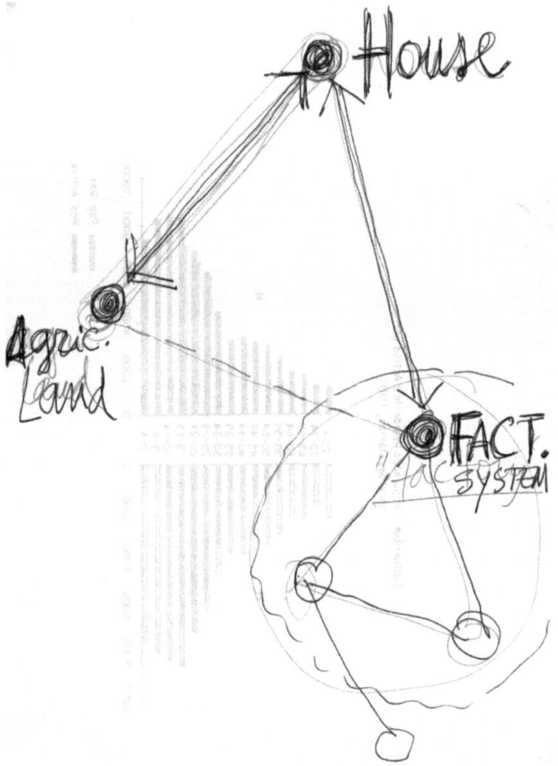

Fig. 20.4 "no title" Source: Portas, N. de Sá, M. F. (1986). Image for the Talk "A diffused redistribution. A case study in the « Media Ave » Region Portugal". Dipartimento Inter-ateneo Territorio, Politecnico di Torino. Reprinted courtesy of N. Portas
"74% of the active population works on local industry, mainly textiles, shoes and clothes manufacturing, but, at the same time, they run small agricultural exploitations that have an important economic role in terms of family subsistence, and also correspond to cultural needs because most of these industrial workers were born peasants and have small ownerships." Source: ibid., pp. 3 and 5

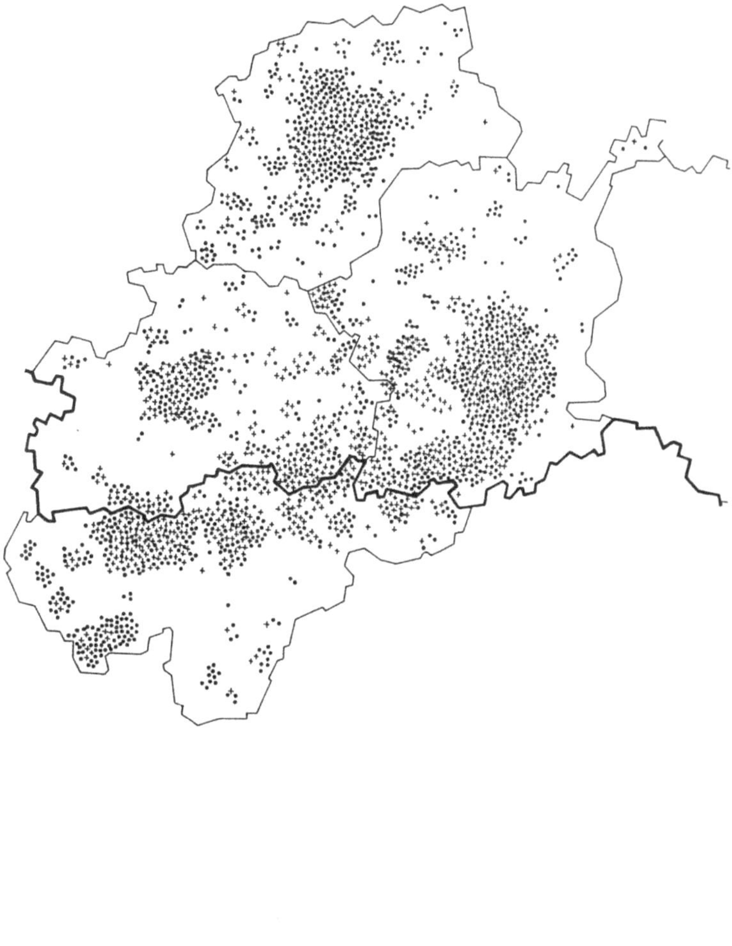

Fig. 20.5 "Location of the industrial units with over twenty employees in the area under consideration. Crosses indicate textile and clothing industries, while black dots indicate other production units." Source: ibid., p. 43. Reprinted courtesy of INU–INU Edizioni, Roma.
"This diffusion model of settlement is accompanied by the diffusion of industry that initially attracted the housing location and now follows the created trend." Source: ibid., p. 5

Chapter 21
Tapijtmetropool (The Patchwork Metropolis), 1989

Willem Jan Neutelings

Introduction

Carlo Pisano
Università degli Studi di Firenze, Florence, Italy
carlo.pisano@unifi.it

Since the early twentieth century and the beginning of the epic of modernism, the Netherlands was assumed as a global planning reference point, being one of the most planned[1] *and one of the most advanced welfare states in Europe, characterised by a pattern of dense urban centres placed in a pristine rural setting. Through many contributions and insights, ranging from the "moral geography"—term coined by Schama*[2] *to connect the long and laborious struggle against the water with the moral inspiration of the Dutch citizenry—till the enthusiastic position of Peter Hall*[3]*—that in 1966 raised the Dutch urban layout as the "right model" for the still-growing world cities—the Netherlands was elevated as a planning benchmark, place of pilgrimage and study trips destination.*[4]

[1]See Dutt, A. K., & Costa, F. J. (1985). *Public planning in the Netherlands: Perspectives and change since the second war*. Oxford: Claredon Press.

[2]See Schama, S. (1988). *The embarrassment of riches: An interpretation of Dutch culture in the golden age*. New York: Random House.

[3]See Hall, P. (1966). *The world cities*. London: World University Library.

[4]See: Astengo, G. (1949). "La lezione urbanistica di Amsterdam. Formazione storica e nuovi ampliamenti". *Urbanistica* 2. Astengo, G. 1966. 'voce Urbanistica'. *Enciclopedia Universale dell'Arte*, vol. XIV, Istituto per la collaborazione culturale. Pisano C. (2018) *Patchwork Metropolis. Progetto di città contemporanea*. Lettera Ventidue, Siracusa. Viganò P. 2018. "Città

W. J. Neutelings (✉)
Rotterdam, The Netherlands
e-mail: info@neutelings-riedijk.com

M. Barcelloni Corte and P. Viganò (eds.), *The Horizontal Metropolis*,
https://doi.org/10.1007/978-3-030-56398-1_21

This context was the sacred field in which in 1989 the young Dutch architect Willem Jan Neutelings, who had just left the Office for Metropolitan Architecture, was called to develop a project for the area of Ypenburg. The site is clearly located just in between two urban areas—The Hague and Delft—but, at the same time, is also part of that complex territorial construction known as the Randstad, where the explosive growth of urban or suburban phenomena has led to a singular blurring of the distinction between the city and the countryside. In this context Neutelings proposed his reinterpretation of the urban form called De Tapijtmetropool *or Patchwork Metropolis.*[5]

Quite often, the conceptualization of a territory, the way in which the ideal link between spatial and socio-political organization is interpreted or even trivialized, diverges from its real condition or its trajectory and we can say, from a general perspective, that the more this territory is idealised and raised to a mythical construction, the more this distance grows.

From this point of view the shocking effect produced by the publication of the Neutelings' manifesto is easily understandable. A few images and a short text were able to discredit the official planning interpretation of the Randstad, depicting its southern part as a continuous field of patches reaching from the North Sea to the Nieuwe Maas river, in which not only the 'periphery'—or the area usually labelled as such—but the inner cities, agricultural areas and residential districts were also transformed into a series of juxtaposed patterns.

Today, the re-reading of Neutelings' work turns out to be an interesting exercise of anachronism, for the Patchwork Metropolis should be considered not simply a sharp critique developed by the "Enlightened Prince of Suburbia",[6] *willing to break up the blacktop of the traditional planning idealism with the jack-hummers of Koolhaasian realism,*[7] *but also a radical abstract model proposing those principles that would rule the territorial transformations in the following neoliberal environment. The late eighties could be interpreted, in fact, as the transition moment where the social democratic Dutch welfare state was replaced by the first neoliberal politics. This passage became visible especially with the supplement of the fourth policy of spatial planning extra, released in 1991 with the acronym "Vinex", often regarded as the starting point of the* post regulative era.[8]

---

moderna e città contemporanea: la Patchwork Metropolis". In Pisano C. (2018). *Patchwork Metropolis. Progetto di città contemporanea.* Lettera Ventidue, Siracusa.

[5]The project has been published with a double title, *De Tapijtmetropool in dutch* and *Patchwork Metropolis* in English, which literally means *The Carpet Metropolis.*

[6]The title of "Enlightened prince of suburbia" was coined by the jury of the Rotterdam-Maaskant prize for the Young Architect in 1991. See Rodermond, J., Uytenhaak, R., & Weeber, C. (1991). 'Jury's report.' In *Willem Jan Neutelings architect.* Neutelings, Willem Jan. Rotterdam: Uitgeverij 010.

[7]See Koolhaas, R., & Mau, B. (1995). *S M L XL* (pp. 958–971 and pp. 958–971). New York: Monacelli Press.

[8]See Balz, V. (2017). *Regional design: Discretionary approaches to regional planning in The Netherlands.* In *Planning Theory, 17*(3), pp. 332–354. https://doi.org/10.1177/1473095217721280.

The way and the scale at which Neutelings looked and represented the territory is extremely similar to the way big companies and big real estate enterprises would have also looked at it in the following decades, developing new housing expansions called, like the mother policy, Vinex. The patches—described by Neutelings as fixed entities, designed and constructed once and for all—have then become the standard formula to implement urbanisation, able to turn the Randstad's lack of coherence into its planning solution—into a breeding ground of prospective projects.

By describing a segmented metropolis, built on variously aggregated patches, Neutelings highlighted many preeminent topics of modernity—the figure of the fragment, the issue of the peripheral condition and the contemporary city's territorial layout—without proposing a "model" for design, a set of guidelines, or an escape route for neoliberal urbanism, but rather establishing, at the most abstract and therefore most general level, the coordinates and dimensions of the design itself, thus defining new limits and new boundaries of the discipline.

21.1 The Transformation of the Southern Periphery of the Hague (from: *De transformatie van de Haagse zuidrand*)[9]

The Randstad as spatial and programmatic Patchwork

Classical urban planning confines itself to the use of two basic elements: green (countryside) and red (built-up area). This reduces the problem of urban expansion to the question of what patches of green can be made red. This view gives rise to an absurd notion of a polarity between city and countryside, a romantic opposition between an arcadian paradise and a megalomaniac metropolis, a red stain sprawling into an endless sea of green.

Unquestionably, this model is no longer adequate for interpreting such a complex system as today's Randstad. A 20-min trip will take a Randstad resident through futuristic oil refineries and brutalist universities, green garden cities and medieval canals, marshy ponds and North Sea beaches, high-tech business parks and furniture megastores, expanses of greenhouses and wide Ruysdael panoramas,

[9]Original title: Urban development study for the Urban Development and Real Estate Department of the Municipality of The Hague, as part of the 1988–89 urban plan.

Source of the text: Neutelings, W. J. Sulsters, W., van Wesemael, P. and Winkler, E. (1989), "De transformatie van de Haagse zuidrand, Stedebouwkundige studie verricht in opdracht van de Dienst Stadsontwikkeling en Grondzaken van de gemeente 's G ravenhage, in het kader van het stadsplan—1988–89 In: Geurtsen, R. Verschuren, P. Kwakkenbos, G. Stadsontwerp in 's-Gravenhage: stedebouwkundig ontwerpen¡. Delft: Faculteit der Bouwkunde, pp. 80–95. Reprinted courtesy of WJ Neutelings rubbish tips, dune reserves, markets, mosques and monumental city squares, opulent mansions and colourful bulb fields.

The Randstad is therefore better understood as a patchwork of spatial, programmatic fragments. In this patchwork metropolis, the opposition between city and landscape is abolished: there is simply one continuous landscape of patterns. Each location in the patchwork has its own specific programme and spatial pattern. It is no longer possible to simply expand: all we can do is to transform existing fragments into new ones.

This model of the Randstad is the spatial reflection of society: not a uniform model, but a patchwork of cultures, races, styles and lifestyles, where the cacophony of ideologies leaves no room for trend-setting avant-gardes. It is a chaos that produces a rich variety of intense experiences, thereby creating the highest quality.

The Southern Periphery of the Hague as Part of the Patchwork Metropolis

In the vision of the patchwork metropolis, The Hague no longer exists as a city, but is simply a collage of fragmented 'Hague' pieces, components of a patchwork of fragments that stretch from the North Sea to the Nieuwe Maas river. Its inhabitants assemble their city like a construction kit. They work in Rijswijk, sleep in Zoetermeer, buy their furniture in Sliedrecht and spend the weekend on the beach at Scheveningen.

The fragments in this patchwork operate in an equilibrium of changing patterns, with no identifiable centre, like stars in the firmament. They are subjected to a forcefield of political, geographical, economic, historical and cultural influences, which give each function its own place in the patchwork.

The Hague's southern periphery cannot therefore be developed as an extension of The Hague, nor as a connection with Zoetermeer or an annexation of Nootdorp. It can only be developed as part of an overall vision for the development of the entire area.

We can conceive of all manner of growth models: concentric expansion, a linear city as far as Rotterdam, a 'finger-city' structure that includes Delft and Zoetermeer, coastal reclamation, a twin city of Delft-Rijnacker-Zoetermeer. However, theoretical models of this kind fail to take account of the existing situation. After all, urban core areas do not occupy an empty landscape, but a completely filled-in patchwork metropolis.

The patchwork itself is the development model, the model of a field undergoing permanent transformation. This doesn't mean that it no longer matters where or how something should be planned. The art is to place new programmes in the existing forcefield in such a way that the equilibrium of the overall patchwork is enhanced—in brief, to come up with a workable transformation model.

Features of the Hague Collage

Fragmentation is a theme in The Hague. The city does not possess any strong structural elements, such as Amsterdam's ring of concentric canals or the development along the Maas in Rotterdam. It is a collage of powerful spaces. In the old city centre, these are the green spaces (*groene kamers*), avenues and squares that serve as city landmarks (Lange Voorhout, Hofvijver, Plein 1813, Nieuwe Parklaan, Malieveld and Haagse Bos). In the new city districts and the periphery, it is the compositions of imposing buildings that form fragments of the city scape (Cultuurplein, Congresgebied, Scheveningen Boulevard, Utrechtsebaan, Plaspoelpolder). These characteristic fragments have a strong linear zone, more or less parallel to the coast. This programmatic layering is a typical feature of the collage structure of The Hague.

A system of connections runs inland, perpendicular to the coast. These are thoroughfare boulevards, which traverse the programmatic layers along defining routes in an almost literal cross-section of the city.

The central amenities in the Hague collage will be shifted outwards. As there is no more room for new large-scale functions in the old city centre, space will be sought elsewhere. These new functions will be grouped into specialized zones, some towards the coast, such as the Congresgebouw area and Scheveningen, but mainly inland, such as the Spuigebied, the Utrechtsebaan and the Schenkstrook.

At a subsequent stage, a peripheral centre zone will be created along the motorways of The Hague's southern periphery. Through its strategic location at the heart of The Hague's fragment patchwork, this location has the potential to develop into an important centre. Favourably located between the old centre and the suburbs, this is the last large site in the conurbation where large-scale programmes can be developed.

Thus, the southern periphery not only becomes an important centre in The Hague collage, but it is also a link in the national service zone that is evolving along the urban motorways of the city's periphery. These ring zones form ribbons of amenities for mass culture, with their own programmatic, architectural and planning laws.

An Analysis of the Southern Periphery of the Hague

Topographical map

The Hague's southern periphery is a zone that contains typical features of a modern urban fringe:

- an arcadian idyll of allotments, sports fields, motocross grounds and recreational parks
- a complex infrastructure of traffic junctions, motorways and marshalling yards

- the loose placement of large-scale elements such as airports, business parks and distribution centres
- a contrasting aesthetic of speed, flowing lines, vistas and skylines
- a symbolic function as an entrance to the city.

Mass map

The Hague's southern periphery is a single coherent area between ragged urban fringes, from the Westland to the Green Heart of the Randstad. It is a continuous zone that can only be developed by means of a carefully planned, comprehensive model: any ill-considered element will work like a dam in a river, causing an unrestrained silting-up of the flow.

Political map

The Hague's southern periphery is not a psychological 'no man's land', but an administrative 'everyone's land'. In economic and urban planning terms, it is a single coherent area, although 10 municipalities—Schipluiden, Wateringen, Delft, Rijswijk, The Hague, Voorburg, Nootdorp, Leidschendam, Zoetermeer, Voorschoten—regard it as their backyard.

Road map

The Hague's southern periphery is a catalogue of severed connections and structures that end abruptly. The absence of good connections between the different parts of the city leads to a mix of national, regional and local traffic on the main routes, with the congestion that this necessarily entails. Moreover, these sparse connections all focus on a single point, like a dead-end funnel into the city.

The Transformation of the Southern Periphery of the Hague

The transformation model for The Hague's southern periphery must not be read as an urban plan, but as a catalogue of interventions to transform that periphery. These interventions are based on the quality of the existing patchwork and on the laws of the new programmes. The various interventions can be made independently of one another, with each one bringing an improvement to the equilibrium of the southern periphery. They have the potential to transform the southern periphery into an archipelago of attractive locations to which future programmes can be linked. This will create new configurations of unique fragments in the Randstad patchwork.

Rapid transit

Public transport will be improved by the introduction of a rapid transit railway system, similar to the French RER or German S-Bahn, to ensure rapid connections between the different parts of The Hague conurbation. It will connect with train stations and motorway slip roads. It will partly follow existing rail routes and, where necessary, new routes—at times either elevated or sunken—will be built along the central reserves of the boulevards.

Motorway system

The A4 links up with the Benelux tunnel. Between Leidschendam and Delft, the motorway has a segregated part for through traffic. On each side is a parallel system of short slip roads that distribute the local traffic evenly across The Hague region. The traffic interchanges link up with the secondary boulevards and with parking garages at stations in the rapid transit system.

Boulevards

The cohesion of the area is reinforced by the development of existing boulevards into a secondary network that links together the fragments, from the North Sea to the Nieuwe Maas river. Lining the boulevards are buildings for commercial, cultural and public functions, creating characteristic urban axes. The introduction of appropriate waves of green lights ensures a good flow of traffic, which moves in flowing fashion at a moderate but constant speed.

Peripheral areas

The urban fringes are supplemented by a modern variant of strip parcelling that effortlessly absorbs existing country estates, sports fields and allotments. The dense pattern draws new boundaries, which heighten the area's contrasts. The peripheral zones form idyllic housing locations on the Vliet or polder, which benefit from the proximity of the city and its amenities.

Linear park

The cohesion of the southern periphery is safeguarded by the creation of one continuous landscape element that stretches from Rijswijk to the heart of Leizo. It forms an anchoring point for future fragments, an outsized point of reference in the area, like the Haagse Bos in The Hague. The enormous circumference of more than 10 km constitutes a wonderful housing area on the edge of the park. This large area also offers room for regional recreational facilities.

Peripheral centre

The central area of the southern periphery will develop into the peripheral centre in the conurbation, part of the national service zone. Thanks to its central location, excellent accessibility and the availability of large sites, it will be the ideal place for implementing large-scale programmes. A high-density strip is planned north of the motorway, while the part at Ypenburg will remain a more open zone with large-scale buildings and reserved areas for future developments.

Motorway locations

The motorway zone will develop into an independent spatial element in the city periphery, a ribbon of building complexes that will strengthen the laws of the motorway. The complexes will be composed as rhythmic noise barriers and city backdrops, as iconic landmarks and monumentalized junctions. The motorway will be raised, covered or tunnelled, with new functions so that it can pass through the urban patchwork unimpeded.

Landscape elements

A number of large landscape elements are used as characteristic fragments in the transformation model. The rolling parkland of Rijswijk, the woods, bodies of water and part of the Ypenburg airstrip, old country estates on the Vliet and the bordered polder roads and greenhouse landscapes of Wateringen and Nootdorp are given new functions as links in the southern periphery.

Phasing and politics

The new layout of the area capitalizes on phasing and political feasibility. The additions to the periphery are autonomous projects within the territory of one municipality. The Binkhorstboulevard is drawn precisely on the boundary between the two municipalities, which means that it can be realized in either municipality. Ypenburg is developed as a single site, with the park section in the territory of a polder municipality and the central section in the territory of two urban municipalities.

Programmatic Details and Land Parcelling

The transformation model creates a framework to be worked out in more detail and interpreted on the basis of future developments. It guarantees maximum freedom for future interpretations, without the loss of a coherent overall picture. Not only are programmes linked to the different locations, but also suggestions for architectural typologies to further exploit the specific qualities of the locations. This gives rise to new high-value fragments, each with their own spatial, programmatic identity. These contrasting identities enhance the quality of the patchwork metropolis as a whole.

Vliet- and Polderstad: socio-bungalows in the shadow of the city

The strips in the fringe areas on the Vliet and polder are filled with 'socio-bungalows', a new type of detached house in the subsidized owner-occupied sector, which can be prefabricated as an industrial product with an abundance of variations. The appeal of this housing typology lies in the combination of the suburban ideal for all and metropolitan amenities close at hand on the boulevards.

Golf city: the third age between the golf course and the *Vliet*

Between Delft and Rijswijk, an attractive area for small apartment blocks in a park landscape on the Vliet will be developed, as a modern variant of the old country houses. These dwellings for retired people benefit from their peaceful location adjacent to the new golf course, where the clubhouse serves as a service centre and golf hotel. The golf park has room for a clinic and a geriatric research centre.

Linear park: family-living amid the green

Along the southern edge of the linear park is a series of four-storey blocks with views of the polder. The blocks are composed of ground-floor flats with gardens onto the park and upstairs flats with large panoramic roof gardens. Where the linear park crosses a boulevard, the blocks are differentiated with neighbourhood amenities.

Heuvelstad: the gold coast on rolling parkland

The rubbish tips of Rijswijk, now transformed into rolling parkland, are ideal locations for the creation of a new 'gold coast'. Along picturesque winding streets stand luxury homes with stunning views of the Hague metropolis, Ypenburg, the expanse of Westland and sunsets over the sea. Golf courses, hockey clubs and the high-end shopping and entertainment centre of Ypenburg are all within walking distance. With a slip road directly onto the A4, a permanent connection with the Randstad is guaranteed.

Commercial boulevard: the modern industrial estate

The northern area of Leizo is organized as a commercial boulevard, a business and trading complex that is organized by sector. Each type of enterprise has separate communal facilities on the boulevard, behind which are extendable chains of offices and commercial buildings. The ends of the chains function as advertising billboards on the motorway.

Science city: modern living in a science park

An area has been reserved in the southern motorway zone for large educational and recreational functions in combination with research and educational centres, as new tourist attractions and employment pools for the region. Free-standing apartment buildings are planned in this science park, with permanent and temporary accommodation for course participants, guest lecturers, tourists and business people.

Ypenburg: living above the peripheral motorway

On the motorway zone are hybrid high-rises with a collage of offices, companies, shops and flats. There is an optimum mix of functions in high building densities, creating a peripheral centre. People live high above the spaghetti junctions, enjoying magnificent skyline panoramas and with every conceivable metropolitan pleasure available in the covered shopping paradise below.

Plein 1999

Partly beneath Prins Clausplein is an open area reserved for large-scale events. All manner of mass events, demonstrations and open-air spectacles can be held here. The area serves as an entrance and parking area for the science park and forms the transition to the commercial boulevard.

Fig. 21.1 "The classical interpretation of the Randstad, green (white) = countryside / red (black) = built-up area". Source: Neutelings, W. J. Sulsters, W., van Wesernael, P. and Winkler, E. (1989), "De transformatie van de Haagse zuidrand, Stedebouwkundige studie verricht in opdracht van de Dienst Stadsontwikkeling en Grondzaken van de gemeente 's G ravenhage, in het kader van het stadsplan—1988–89", in: Geurtsen, R. Verschuren, P. Kwakkenbos, G. Stadsontwerp in 's-Gravenhage: stedebouwkundig ontwerpen¡. Delft: Faculteit der Bouwkunde, p. 82. Reprinted courtesy of WJ Neutelings

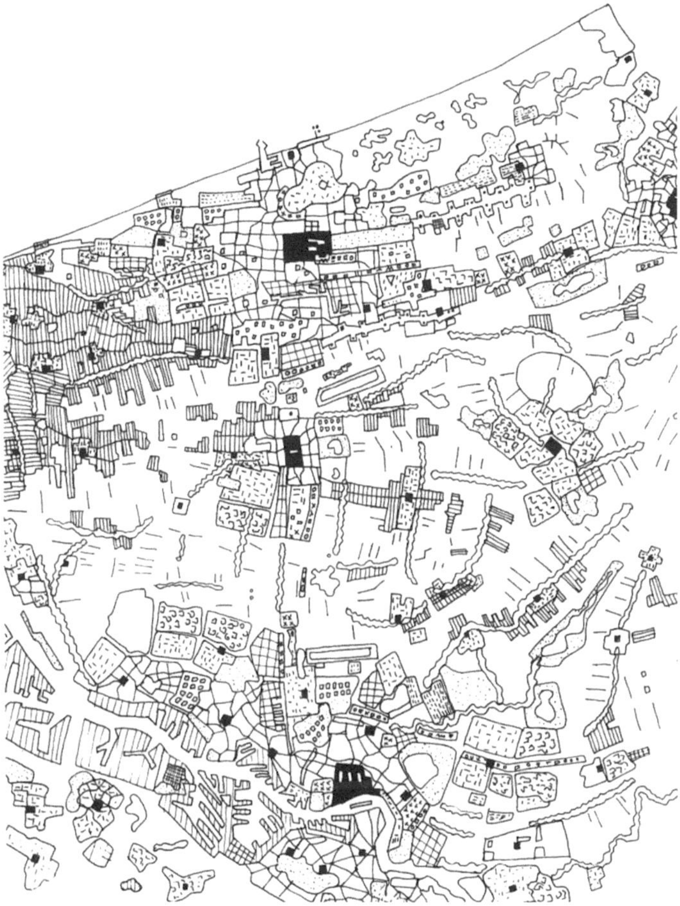

Fig. 21.2 "The patchwork metropolis, the Randstad as a spatial, programmatic patchwork". Source: ibid., p. 82. Reprinted courtesy of WJ Neutelings

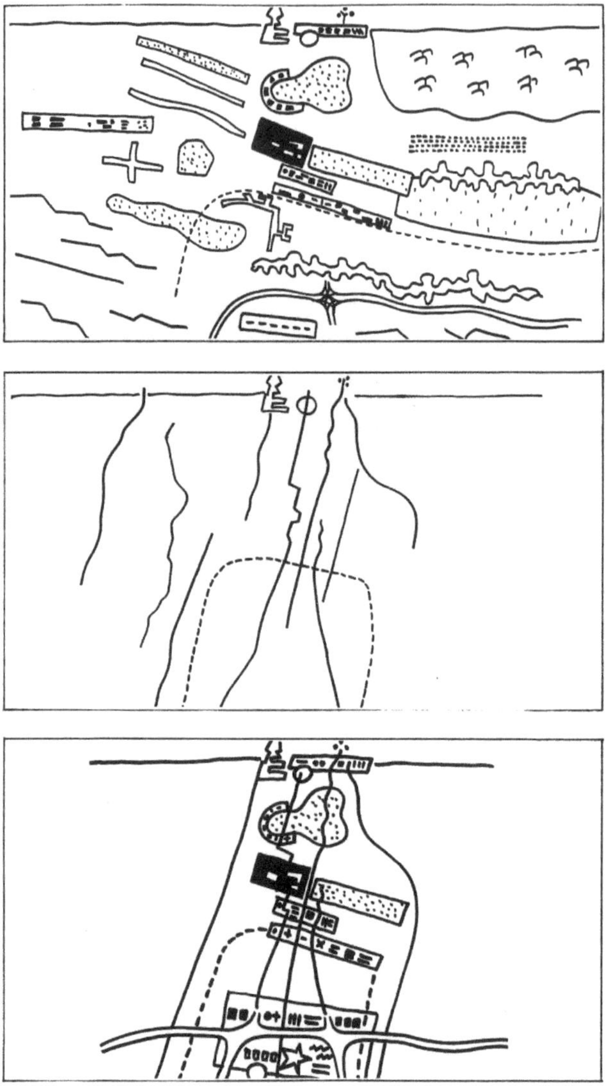

Fig. 21.3 Illustration of the features of the Hague collage. "Fragmentation is a theme in The Hague. [...] It is a collage of powerful spaces. In the old city centre, these are the green spaces [...], avenues and squares that serve as city landmarks [...] In the new districts and the periphery, it is the compositions of imposing buildings that form fragments of the city scape [...]. These characteristic fragments have a srong linear zone, more or less parallel to the coast. This programmatic layering is a typical feature of the collage structure of The Hague.". (see part "Features of The Hague collage"). Source: ibid., p. 84. Reprinted courtesy of WJ Neutelings

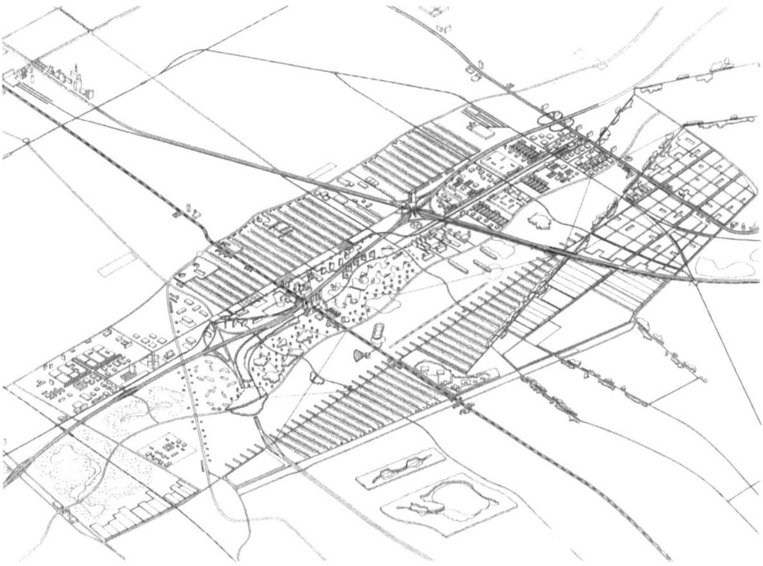

Fig. 21.4 Illustration of the programmatic details and land parcelling (see part "Programmatic details and land parcelling"). Source: ibid., p. 91. Reprinted courtesy of WJ Neutelings

Fig. 21.5 Illustration of (counterclockwise from upper left): Vliet- and Polderstad, socio-bungalows in the shadow of the city, Golf city, the third age between the golf course and the Vliet; Ypenburg City, living above the peripheral motorway"; Plein 1999, Source: ibid., pp. 92 and 95. Reprinted courtesy of WJ Neutelings

Chapter 22
La Città Diffusa (The Diffuse City) (1990)

Francesco Indovina

Introduction

Martina Barcelloni Corte
Faculté d'Architecture, Université de Liege, Liège, Belgium
mbarcellonicorte@uliege.be

Soon to become a fundamental reference for researchers and urban planners of later generations, the neologism città diffusa, *was coined at the end of the '80s by Francesco Indovina and his research group to describe what, in the area of the Central Veneto (North of Italy), was starting to be perceived as an unprecedented form of city.*

The juxtaposition of the terms 'city' and 'diffusion' was unusual for the time, if not to indicate negative externalities of urban growth. The abstract ideal of the compact "European city", had in fact led theorists and researchers to refuse the dynamics of urban diffusion and see it mostly as a problem that required solving. With his work and consistently with the research brought forward earlier by Samonà and the school of Venice, architect and urbanist Francesco Indovina tried to contrast this distortion, attempting to analyse the process underway, without prejudice.

Published in July 1990 by the Territorial Department of Economic and Social Analysis (DAEST) of the Istituto Universitario di Architettura di Venezia *(IUAV), the study* "la città diffusa" *renders the first results of the research* "Ipotesi sull'area centrale veneta. La città diffusa" *(1987) coordinated by Francesco Indovina and*

F. Indovina (✉)
IUAV University, Venice, Italy
e-mail: indovinaf33@gmail.com

M. Barcelloni Corte and P. Viganò (eds.), *The Horizontal Metropolis*,
https://doi.org/10.1007/978-3-030-56398-1_22

funded at 60% by the Italian Ministry of Education. The research study was implemented together with a small group of scholars from the University of Venice: M. Torres, L. Vettoretto, M. Savino, M. Sernini and F. Matassoni, who authored the different essays within the book, all aimed, via different methods and sources, at verifying the same hypothesis. What was highlighted throughout the research project, was the capacity of a territory–extensively infrastructured for agricultural production–to absorb an unanswered demand for city.

Through both the analysis of broad dynamics and specific case studies, Indovina and his team were able to highlight how the exiting of population, activities and services from the compact city and their relocation in the broad and more inexpensive agricultural territory, was giving rise to a "true and proper city" of an unprecedented form and size, produced by "emerging needs that had not found a response in the consolidated city". In other terms, and in contrast with the usual readings of urban phenomena, territorial dispersion was read as a particular and not necessarily negative response of the agricultural territory to the urban crisis that manifested itself in the preceding decades. The exiting of the population from the compact centres was in fact the outcome of an unsatisfied demand for "urbanity" that, affected by the changes in the production process, pursued a search for advantages for single families and businesses.

Through the different contributions, the research study described how, gradually, the abandonment of sizeable quotas of agricultural activities, the modification of the production processes, the changes in lifestyle, the modification of reference models, the increase of the costs of the concentrated city, the modification in the organisation of retail trade and the creation of large leisure facilities, first produced the urbanisation of the countryside and subsequently the formation of the città diffusa *itself. The actual changeover to the* città diffusa *true and proper was recognised by the authors in the enhancement of the inhabited territory with a system of services that the settled population would gradually "use in the same way as that of the compact city". The availability and use of these facilities would turn the dispersed urbanisation into a Metropolis, a self-organized process of enhancement that, thanks to a system of functional and social objects and relations, acquired ever greater "urbanness".*

The città diffusa *was seen to be an anomalous Metropolis since it didn't have its traditional physical and morphological characteristics, while preserving the functional and social ones; broadly equipped with everything that was private but underequipped regarding the public sphere, a "Metropolis that "preserved the drawbacks and benefits of the old one, while adding its own drawbacks and benefits".*

Indovina's research on Veneto's città diffusa *will play a crucial role in launching an important season of studies–throughout the following decades–on the new forms of the contemporary city, both at the national and at the European scale.*

22.1 The Diffuse City[1]

Foreword

This research sets out to assess the operative efficacy of the concept of the diffused city: to find out if, in substance, this concept can help us to appreciate and interpret a specific phenomenology of organization of the territory. The research investigates the form assumed by the organization of the territory in the central Veneto and hypothesizes that this "form"—not morphological but of spatial organization—can be interpreted by means of this concept.

The focus on central Veneto is not meant to assert that such territorial phenomenology manifests itself only in this part of the country, but it does indicate that it seems to be possible to observe it here in one of the places in which it has taken on a widespread and clearly visible character.

The starting point is the observation that remarkable transformations of the settlement phenomenon have taken place, and that these transformations require interpretation precisely because they do not appear to be a sort of expansion of the previous phenomenology, but almost a change of state. A situation that imposes verification on different levels: that of the real context, in order to measure the transformations that have happened and to identify their "causes"; that of interpretation of territorial dynamisms; that of interpretative concepts.

This note has the aim of putting into order the considerations and hypotheses that have prompted and prompt empirical exploration and field research.

Transformations of the central area of the Veneto

While empirical research can make it possible to measure the transformations that have taken place in the central area (over the last fifteen to twenty years), it is worth examining, even in a summary manner, the directions taken by those transformations.

It can be asserted that the territorial structure of the central Veneto area, as it presents itself today, is undoubtedly different from that of the past, giving rise to a new phenomenon. We are looking at something different from the previous territorial structure, a change that is not only quantitative but also qualitative, and the concepts elaborated in the past no longer seem pertinent. Whereas the adjective "diffused" could previously be applied to urbanization, today a new term is needed. We can tentatively define this new territorial structure with the term "diffused city."

[1]Excerpt from: Indovina, F. (1990). *La Città diffusa*. In F. Indovina, F. Matassoni, M. Savino, M. Sernini, M. Torres, & L. Vettoretto (Eds.), *La città diffusa*. Venezia: Daest, pp. 21–43.

In a certain sense, the diffused city comes in the wake of "diffused urbanization," but the two phenomena appear to be completely different on a territorial level and an economic-social level, representing different stages of organization of space, as a result of the reorganization of socio-economic processes. What we are seeing, then, is the manifestation of a truly new phenomenon that calls for new conceptual elaboration.

This need appears to be urgent because the diffused city (or whatever one wishes to call this new structuring of space) tends to become a recurring mode of organization of the space of our country. The phenomenon resembles, though only on the surface, that of the North American suburbs (a fact that does not help in its interpretation); or, if you will, the diffused city might be seen as the Italian (European?) response to the problems that have generated that very particular form of organization of space on the North American continent.

In absolutely schematic terms, and in a way that may seem more allusive than descriptive, we might say that the forms of organization of the territory in question (but probably also of many parts of the region and the country) have passed through three phases:

- cities set into an agricultural space;
- cities surrounded by urbanized countryside (i.e. countryside with scattered residential and other construction);
- a single large city that incorporates certain zones of countryside.

  Each phase refers to a specific cycle of socio-economic development and constitutes the composite result of strategic and "spontaneous" activities; the phases also correspond to different modes of reaction to the emerging of needs of a productive type or the rise of new social aspirations.

  However, it is important to emphasize that what has been stated thus far does not imply the prospect of a fixed or predetermined path of transformations in the organization of space. We are not saying that each form of territorial organization must necessarily pass through the three phases indicated above (stages of a preset path); nor are we asserting that a given evolution of the economic structure will necessarily lead, in all situations, to an identical transformation of the settlement system. We have simply set out to indicate a probable path, which it is possible to reconstruct in the area addressed by the analysis.

Several socio-economic connotations of the phenomenon

The economic advance of social strata previously connected with agricultural activities, and their abandonment of the primary sector in favor of the secondary sector, has prompted these strata themselves to give visible form to this improvement and change of culture (in the anthropological sense) through the "home," nearly always single-family, self-organized (often self-built), located either in areas belonging to homeowners (or relatives) or in agricultural areas with low land prices. In other words, the new is wed with an ancient way of thinking: there is a change of typology, but the culture of "stuff" continues to prevail.

The resulting urbanization thus seems to be scattered, lacking in an overall design, and not supported by an adequate grid of infrastructures and services. It branches out from existing settlements—older or newer, larger or smaller—to impact the territory as a whole, without following particular axes, not even spreading like an oil slick, because very often it is anything but seamless. We are looking, that is, at the phenomenon of the "edified" countryside, known as "diffused urbanization," typical of certain regions and particularly evident in the Veneto.

To a very limited extent, the phenomenon of the self-built or in any case self-organized home is presented as an act of "emancipation," so to speak, from paying rent; most substantially, it displays an improvement of economic conditions (only in a later phase does the aspect of "release" from the condition of being a tenant also emerge, while it is only marginal in this phase). The phenomenon of "scattered residence," then, can be seen on the one hand as the translation under new economic conditions of a rural culture focused on "stuff" (which emblematically and also traditionally takes the form of home ownership), while on the other it expresses the first symptoms of urban discomfort, above all that of the cost of housing. Furthermore, the settled population comes to a great extent from the place itself: these are not so much processes of immigration as modifications of habitational distribution within populations already residing in a given territory (municipality, area, and so on). In substance, the population of the zone itself, due to economic advances, gives rise to a different organization of its own settlement.

This relocation of edification and population in the territory, furthermore, also due to its rural cultural matrix, has not expressed a pressing "demand" for services. Precisely due to this character and due to the absence of "demand" for services, this phase of diffused urbanization takes form at a low level: it is lacking in basic services, and even the circulation infrastructures are inadequate, or far from meeting any standard. There-fore diffused urbanization not only consumes large portions of territory, but also acts as a powerful factor of spatial impairment.

Inside this situation, on the one hand, the connection between residence and productive activity develops, while on the other there is a widespread rise in small productive activities. The ongoing processes of restructuring of large business "liberate," so to speak, a trained work force that is forced to activate itself in an "autonomous" way, oriented either towards work in the home or the formation of small businesses. These phenomena of productive growth and diffusion are facilitated by the "productive decentralization" implemented by large companies, which opt to relocate production in spatial contexts where the supply of labor is overabundant with respect to the institutionalized demand (a phenomenon that has impacted various geographical areas of the country, including the Veneto).

It should be pointed out that at first the settlement structure, due to its functional characteristics, takes form in islands: the interconnections between the individual settlements seem to have little importance. Enclaves take form, often based on relationships of kinship, that interact with the surrounding territory in a limited way (with the exception, obviously, of mobility and work relationships). Even routine commercial transactions are reduced, also due to widespread agricultural production for self-consumption; the utilization of collective services diminishes (apart from the fact that they are lacking) due to a choice based on economic convenience as well as cultural position, favoring forms of mutual aid inside an extended family; the utilization of services on a higher level is also reduced, due to lack of economic wherewithal and a lack of supply.

In continuity with this initial phenomenology of territorial settlement, it is possible to identify a sort of "second wave" that has more direct origins in dissatisfaction with the "city." It is difficult to determine the constituent elements of this dissatisfaction without appropriate in-depth study; in particular, it is hard to know if this discontent can effectively be attributed to the overall "urban condition," or instead to certain specific shortcomings of individual cities. In substance, the weight of rejection of the city as such with respect to the weight of specific inefficiencies of the city, we might say, is unclear. Obviously this distinction is not just nominal, but would indicate a different attitude regarding the urban phenomenon. Of course it is always possible to assert that the two stances coexist, which is true; but the difference in their weight is no minor issue. Indirect clues would seem to indicate that the influence of the shortcomings of the city is greater than the rejection of the city as such.

In particular, the idea is to take the problem of housing as one of the key questions that has influenced the choices of families and determined the relocation of the population in the territory. The supply and cost of housing (as the main indicator of the overall cost of living) constitute non-marginal elements of the shortcomings of the city. It is worth emphasizing that this is a factor of coercion that is mystified when it is asserted that the positioning of the population in the territory corresponds to the urge to take advantage of greater freedom of habitat choice. Furthermore, it can be currently observed that the supply of housing offered by the concentrated city increasingly fails to correspond to demand in terms of quantity, quality and cost.

The phenomenon can also be described in the following way. The social middle class has developed a "conception of living" (or a model, we might say) that does not find fulfillment in the concentrated city. In fact, the housing supply that can be afforded by this social stratum does not correspond to that conception of habitation, which would instead be satisfied by supply that is not accessible from an economic viewpoint. In this situation the differential between the price of housing offered in the context of diffused urbanization and housing available in the concentrated city plays a key role in determining population movement. Nevertheless, what takes form is not a factor of choice, but a forceful element of constriction set by certain habitational models. In other words, segments of the city's population find themselves forced to move into the urbanized territory in order to achieve their model of habitation. If, as underlined above, housing represents the main factor of the overall cost of living, in the diffused territory living can be less costly than in the city, with differences of "quality" whose appreciation calls into play cultural problems, life choices, or ideological positions regarding a reality to which inhabitants are subjected.

At this point a short note is needed on habitation models and above all on their modification. The way individuals and families relate to the city is neither a constant nor a "natural" outcome; instead, it is a cultural and political outcome. In a given phase (a phase of "policy") this relationship has above all had a collective connotation: it was the city, as a collective possibility of interrelations and an endowment of collective services, that formed the basis for the concept of living. Living as an element of relation between inside (the home and the individual) and outside (the collective services of the city, and the other people); the qualification of habitation as the relationship between needs and the services offered; collective services as a "right" connected with residence. A phase, that is, with a central social focus. When this phase declines (the causes are observed, not analyzed), so does the concept of habitation: it focuses intensely on the home, while expectations regarding collective services subside, which callously tend to be replaced by "market" services. For the purposes of this discussion, what seems fundamental is the shrinkage of the concept of habitation to the mere "home" and the weakening of the relationship between habitation and community services. It is clear, then, that the relocation of the population, within certain limits (this is not a matter of an "absolute" phenomenology, but of partial changes, while the changes of sensibility seem to be more consistent), tends to follow new, different "rules."

It should be pointed out that in order for the above-mentioned price differential to play a decisive role in settlement choices, the diffused urbanization has to possess certain characteristics of an urban type, however modest. The new wave of population to be settled, in fact, has gone through urban experience that has shaped its demands and its needs. Though the concept of habitation has changed (according to what we have outlined above) the urban experience still has its influence; even in relation to a real or ideological "rejection of the city" certain habits die hard.

In substance, the passage from diffused urbanization to the diffused city is simultaneously quantitative and qualitative, where quality, as we shall see, is also induced by quantity.

Social typologies and structural processes

Analysis of the territorial spread of the population reveals that the social type of the population tends to be modified over time, also in correlation to the territorial "origin" of the population itself. Though aware of the irrelevant character of generalizations, we can still identify different phases of the phenomenon of territorial diffusion of the population in the area being considered:

- first phase, involving the "local" achievement of better living conditions. The achievement of higher economic levels prompts the social strata involved to seek better habitat conditions in their same area of origin. This phenomenology may also have led to a relocation of the population within the same municipality, with abandonment of the "center" to settle in the urbanized countryside (urbanized though with the short-comings outlined above);
- second phase, where "borderline" immigration is prevalent, still from restricted areas but also from the "city," of social strata of modest economic means. These flows tend to be attracted by and to reinforce processes of diffused urbanization, also due to their limited economic possibilities;
- third phase, where there is prevalently a flow of immigration of urban origin, of the middle social strata. A phase, that is, marked not so much by a phenomenon of attraction as by one of "flight" from the city.

These phases are obviously characterized by a positive dynamic in the area of housing supply. While at the beginning there is a sort of identification between housing supply and demand, in the sense that the demanding subjects are themselves the "promoters" (and in frequent cases self-constructors), later we can see the importance of institutional supply, be it public, private or cooperative. It is precisely this new supply that determines modifications of the settlement type, also for objective reasons connected to the logic of the suppliers. While on the one hand, in fact, the model of diffused settlement is not questioned, on the other factors of concentration and intensification arise (as, for example, in the PEEP low-cost housing programs).

The quantity of settled population and the different social experience of the new immigrants generate, in time, growing demand for services (public and private). In other words, the attraction of a "place" is also determined by a minimum of available services (we will not look into questions of the "threshold," both because they are complex and because phenomena of attraction do not always have a connotation of optimization). Furthermore, local administrations, to achieve part of the objectives that justify their "existence," to gain the consensus of a more demanding population and to attract new residents to expand the size of the city (numbers are always seen as an expression of "power"), engage in a policy of incrementation of infrastructures and services.

This is the path, though within the limits of a generalization that does not reveal "local" characteristics, through which small and medium settlements that have expanded through diffused urbanization begin to take on connotations of an urban type (to the extent that this can be said of a situation that is lacking in two of the particular features of the urban condition: intensity and density).

Nevertheless, the above observations should also be correlated with certain structural processes: the connection between structural phenomena that impact the production sector and processes connected with residence (as we have seen) is precisely what can supply indications to explain the motivations behind processes of relocation of the population.

Looking at the mechanisms that have impacted the production sector (and still taking the limits of generalization into account), we can observe an initial phase in which the most influential phenomenon seems to be restructuring, while relocations of production seem to take on greater importance later (though the two phenomena cannot be reliably separated).

As we know, the reasons behind processes of relocation are multiple; some are determined by particular market conditions (raw materials, potential outflow, labor, etc.), others by technological opportunities; some are activated by processes of "attraction" and others by those of "flight." They almost never exist in a "pure" state, which further complicates the investigation. So what is outlined below should be taken for its value as exemplification.

Industry located in the city, through the effects of urban growth, winds up being "surrounded," as if the city were trying to suffocate it. On the one hand, production becomes more costly (due to restrictions, administrative constraints, congestion, etc.), while on the other the process of value incrementation of areas (formerly peripheral, now central) can make relocation of industry elsewhere economically advantageous. Therefore processes of productive restructuring and corporate reorganization are combined with processes of relocation in the urbanized "territory."

The migration of productive activities, predictably enough, also causes flows of movement in the population (obviously in proportion to size). This happens in two ways: on the one hand, a part of the staff follows the company to its new location (moving as a result of the movement of the productive activity), while on the other the relocated company generates demand for labor, which attracts population if it is not locally satisfied (and generally this is the case).

It should also be observed that the relocation of productive activities in the "territory" is followed by the relocation (or the new formation) of businesses that provide services to businesses. Obviously this does not mean all the service activities, but simply the most "banal" so to speak, and those that do not need to interact with other functions of an urban type. But we will return to this point later on.

Both productive and service businesses bring about a modification of the fabric of local relations, both internal and towards the outside, while also causing social alterations that tend to have an impact on the "compactness" of "local societies." What seems to have a certain pertinence, at first glance, is the fact that the location of these businesses is also of a diffused rather than a concentrated type.

Provision of services and urban character

Due to the line of reasoning that is emerging, it seems important to emphasize that the set of processes observed results in the involvement of a sizeable population in the diffusion process. The quantitative aspect, as we shall see, is not a negligible factor: as a whole, the population that resettles in what we have called the diffused city is numerically consistent, equal to that of a large city, which becomes decisive to explain the modifications of territorial quality.

We have already observed how the encounter (and at times the clash) between the policies of local institutions (at the level of urbanism and services) and the "demand" asserted by a growing population that is modified in social and "experiential" terms leads to a growth in the supply of infrastructures and collective services. Obviously the phenomenon does not appear in a homogeneous way, from either a spatial or a sectorial viewpoint. Furthermore, the phenomenology regarding the problem we are examining here is not singular: there are cases of services "diffused in the territory" together with cases of spatially concentrated services (therefore with the formation of specialized areas, which are almost never inserted in traditional urban contexts), for which the supply of circulation infrastructures provides and facilitates access. The recurring typology seems to be a mixture of concentration and diffusion, but the rule is the same: to give rise to a system of offerings for the diffused city, for a population located in a wider territory of lower intensity.

This greater provision of infrastructures and services (though still substandard) begins to give an urban tone to the settlement. It should be stated that the diffused city already has urban connotations today, but remains sub-standard, so to speak, with respect to the concentrated city. Full achievement of a city effect, in substance, still re-quires infrastructural investment and the supply of ulterior services.

Private services deserve some degree of consideration: we can notice a certain lack of the type of services defined as urban, but—and this is the important thing—this typological shortage does not correspond to a functional shortcoming. This constitutes another particular characteristic of the phenomenon being examined. To express the matter in a synthetic and schematic way, we might observe that situations of diffused urbanization to not encounter, in the private sector, services of an urban type, but instead seem to leap directly to metropolitan services. We will indicate the conditions that give rise to this leap below.

The fundamental characteristics of the city (as we are accustomed to understanding it) include high density and intensity. The population settled in the context of diffused urbanization, obviously in circumscribed areas, while taking on a considerable size (that of a big city) does not have that connotation of density and intensity that is associated with the traditional (compact) city. Without delving into the complex question of "proximity," we can agree that in operative terms high density and intensity can be interpreted as the possibility for a large population to "easily" access every point of a given territory. In a big city the numerous population has the possibility of easy access (in physical terms) to every point of the city and above all to every point of its "neighborhood." Clearly this is a highly differentiated accessibility. On a technological plane we can distinguish between pedestrian and mechanical access (according to different means of transport); from an administrative viewpoint there can be limits; from an economical-social viewpoint situations of exclusion can arise, etc. In any case, the point of view that prevails in considerations regarding the (compact) city is that of accessibility measured in terms of multiplicity of opportunities and of proximity. From this stand-point, the difference between the traditional city and the diffused city reaches its maximum: in the latter, in fact, the population is strongly penalized in terms of physical accessibility to every point. Nevertheless, the difference between these two city "forms" can be reduced when there is a greater inclination for movement (nothing is right around the corner) and greater accessibility when measured in terms of time. High mobility and easy accessibility, in substance, erase the effects of the absence of density and intensity.

The increasing spread of the private automobile but also, over time, the modification of its use (its transformation into a sort of "shoe," in the sense that one does not walk barefoot), have enormously increased the inclination towards mobility, on the one hand, while the "accessible territory" has expanded as a result, on the other. This fact tends to transform a diffused population into an intense market reservoir, which as such becomes appealing for the offering of certain private services. Increased mobility, that is, expands the spatial confines of the possible market reservoir.

This is not the place for discussion of whether the increase of automotive mobility is an independent or induced, freely chosen or coerced phenomenon; here we can simply observe that without growth of mobility based on private vehicle ownership the diffused city could never have happened. It comes into being and lives because it is a city of motorists.

Getting back to the problem of services, we can point out that the appeal of offering a "private service" depends on the existence of a suitable market reservoir, or of a suitable quantity of population to be "served." It apparently makes no difference if the market reservoir is composed of a concentrated mass of population or by one that is not concentrated but has a high inclination towards mobility. But this is only apparent, because in spite of everything a situation in which the "service" and its "market reservoir" both exist in a limited area is not equal to the case in which both of them exist in a much larger area.

In the specific case of the diffused city, it can be hypothesized that it is precisely the mass of population settled in a diffused way combined with its inclination towards mobility that determines the appeal of providing a private service. Nevertheless, that private service, in order to fully exploit the market reservoir, must have some specific characteristics.

In particular, such services have to be easily accessible and highly attractive ("showy," we might say), capable of canceling out—due to their great appeal—both the (psychological and physical) "distance" from potential customers, and the competition of the compact city. In this situation, taking the cultural context of the moment into account, a service on an urban level does not seem adequate: its potential offering is far greater than the demand of the surrounding population (limited area), while its "appeal" is not great enough to attract "distant demand." Instead, a service on a metropolitan level programmatically addresses a much wider market area, and in this sense it is equipped (it has the image) to attract (arouse) distant demand. Therefore this is the type of offering that can more easily be located in an area of numerous but not dense population, since it uses attraction (and therefore the mobility of customers) to compensate for the lack of density of the potential market.

It should also be stated that such metropolitan services, even when their market target is identified as a cluster of high intensity (a metropolis, or a least a big city), generally find it advantageous to locate "outside the city," precisely to boost accessibility for an almost totally motorized consumer, and to exert maximum potential appeal (accessibility is seen here as a set of territorial conditions that permit the use of the service on the part of a significant mass of customers, meaning not just access roads, but also parking areas, etc., conditions that are increasingly rare on an urban level).

The insertion in the context of diffused urbanization of services for persons of a metropolitan type (shopping centers, hypermarkets, large specialized dealerships, large entertainment or sports facilities, etc.), productive activities, business-to-business services (consulting, data processing centers, draftsmen, etc.), sales outlets connected directly with production (or with apparent production), and therefore specialized points of sale (for example: Just Lamps, 100% Brides, etc.), triggers a different configuration of low-density urbanization.

New territorial hierarchies

This different configuration of low-density urbanization is what has been called the diffused city. Therefore this territorial phenomenology stands out for:

- a sizeable mass (comparable to a city, in other words) not just of population but also, at least partially, of services and productive activities;
- a scattering of that mass across a vast territory; so vast as to not possess, as a whole, phenomena of high density and intensity. Obviously this does not mean that there cannot be individual "points" of high density; it simply means that the spatial configuration does not give rise to significant phenomena of density and intensity of an urban type;
- a high level of connection between different points of the territory, or namely a territory that has multiple connections of a horizontal type (infrastructures) capable of guaranteeing the possibility of very extensive mobility.
  At this point it might be useful, though in an extremely synthetic form, to identify some of the questions raised by such a spatial configuration. This can make it possible to delve deeper into certain connotations of the diffused city.
  From what has been outlined above, the difference that can be seen between the diffused city and a hypothesis of metropolitan area should be evident; a difference that is completely concentrated around spatial hierarchies.
  The term «metropolitan area» indicates (generally) a territory with a high level of hierarchy: there is a center (the large city) and a series of small and medium satellite cities (diffused in the surrounding territory) that are linked to it by hierarchical lines. A territory, as they say, marked above all by vertical connections from the periphery to the center of the territory. The existence of these connections of a vertical type is what makes it clear that we are looking at a hierarchical

situation; which means that the location of activities and services, the relocation of the population, the overall structure of the urban offering, are ordered in a ranked way: each service is positioned in space in the point that corresponds to its ranking (from the lowest in the periphery to the highest at the center).

On the other hand, as we have seen, the diffused city relies on a spatial organization marked by abundant horizontal connections. Just as the vertical connections point to a hierarchized territory, so the horizontal connections indicate a "non-hierarchized" territory. We can make the provisional assumption that the diffused city, precisely because it is marked by horizontal relations, is characterized by an absence of spatial hierarchies.

The multi-directional nature of flows (not just of persons, but also of goods and information) typical of the diffused city (which in this sense resembles the city) contrasts with the mono-directional nature of the metropolitan area. Obviously we are taking the situation to extremes. In fact, it would be possible to point out that horizontal and multi-directional flows are also multiplied in the traditional metropolitan structure, attenuating the hierarchical element or at least conveying the sensation of its attenuation. Nevertheless, this seems to be a marginal issue with respect to the line of reasoning that is unfolding.

To prevent misunderstandings, the absence of hierarchies of the diffused city should not suggest the idea of a sort of "self-sufficient territory," nor of a space exempted from the overall context of territorial hierarchies. The diffused city enjoys and is subject to general territorial interrelations: it is inserted in the context of territorial rankings. It is possible to clarify this observation by making reference to real phenomena.

Observing the flows of population movement based on different motivations in the diffused city, we can notice that the horizontal flows are overlaid by existing vertical flows, above all for employment, consistently aimed towards certain polarities. Furthermore, in time these flows can be increased precisely due to the growth of the population, which moves, as we have seen, from the "city" to the diffused city, but very often maintains its original employment in the former.

The phenomenon can also be observed in the different phases of spatial structuring: while in the phase of diffused urbanization there were closed enclaves from which a movement of commuting towards external polarities for reasons of employment was generated (a unidirectional, hierarchical movement), in the phase of the diffused city that movement, which remains, is very often joined by an "internal" movement, again for work (new opportunities that arise in the diffused city, workers who have "followed" a company that has moved, etc.). The result is a multi-directional character of the movements motivated by employment.

The same can be said of all the other types of movement. In their sum, the traditional and mono-directional movements are attenuated, in the sense that their relative weight is reduced.

Precisely this phenomenology, described here in terms of mobility, leads to the con-figuration of a multifaceted territorial reality. It is possible, for example, to identify zones that are configured as "dormitories" for the labor force employed elsewhere, but at the same time have a character of urban complexity (though often of modest quality and, above all, low intensity). This is precisely because the constructed territory is placed over a fabric of historic urbanization where "centers" traditionally of modest size (and lacking in other attributes) become important nodes in the whole of the diffused city. It can moreover be noted that the "points" of location chosen by the large activities of service or production do not bring about phenomena of aggregation, but of scattering, which while they lead to negative results in terms of land consumption, territorial debilitation, shortage of infrastructures and increased mobility, nevertheless also prevent processes of congestion. This multiplicity of configurations within a given framework is what appears to be of remarkable interest.

The conception of the city envisions a population that performs its functions (production and reproduction) mostly inside the city. Taking for granted that the process of urban organization necessarily gives rise—or so it is said, at least—to spatial concentrations of those functions, it has been possible to observe that elements of spatial specialization (for example: dormitory district, business district, etc.) represent powerful factors in the reduction of quality of the urban context. Nor is the question simplified by observing that very often the processes of specialization described in terms of functional connotations are actually the result of the social structure.

In the diffused city, for all the reasons described above, on the one hand processes of spatial specialization are less evident, while on the other, nevertheless, urban complexity seems to be diluted, so to speak. If this is the case, we are forced to look at certain phenomena of territorial specialization in a different way, in the sense that while in the concentrated city they appear to be "shortcomings," in the diffused city they do not necessarily have a negative impact on urban functioning and conviviality (or at least they seem to have less impact). In substance, the diffused city generates hierarchies and specializations within itself; nevertheless, in their intensity, "form" and organization, they seem less negative than their counterparts in the concentrated city.

The theme of territorial hierarchies, however, does not only have to do with the relationships inside the diffused city, but also with the relations of that city with the rest of the territory or—in the case in which the diffused city contains a "traditional" concentrated city—the relations between it and the rest of the diffused city itself. From this standpoint, it does not seem acceptable to call into play a sort of demise of spatial hierarchies.

While on the one hand there is no doubt that the concentrated city, in any context in which it is located (taken per se, or seen as part of a metropolitan area, or part of a diffused city), "yields" something, on the other hand just what is yielded needs to be analyzed. We might say that the services (for persons and also for businesses) that move into zones—which in a generic way we will call "extraurban" (especially in zones of such widespread urbanization as to qualify for the title of diffused city)—are for the most part banal, while the concentrated city continues to be the location of the more exclusive services. The same can be said of productive activities: the concentrated city expels the more mature activities, while it maintains or attracts advanced productive activities (those with greater added value).

We are faced then with new forms of spatial hierarchy that can be described, in an extreme and synthetic way, by means of the following formulation: the functions that hierarchize space in the present historical phase are not those that are pertinent to large masses (of consumption, employment, production, etc.), but are instead the more innovative functions capable of determining processes of "control" (over population, economic mechanisms, the spread of innovation, the financial markets, cultural production, levels of quality of life, etc.). In other words, one of the fundamental characteristics of the traditional function of the city is accentuated: the concentration of "power" in the various, changing forms that power assumes. If certain characteristics of an urban type spread into the territory, giving rise—in keeping with specific connotations—to the diffused city, in some cases, and in others to forms of diffused urbanization or compromised territory, the "production" of urban characteristics that confirm the ranking of space is not depleted. The resulting territory seems to be structured according to new hierarchies, taking into account the fact that in this new context the old "factors" of ranking are no longer relevant for the new activities (of production and service), nor has the location of traditional functions (of service and production) ceased to be capable of generating spatial hierarchies.

The diffused city, as described above, is not exempt from the process of spatial hierarchization, and therefore as whole (like any other city), regardless of its administrative subdivisions, it is positioned in a set of spatial hierarchies of a higher order. More-over, inside it—though in a less accentuated way than in the concentrated city, precisely due to its genesis—it presents a hierarchical articulation of an urban type, attenuated by high levels of mobility and accessibility (which generate other problems but attenuate exclusion).

The problem of the borders and the variables at play

At this point, however, a crucial question emerges that has always been substantially implicit: that of the borders. The borders of the concentrated city also constitute a theoretical and practical problem, but from a simplistic viewpoint the administrative structure takes care of the issue: the administrative borders define the territory of the city.

In the diffused city the question is posed in a different way: the solution of administrative boundaries doesn't work. In this case precisely the opposite is true: the definition of the "borders" of the diffused city is what makes it possible to determine which administrative units can be considered as belonging inside the diffused city itself. But there's more: the definition of the borders is also decisive for the recognizability of the phenomenon.

The definition of the borders, in this case, is not a priority of the research on the connotations (and the very existence) of the diffused city, but represents an area of research in its own right.

While the area of the diffused city can be identified as an impression, only in-depth study of relations can define its boundaries with the necessary precision. It should be observed, in fact, that due to the connotations of the diffused city it can also remain submerged as a result of an error in the definition of its boundaries. Obviously this submersion is meant in the sense of the plane of identification, because on a factual plane the city is active, regardless of any "recognition."

Regarding the variables called into play, both to define the borders of the diffused city and to identify it, the following are rough indications for an initial approximation of the problem. The research itself will enrich the context of the hypotheses and the analytical variables to be considered:

- the relocation of the population, obviously, is one of the main variables. It will be important to analyze this phenomenon over time, identifying models of settlement behavior, also in relation to housing supply;
- location and development of productive activities. The phenomenology that seems to be most pertinent is that related to exogenous and endogenous growth, both regarding migrations or relocations, and regarding the "new" opportunities offered in and by the diffused city;
- location of services for businesses, according to their character of role and function;
- location of services of a non-metropolitan type for persons, i.e. identification of the "urban" structure of individual points of the diffused city;
- location of services of a metropolitan type. It is hypothesized that these presences can represent one of the constituent elements of the diffused city;
- inter-relations and mobility of the population;
- infrastructuring;
- what it is generically possible to call "ways of life."

  What is outlined above, obviously, constitutes only the indication of areas for development of the analysis. The specific content of this analysis can only be identified in relation to the hypotheses of interpretation previously put forward and eventual new hypotheses suggested by the analysis itself.

Can the diffused city compete with the concentrated city?

We can conclude by raising a question for which the research might provide some initial indications: on a par with the concentrated city, can the diffused city be a pole of attraction for those activities whose presence determines a "high" position in the over-all territorial hierarchy? Or, to put it another way: can the diffused city compete with the concentrated city in terms of appeal for activities of innovation and "power"? This question does not seem to be without interest, if we consider the fact that what is at stake is the structure of the organization of the country, as it has historically been "constructed."

Finding simple answers seems to be no simple matter, also due to the implications of a general order posed by the question, and due to the lack of specific, detailed analysis. Certain notations, in any case, can be put forward to indicate the aspects on which to concentrate first.

The indications, precisely because of their character as such, can be listed in a schematic way, point by point:

- the first element gleaned from experience indicates that the power of attraction of the diffused city for those activities that can be identified as the expression of "present power" is very low. While the territory of the diffused city has appeal for the location of mature activities and banal services, etc., it has very little appeal for activities of innovation and advanced technology, for financial centers that are not "service centers," for cultural and scientific activities on a higher level, etc.;
- the lack of importance of traditional factors of choice of location for this type of innovative activity seems to diminish the potential of the earlier poles of attraction. However this does not mean that there are not other factors of location appeal. It only means that the old ones have been replaced by new factors, where the latter seem to be more concentrated with respect to the former. It is a matter of "structures" (universities, research centers, libraries, etc.), but also of "residential preferences" of high-level technicians (who prefer the concentrated city). The hypothesis could be made that these are factors of strong "sedimentation" that cannot, that is, be easily recreated, and for this reason are more concentrated in a few points than were their predecessors;
- finally, it should be observed that on the one hand innovations in telematics do not yet have a "diffusive" character, so in contradiction with their "essence" they produce concentration, while on the other they are not able to replace the interrelations among subjects with high power content (of whatever type).

It seems possible to state that only a "territorial policy," in the general sense, that takes stock of this new spatial configuration to transform it from a prevalently spontaneous outcome into an overall project of spatial configurations (in keeping with options that have yet to be defined), could put the concentrated city and the diffused city on equal footing, within certain limits. This would be a project of qualification that could make the diffused city emerge from its "residual" role, which cannot be otherwise eliminated. While not even a suitable territorial policy could eliminate what we have called strongly sedimented attraction factors, it could nevertheless reconstruct a less unbalanced territorial hierarchy, paying greater attention to non-reproducible resources, and operating in a less selective way on the social plane. In this sense, the diffused city could play a balanced and balancing role in the organization of space, both in terms of (innovative) production and the social quality of life. But then the diffused city would not be simply the result of the real estate process, but a project of reorganization of the territory.

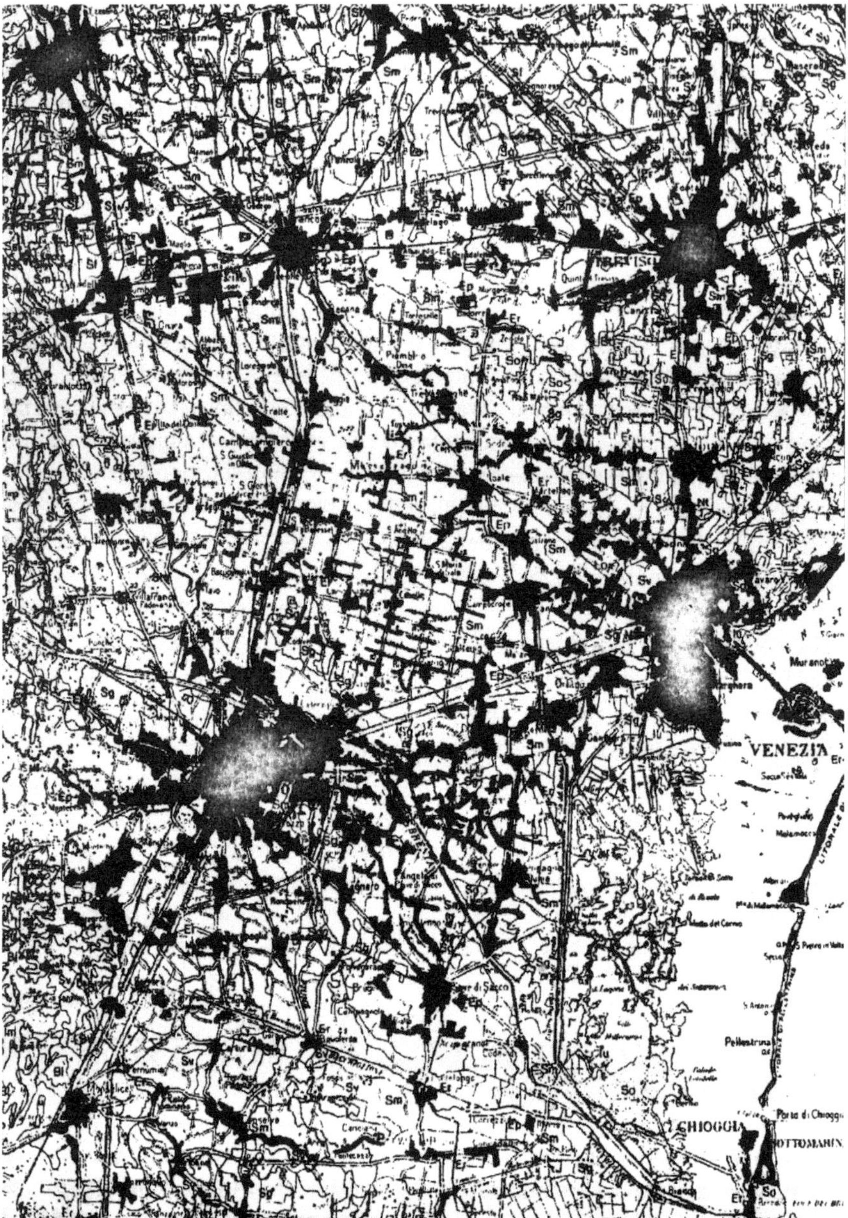

Fig. 22.1 Urbanized area between Padova, Venezia and Treviso. Source: Sernini, M. (1990), "Appunti di ricerca intorno all'urbanoide in sviluppo tra le citta' di Padova, Venezio-Mestre, Treviso". In Indovina, F. Matassoni, F. Savino, M. Sernini, M., Torres, M., Vettoretto, L. *La città diffusa*. Venice: Daest p. 59

Fig. 22.2 "Delimitation of the examined territory and identification of settlement typologies areas. Grey: area of cities; white: areas of minor isolated centers; dotted: area of linear settlement". Source: Torres, M. (1990), "Città concentrate, citta' reticolare e centri isolati nell'area Padova-Mestre-Venezia-Treviso". In ibid., p. 75

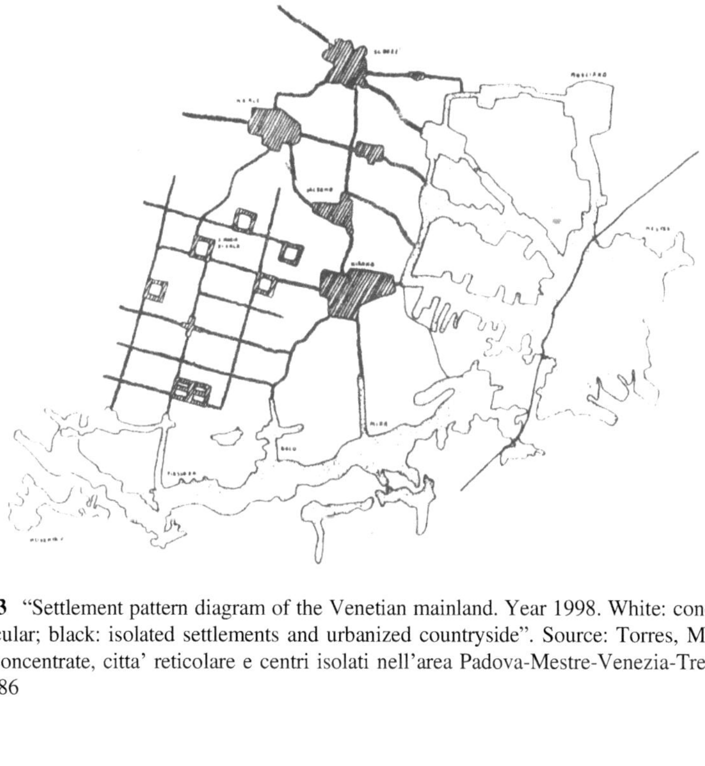

Fig. 22.3 "Settlement pattern diagram of the Venetian mainland. Year 1998. White: concentrated and reticular; black: isolated settlements and urbanized countryside". Source: Torres, M. (1990), "Citta' concentrate, citta' reticolare e centri isolati nell'area Padova-Mestre-Venezia-Treviso". In ibid., p. 86

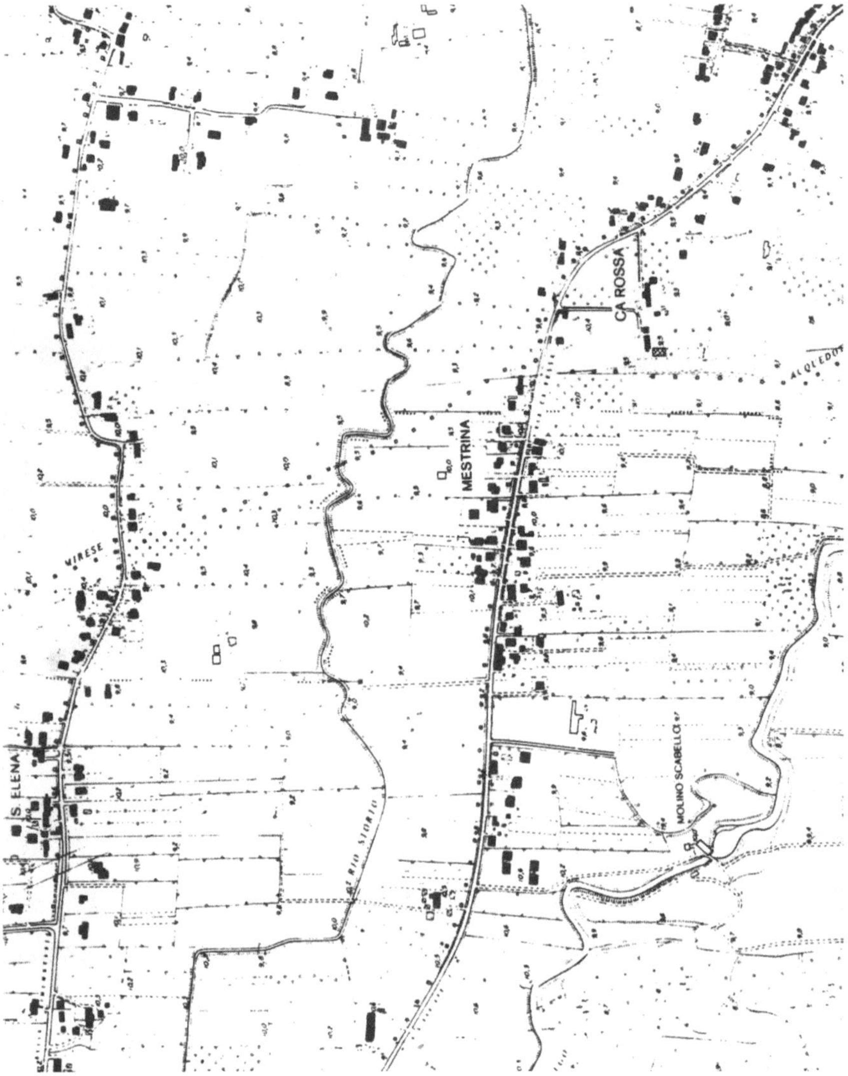

Fig. 22.4 "The parallel linear settlements of S. Elena and Mestrina". Source: Matassoni, F. Savino, M. (1990), « Alle radici della città diffusa: l'indifferenza localizzativa e la pianificazione mancata ». In ibid., p. 118

Fig. 22.5 "The diffusion at Veternigo on the traces of the roman centuriatio". Source: ibid., p. 120

Chapter 23
The Periphery (1991)

Bernardo Secchi

Introduction

Tommaso Pietropolli
Laboratory of Urbanism - EPFL, Lausanne, Switzerland
tommaso.pietropolli@epfl.ch

Between the early 1980s and the mid-1990s, Bernardo Secchi published regularly editorials in Casabella, *at the time directed by Vittorio Gregotti, contributing to characterize the magazine as a relevant place in which to reflect on the making of the urban Project.*

Secchi's contribution for the issue 583 (October 1991) "La Periferia" reflects on a topic that was increasingly becoming evident for those involved in interpreting and designing urban territories in Europe,[1] *namely that of the abandonment of vast industrial and technological areas, often located in peripheral positions and that in those years started to be redeveloped through punctual and poorly organized projects. It could be said that Secchi, using this evidence as a pretext, questions the transformation of the European city-territory, reading differently its spatial organization and rethinking the construction of its Project, that evidently could not refrain from considering these newly available urban materials.*

The text is rich and dense with reflections, many of which could still today be considered as unexplored fields for research; however, some of the author's intuitions more than others clarify the close link existing between this text and the

[1]In the same years, Secchi was working on several urban plans, for instance that of Jesi (1983–1988) and Siena (1984–1990) and was teaching at EAUG in Geneva.

M. Barcelloni Corte and P. Viganò (eds.), *The Horizontal Metropolis*,
https://doi.org/10.1007/978-3-030-56398-1_23

research on the Horizontal Metropolis.[2] *Three main questions emerge from "La Periferia", all differently contributing to the construction of the research on "horizontality" both in methodological and conceptual terms.*

The first is a question of spatial configuration. The periphery is a relative space, it is not absolute; it refers to a centre, being this real or assumed. Without a centre the periphery cannot exist. Starting to observe the "homogeneous or coagulated mesh" of urban materials from the inside, Secchi realises that other spatial configurations are possible, once we accept that different hierarchical organizations of space are possible, and thus opens to a different reading of the peripheral spatial capital.

The second question is linked to the tools that the architect must equip himself with, in order to understand changing territories. If it is true that Secchi remarks a problem of "pertinent description" for the contemporary periphery and—to a certain extent—also one of aesthetics related to the same territories, the descriptive approach that he proposes should be considered a highly projective move. The search for new gazes and descriptions, fundamental to better understand and re-conceptualize territories that are becoming gradually illegible,[3] *can help to develop new imaginaries and avoid the limitations of a biased reading.*

Design is at the centre of the third question. Working within a fragmented spatial condition, in order not to accept a paratactic construction of the territory, in which different materials and projects approach one another without systemic coherence, Secchi invites–aware of the difficulties–to reconsider the capacity of the project to elaborate "open structures". Organizing and structuring the various urban projects in a systematic way, allowing the new "urban ecologies" under formation to develop completely, the project of these vast portions of territory might enable the construction of new and different conditions of territorial habitability, for which different conceptual paradigms need to be developed.

Today, in the light of a general impoverishment of the Project as federator of shared intentions between the different actors of urban and territorial development, Secchi's hypothesis of organizing design around "figures"—horizontality being one of these—able to guarantee clarity and unity for the Project itself, beyond the complexity of the single case, seems to be particularly relevant. Moreover, in the frame of a debate overcharged with ideologies against or in favour of urban peripheries and scattered urban regions, reasserting the need of a reflection on the Project for these spaces, starting from the urban materials themselves, is a fundamental, methodological legacy of Secchi's text.

---

[2]Bernardo Secchi and Paola Viganò explicitly worked on horizontality in occasion of the vision for Brussels 2040, but reflections on horizontality could be traced in their work through the explorations of different figures and concepts, in occasion of different research activities and projects: from porosity in the case of the Grand Paris, to the concept of isotropy in the case of Veneto region, and lastly until the figure of horizontality in the case of Brussels.

[3]Due to the accumulation of layered and not only geographical rationalities, mainly related to territorial physical structures (topography, hydrography, infrastructures), but also to the different spatial structures of which different economic and social models have been used over time.

23.1 The Periphery[4]

All over Europe, the focus is turning to the urban peripheries, for different reasons in different situations. There is an awareness that the peripheral zones are those in which the greater part of the social problems left unresolved by the redistribution policies of the last few decades are still concentrated; that these are the zones in which to come to terms, in the words of Ernst Bloch, with elements from the past that have not yet been "indemnified." There is the sensation that these problems are being exacerbated by the recent wave of immigration from outside of Europe, by the new forms assumed by social multiplicity and heterogeneity. The fear is that the situation could explode into new urban conflict, as already sporadically occurs. The periphery prompts a new type of reflection on society, its behaviors and changes, and our capacity to represent, comprehend and predict them.

Above all, it has been the periphery that during the last decade has generated a new theme for urbanists, unpredictably wide and profound in scope: that of the abandonment of vast industrial areas, of major technological resources, significant parts of the fixed assets of the industrial city. The periphery appears today as the site of a withdrawal and, at the same time, an opportunity for a new project. And this prompts us to reflect on the design of the city.

It is again in the periphery, however, crossed by thousands of boundaries, that every project and every policy funs up against the perplexities and contradictions of a state mechanism eternally searching for a "stable form" that corresponds to the morphological arrangement of the society, the economy and the territory. Today, the periphery strongly resists any type of institutional engineering; it refuses to be attributed, to be included within new conventional boundaries; it is discovering its own local character and its own kaleidoscopic identity.

Discussed until recently mainly as a place of an absence to be filled, a lack of social services, green areas, collective spaces, order, quality and meaning, contrasted with the strongly connoted space of the historical city or the rural landscape, the periphery has become the place of indictment regarding modern city planning and its claim to operate as a philosophy of history, a tool of human emancipation. This is also an indictment of architecture, sloppy technique, speculation and real estate ownership, and everything that gravitates around them, of the conflict of partial interests with the common weal. The accusation extends to the administration, its incapacity to set priorities, to put the problems into order and to solve them in correct ways.

The different European situations, the peaceful and civil peripheries of Geneva or Bologna or Zurich, the outskirts of Frankfurt or Amsterdam or Stockholm, Paris, Madrid and London, or Toulouse, Naples and Palermo, today raise issues that are less than reassuring. The cause behind the issues and problems of each situation would not seem to be the per capita income of the country, its level of economic development; nor the institutional and administrative context, the functioning of the

---

[4]Source: Secchi, B. (1991). La Periferia. *Casabella*, 583, 20–22.

political system, the specific system of land management, the specific history of urban planning and its experiments, the size of the city, its status as a capital or a less important center. Each example seems to invalidate or at least to render seriously insufficient every one of these explanations that have been traditionally, and perhaps too hastily, called into service.

Prior to any plans or projects, today the periphery needs pertinent descriptions and specific explanations.

It might be worth starting with the blurriest images, going against the traditional approach and reaching the periphery from the "outside," from where the territory begins to be inhabited by populations no longer connected to rural ways of life, but scarcely resembling the "blasé" stereotype of the metropolitan man: "ni indiens, ni urbains," André Corboz would say. In vast regions of the center of Europe we realize we are immersed in urban space without being able to connect it to a center, without managing to attribute its formation to the explosive expansion of a capital. It is very hard to find the boundaries or to estimate the size of this immense urban space. It appears to us like the result of a flood: a homogeneous "mush" of single-family houses on lots, in which certain more substantial "lumps" are floating, chaotic clusters of incoherent material. Peripheries that swallow up, almost concealing and erasing, the framework of the historical cities that so deeply imprinted the geography of W. Christaller and the spatial analysis of the 1960s. Among the different "lumps," in disorder, like other coagulations, there are semi-solid "plaques and filaments" made of a different material: the towers of the new office complex, the big sheds of industrial zones, the "market streets." Enclosed by all this there are large green spaces, pieces of countryside absorbed in the urbanized expanse; agriculture that manages to resist constructed development in certain parts of Europe; areas awaiting future urban expansion, already invaded here and there by individual objects, in others. Anyone who flies over Europe, or just performs that mental exercise by observing maps, cannot fail to be amazed.

I propose examining the "lumps," the harder, more solid parts, from the viewpoint of the enormous apparently homogeneous expanse of houses on lots, and to avoid superimposing on that expanse an image and a vocabulary formed in other eras, when the city-lumps could perhaps still be conceived within a "rank-size rule" as "central places" surrounded by a "periphery" of residences, with the "void" of the countryside between one settlement and the next.

Born from the "lumps," as their "illness," "excrescence," "metastasis," as a "place into which everything that is residual and temporary is poured… symbol of rejection and attempt at the same time… foam that seethes along the walls of the city… becoming a tide and then a flood" (Le Corbusier), the periphery and, beyond it, the enormous expanse of extensive edification of space to which I am referring have now become the city. What I suggest is to recognize a "stable form" here; and to instead consider the solid, dense and complex parts, the "lumps, filaments and plaques" of the polycentric city, the historical cities and their immediate outskirts, the districts of the public city, the market streets, the industrial zones, as socially, functionally and morphologically unstable parts, for which new roles and possibilities can and must be grasped, through the project.

It is certainly difficult to connect quantitative estimates to these images. Every term, when inserted in propositions of this type, inevitably takes on a fleeting, metaphorical or unbearably conventional character; it is reductively referred to definitions, classifications and subdivisions that are not well suited to operations that attempt to say something more than what has already been said by the gaze. But it is undoubtedly safe to say that a large part of the European population, a percentage that is growing over time almost everywhere, lives inside the great expanse of extensively developed territory, i.e. inside a space that is not precisely city, but is also not countryside.

Subdivided into typologically, morphologically and socially recognizable parts as in the Belgian "green periphery," a "collection of standardized prototypes randomly scattered" as in France, "a place of planning subterfuge that fills the landscape with constructions of banal architecture and bad taste" as in England and Holland, in which we can observe the "ardent desire to be comme il faut and to follow fashion," as in the new German suburbs (Marcel Smets), an area which seems to host a concentration of the pretentious, the vulgar, but also of typological invention as in the metropolitan areas of Brianza and the "diffused city" of the Veneto, this "improvised, eclectic and libertine" part of the European city has usually attracted the moralizing criticism of the urbanist who observed excessive land consumption, and the more polite criticism of architects, who point out the ignorant use and contamination of technical and linguistic materials derived from the widest variety of sources. I would propose to observe the periphery with greater "pietas."

From the outset, the formation of the periphery has been associated with an enormous increase in physical and social mobility, made possible by modern means of communication: first the railroad, then the automobile. Gauting, Wannsee, Meudon, Twickenham: Rudolf Borchardt and Seebohm Rowntree, at the turn of the century, observed the phenomenon with different gazes, but discussed it in terms that are remarkably similar to those we use today.

In Italy, it is above all in this part of the urbanized territory that we can find—in Lombardy, the Veneto or the Marches—the "strapaese" of Giuseppe De Rita, "the parts of the society many felt should be modernized, considering them almost Vendean" … "the submerged economy of small businesses" … "the 'strapaese' that has changed itself, with unexpected vitality and outside the magisterial canons of the elites." In other parts of Europe this part of the territory often contains the heirs to a working-class population, who through fierce petit bourgeois individualism have redeemed themselves from the past, fulfilling long cultivated aspirations in the present.

The recent "Social Survey in Lombardia" and similar studies conducted in other European situations show that inside the apparent homogeneity of this urban space it is possible to find "new polarizations, new inequalities that the image of a 'post-industrial,' 'plural' socio-territorial reality runs the risk of concealing" (Antonio Tosi), "new forms of marginality and poverty," as well as new wealthy and exclusive "minorities." It is increasingly difficult to interpret their territorial distribution on the basis of the main rules of a traditional order. The descriptive efficacy of indicators like the center-periphery distance decreases over time, while the importance of the level, value and quality of housing, or of "ecological" factors positioned without any connection to the previously existing grid of urban centers, increases. Inside this area we can detect kaleidoscopic systems of economic and affective relations that in "a 'modern' organizational context, alongside and in interaction with large-scale networks and formal processes," assign increasing value to local and informal relations over time. Based in a non-secondary way on behaviors of mutual support that seemed to have vanished on the urban scene, or obversely on behaviors of identification-exclusion typical of "minorities," they lead to consolidation of short-range family networks that partially replace the organization of the "welfare state," and to the formation of the elitist district, isolated and protected from those who do not belong to the "minority" that lives there and identifies with it.

A more careful observation of the morphological characteristics of this vast area demonstrates, though hidden by new subjects, new activities and lifestyles and the new structure of consumption, the importance for much of contemporary society of historical continuity and habits, a theme dear to Odo Marquard.

Continuity and innovation, conservation and change seem to coexist and intertwine in these areas as in few others; in no part of the European territory has the past been so thoroughly erased and, at the same time, so completely conserved. The persistence of the memory of places, of place names of reference, of the irrigation ditch and the row of trees that marked property or crop lines, and their transformation into streets, lanes, walls, alignments; the growth of the house with the family, the yard and the vegetable garden, the division of tasks and time: the importance of continuity and habit is also expressed in these ways.

On maps of Brianza we can still see the minimal rationality of the old subdivision of the fields; it is even clearer in the Roman "aggeratio" of the "diffused city" of the Veneto, in the "pavillonaires" of the suburbs of Geneva, or the "banlieue radieuse" of Belgium… In the system of persisting and permanent features social practices and processes are also though not above all represented, which have been able to economically transpose their efforts, innovating without creating a "tabula rasa." One of the possible histories of social emancipation is portrayed, in which we can recognize "traditions, customs, habits of knowing and of acting" firmly rooted a large portion of the European populations. An emancipation through private initiative that is ambiguously opposed to the hypothesis of an emancipation by collective means that characterizes the program of research and experiments of modern urbanism.

In disorderly ways, the apparent homogeneity of this immense urbanized space is interrupted by "lumps," points of denser concentration, consisting of a chaotic collection of incoherent urban materials. Volumes and areas of different size, form, appearance, material consistency; witnesses to paratactically associated human and design experiences: the apartment house, the factory, tract housing, parking, the school, the service station, the church, another factory, the farmhouse, the supermarket, the office building, more parking, a leftover lot, a playing field, the workshop, the large industrial plant, the bus depot, the gasometer, the urban park, the sports facility and the dump. In a wide variety of literary forms, lists of this sort have become the conventional form for the description of a “world of objects” that until a short time ago we identified with the terms “periphery” or “suburb.”

This is the part of the European city in which historical continuity has been interrupted in the most evident ways. The part which, abandoned by the subjects and intentions that built and crossed it, inexorably slides, like our past, into the realm of the incomprehensible. Only archaeological digging could grant order to the accumulation of objects and experiences, to put into sequence the remains of the outskirts narrated by Pasolini, of the working class suburbs and villages, the developments of the subsidized housing institutes amidst the gasometers and rail yards, the wholesale markets, the prisons and hospitals expelled from the city center and now abandoned. Only archaeological digging could manage to reveal the sequence of the garden cities and developments in which the research program of modern urbanism has depicted itself, and those—beside them—in which it has displayed its relationship with unbridled real estate speculation. Visiting Praunheim, Hellerhof or Westhausen today, in the subsequent disorderly edification, or locating the large fragments of the Roman “public city” in the sea of illicit construction, means taking an embarrassing voyage in the history of human experiences and ideas. It is often appalling.

One is appalled by the inability-impossibility of most of these experiences, based on a program that was supposed to be demonstrative and cumulative, to become morphogenetic factors, in the widest sense of the term; generators of a new “compositional” rule of new and ancient materials, activities and urban subjects.

Those who visit the peripheral zones of modern urbanism today, the zones of the period between the two wars or those of decades closer to us, can recognize practices and processes in which an enormous quantity of social and figurative imagination is still represented. They can also observe the preference for many of these developments today, which are sought out, purchased, restored, cared for by old and new inhabitants, and by the more advanced, intelligent municipal administrations: the Lyon of Tony Garnier, the Frankfurt of Ernst May, the Stockholm of Uno Ahrén, the Geneva of Maurice Braillard, the QT8 of Piero Bottoni. The bewilderment takes over when we wonder what factors have prevented this city from fully developing, not granting it the time to be assessed, appreciated and understood; factors that have literally “pulled the rug out from under” and prevented the rise, in the strict sense of the term, of large “territorial projects” like Romerstadt or Berlin-Britz; that have combined heterogeneous and obtuse material without any order, which have replaced imagination with an encoded procedure, making the project of urbanism into a sort of “Parallel Action.”

This part of the European space, this “world of objects” that in ever more disturbing ways fills contemporary figurative culture, represents one of the most salient characteristics of our time. Here, the complexity of our society rebels against “any reason that claims to understand, judge and dominate it, against any basis that attempts to indicate a path, hailed as a liberation from tyrannical models; the Babel of languages is praised, words released from all meaning, the otherness of details that cannot be traced back to universals. The decline of the basis, of any unitary reason, is undeniable; what bears discussing is whether this is always a joyous decline" (Claudio Magris).

Encapsulated inside the clustering debris of edified materials of the periphery, the walls, the pavements, the gardens of urban centers laden with history are gradually emptied of their inhabitants, activities and roles. Literally gutted inside their buildings, they have become increasingly "abstract"; in many of their parts, they have fully consummated the split between the cumulative character of the past and the provisional character of the present, between the form of materials in which "traditions, customs, habits of knowing and acting" are deposited and the function now performed inside each building. Those who visit the European historical city today in an aware manner remain bewildered: the working-class apartment has become an elegant residence, the large hotel a bank headquarters, the noble palace houses the functionaries of an international agency, the old bourgeois districts now occupied by recent immigrants, make the city resemble a "trompe-l'oeil" on which our disquiet that has become historical sensibility is exercised.

The populations that live in the diffused city or the periphery like these simultaneously unitary and complex spaces. They grasp their "collective" character, and they flock to them in droves, on a daily, weekly or seasonal basis, to take a walk, to go shopping, to see and be seen. All this generates a rapid slippage of these places toward something "other," toward a reduction to the status of a bazaar, with connected services and parking.

A description is perhaps already an interpretation and a project: what is taking form in Europe is a new urban "ecology," in the sense of the term used by Reynar Banham; actually a multiplicity of new ecologies, layered and intersecting. They have not yet found their "representation," the urban and architectural project that adequately represents them and permits them to completely develop. This is the aspect which, more than anything else, should prompt us to rethink the entire urban phenomenon in Europe. It is no longer just an expansion of the city of the nineteenth century. The abandonment of large industrial areas, of large infrastructures, large technological plants, of the spatial structure itself of the city of the 1800s thanks to which the contemporary city, until now, has been able to survive, is an indication and a late symbol of this fact. The ways in which, all over Europe, these large areas are subdivided, cut into pieces, ground up as a result of procedures without any real content, is an indication of the total absence of ideas in which this parting takes place.

Yet the discontent regarding the "world of objects" in its concretely achieved guise is widespread. To a great extent, it is also highlighted by important social practices: the flight towards the privacy of the extended city and the use of historical space, of its character that is "designed" and "vague" at the same time, precise in its forms and open to the widest range of uses, for the celebration of everyday collective rituals: shopping, entertainment, recreation, expression of opinions.

Perhaps what we need is a new overall project; not a series of projects, in succession, but a single unified project that selects and at the same time defines, on different scales, the relevant themes, embracing diversity and singularity without repressing them, but giving them a complete meaning. A project, above all, that has the courage to subject itself to the falsification of the future. The center of this comprehensive and unitary project lies perhaps in the solid and semi-solid parts, clustered or in the process of doing so, between the historical centers and the zones of suburban housing, in all their instability, that is their possibility to be "other" from a morphological, functional and social viewpoint; to build the "structure" of the entire urbanized space.

It is futile to expect to modify the character of inhabitable space in the contemporary city through zones of dubious "high quality" distributed in physical and social space either randomly or in keeping with corporate or political-administrative logic. Precisely the necessary marginal, punctual, singular character of everyday intervention should bring us back to imagining large "frames," large open structures in which to define the singularity of existing materials: the enclosed scrap of countryside, the zone of tract housing, the commercial drive, the market street, the business center, the industrial zone, the quarter of the "public city," the freight yard, the parking lot, the sports facility. We have to imagine interventions that will complete the frame, slowly modifying it, bending it until it takes on new meanings, until a new inhabitable space has been built.

I know that there are many difficulties, but I also know that all this is concretely possible, and partially desired; and that in certain parts of Europe new policies are making concrete strides in this direction.

Fig. 23.1 Piano Regolatore of Bergamo, Bergamo est (extract). Interpretative map of the productive activities (operatig activities in black; discontinued activities in white).
Source: Secchi, B., Gandolfi, V., 1994, Bergamo. Progetto del Nuovo Piano Regolatore Generale, Bergamo, Comune di Bergamo. Published also in: Secchi, B. (1994). *Tre piani, La Spezia Ascoli Bergamo*, C. Bianchetti (Ed.). Milano: FrancoAngeli, p. 139

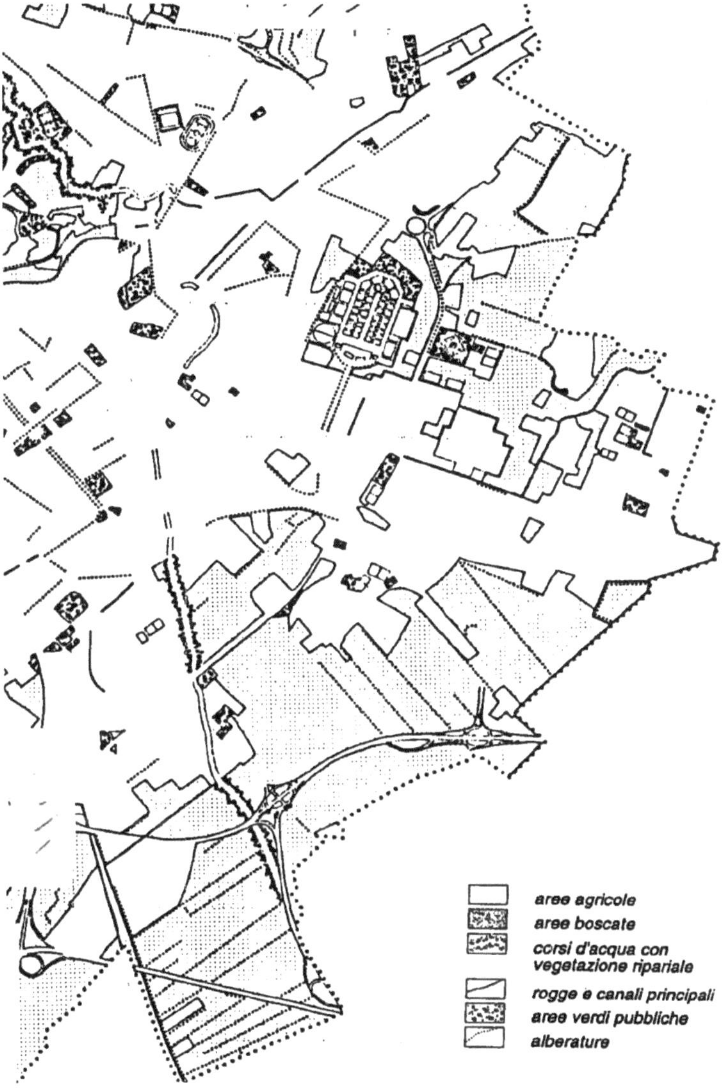

Fig. 23.2 Piano Regolatore of Bergamo, Bergamo est (extract). Interpretative map of the "landscapes within and around the city".
Source: ibid., p. 147. Reprinted courtesy of IDEM

Chapter 24
La Ville-Territoire (The City Territory) (1991)

André Corboz

Introduction

Marine Durand
Lab-U (EPFL), Lausanne, Switzerland
marine.durand@epfl.ch

When André Corboz used the well-known metaphor of the "palimpsest"[1] *to describe the long-term process of stratification, accumulation but also modification of the territory, he already rejected the old antagonism between city and countryside, stating that the foundation of any project must take hold form the territory itself, which needs to be well investigated and understood in its historical and physical depth. With the article "Towards the City Territory" published later, in 1990, André Corboz warns his readers of the danger of the widespread refusal for this new urban condition. Built around a brief but convincing historical reconstruction of the evolution of the city, the article denounces the inability of urbanism to recognize the emergence, development, and above all the extent of this new form of city. An inability of this kind, in the opinion of the author, impedes the understanding of this new condition and—consequently—the adoption of the necessary ways and means to guide the transformation of the same.*

Except from the article's title, the term of "City Territory" isn't used in the text, but André Corboz clearly refers to the figure of a city that is "coextensive" with the whole territory to describe the contemporary urban condition in Europe and more specifically on the Swiss Plateau, especially when referring to Meili's proven prediction of a "Großstadt". After having developed the metaphor of the "Palimpsest" and the figure of the "City Territory", Corboz will work and develop

[1]«Le territoire comme palimpseste» in *Diogène* 121 Jan-March 1983 pp. 14–35.

M. Barcelloni Corte and P. Viganò (eds.), *The Horizontal Metropolis*,
https://doi.org/10.1007/978-3-030-56398-1_24

the metaphor of "hypercity" in 1997.[2] *As happened for other authors collected in the second part of this book, taking hold from the analysis of a specific territorial condition, the Swiss Plateau, André Corboz will propose a general theory to describe the condition and transformation of "human settlements of wide dimension in the Western world".*

It is interesting to note that, in all the three texts mentioned above, the author refers to J.J. Rousseau's portrait of "Switzerland as one big city", outlining Rousseau as a visionary, closer to reality than the contemporary Swiss people (but we could say Europeans) themselves. If the current mental representation of the city can no longer be deemed valid, and if categorisation—in terms of city and countryside—is gradually yet inexorably becoming obsolete, a completely new awareness and imagery needs to be triggered.

24.1 Towards the City Territory[3]

If current trends within the urban phenomenon are generally perceived as a loss of identity, this is partly due to the fact that our idea of the city has remained stuck in the pre-industrial era. Before the revolutionary age, cities were distinct from the countryside both physically (thanks to their walls) and legally (almost everywhere, country dwellers were their subjects). Three factors working in succession helped to dismantle this opposition: advances in artillery, explosive population growth and the transport revolution.

The first of these developments led many cities to demolish their fortifications, creating new cities without any defensive system. This transformation started in the second half of the eighteenth century and reached its peak in the nineteenth century. Now, planned districts provided a smooth transition between the urban core and the suburbs. From this time forward the city was structured in two zones: the centre and the periphery—the latter constantly expanding.

Thanks to the combined effects of improved nutrition, the disappearance of certain endemic diseases, and medical advances, populations boomed—their growth rate was unprecedented. For complex reasons, the rural population emigrated to urban areas where it would constitute industry's "reserve army". This trend accelerated after 1900, reaching spectacular proportions after the Second World War.

The creation of public transport systems, and the railway in particular, made it possible for the home and the workplace to be physically separated; for an

[2]In 1994, the idea of hypercity in analogy with hypertext is shortly evoked in his text "Apprendre à décoder la nébuleuse urbaine" in *Les Cahiers de l 'Institut Art et Ville*, 8. 1994. pp. 4–12.

[3]Source: Corboz, A. (1990). *Vers la Ville-Territoire*. In M. Haller, H. Hauser, & R. Zaech (Eds.), Ergänzungen, Ergebnisse der wissenschaftlichen Tagung anlässlich der Einweihung des Ergänzungsbaus der Hochschule St. Gallen (pp. 631–635). Bern and Stuttgart: Verlag Paul Haupt. © English translation: Susan Mackervoy/Ros Schwartz Translations, London

increasingly large section of the population it became possible to work in the urban centre while living on its periphery.

Other factors also contributed to this general trend—the doctrines of the Physiocrats, for example, and above all their laissez-faire development policy, with its accompanying use of urban land rent as the primary mechanism driving the physical transformation of urban areas.

As a result cities saw explosive growth, sprawling into their surrounding areas and becoming conurbations. The changes were not just observed within the cultural sphere, they had also been anticipated to some extent. The idea of a city coextensive with an entire territory appeared as early as the seventeenth century: in 1676, Gabriel de Foigny [wrote a utopian novel, The Southern Land, whose] protagonist, Jacques Sadeur discovers a continent divided into 15,000 squares, each comprising 16 districts with 25 houses inhabited by 16 people. Simon Tyssot de Patot picked up this idea in 1710, in his imaginary account of a country without towns or cities, where the population is spread evenly across the whole territory, organised according to a strictly regular grid pattern of canals and roads.

America's Land Ordinance of 1784, written by Thomas Jefferson, proves that such visions were not merely the ironic or deranged products of a systematic mindset: it calls for the Union's future territory, west of the 13 founding states, to be divided into identical squares, in order to distribute the future population evenly across the whole of the territory. Jefferson, anti-urban and a Physiocrat, wanted an agricultural, Rousseauist republic—and this is what happened: today, 78% of the United States is indeed divided up according to this regular grid principle.

An important but little-known text—relating to a very different sphere, but viewed from a similar angle—casts a completely new eye on the relationship between human settlements and their geographic base: "The whole of Switzerland is like one large city divided into thirteen districts, some in the valleys, others on the slopes, others on the mountains (…). There are districts that are more or less populated, but all are sufficiently so to indicate that one remains within the city." Date: 1763. Author: J.-J. Rousseau. From our perspective 18th-century Switzerland was positively idyllic—which makes this urban metaphor all the more surprising.

Shortly after this, various urban projects came into being which aimed to establish a continuity between town and countryside. The fourth plan for Carouge —the town created by Piedmontese architects in the current canton of Geneva— from 1781, for example, and the second plan for the Royal Saltworks at Arc-et-Senans, dated 1774 but actually produced after 1790 and only published in 1804.

Nineteenth-century theorists would pose the same problem on a very different scale. In 1873, the Catalan Ildefonso Cerdà envisaged the "territorialisation of the city" following a formula he himself had proposed in 1867: "Ruralise the city, urbanise the countryside: replete terram!" In 1882, Arturo Soria y Mata of Madrid introduced the idea of the linear city, a 500-m-wide urbanised strip of unspecified length that would span continents, connecting existing cities. In both cases the aim was to combine the best aspects of urban and rural life while removing the disadvantages specific to city and countryside respectively. The same principle informed the English concept of the garden city, introduced by Ebenezer Howard in 1898; with their controlled size and their balanced mix of activities, garden cities were intended to form a network covering the entire territory, taking the place of urban agglomerations.

Another fundamental aspect of the modern city emerges in the work of various theorists towards the mid-eighteenth century. Confronted in Rome initially and then in Paris by the problem of large-scale urban development, they abandoned the old ideal of intervention via tabula rasa and the imperative of regularity that justified this approach. The city had become too complex to permit anything but partial interventions, minor adjustments, simple alterations, adapting to the needs of what had now become a polycentric system. From the works of Lione Pascoli (1733) to those of the abbé Marc-Antoine Laugier (1753 and 1765), via the famous pseudo-archaeological reconstruction of Piranese's Campo Marzio (1762) and the famous engraving in which Pierre Patte grouped together the monuments planned for Paris to the glory of Louis XV (1765), there was a tremendous ferment of ideas and morphological proposals attempting to give form to the urban irregularity.

Yet the city was not just complex—it was dangerous, too. In 1769 the same Pierre Patte published a treatise in which the siting of noisy and foul-smelling activities, hospitals, cemeteries and prisons, waterworks, drains, waste disposal and public lighting are for the first time regarded as an integral element of the architect's concerns. This development marks the birth of sanitarianism, the discipline that would dictate, via the principle of zoning, most of the urban innovations of the following century.

The city progressively ceases being an artefact that celebrates the sovereign's glory, becoming instead a system of economically rational functions and channels: no longer a static space for parades but a network ensuring efficient supplies and services, public health and its citizens' safety.

Not until the twentieth century would geographers describe the dominant features of the new urban reality which, building on these various foundations, became established from around 1850. First of all they noted merging phenomena, among which the cases of Roubaix-Tourcoing and Elberfeld-Barmen are the most well-known (in the second case the new entity was renamed Wuppertal in 1930); in 1915, Patrick Geddes coined the term "conurbation" for ensembles of neighbouring towns forming a system; in 1961 the term "megalopolis" was used to describe the loose urban conglomeration extending from Philadelphia to Boston as a single integrated complex. Today, the Swiss Plateau constitutes a "Großstadt"—a metropolis—in terms of its formation: communications of all kinds within the area extending from St. Gallen to Geneva exhibit a greater intensity than those connecting the same ensemble to neighbouring areas.

These various considerations show that the urban transformations shaped by the Industrial Revolution were accompanied by an ongoing discussion of the process—of which many more examples could be cited, especially in the social sphere. And yet, because urbanism is not part of the general culture, this transformation was only perceived more widely after the fact, and most often with dismay. This process of accelerated territorial transformation would, it was asserted, threaten the identity of cities (but do cities really still exist?), of races, even of nations.

Certainly no-one enjoys witnessing the brutal urbanisation of particular landscapes, the destruction of coastlines by the hotel industry, the conversion of a historic local site or feature into a caricature of itself to appeal to tourists. Yet globally rejecting the zones that currently form Europe's urban backbone is dangerous because it stops us imagining what the new urban condition might be and prevents us creating the resources we will need to guide its evolution. What it is absolutely vital to understand is that our traditional mental representation of the city simply no longer applies.

At a time when the media now almost exclusively define our lifestyles based on uniform sociocultural stereotypes, the city/countryside opposition is losing its meaning. The city as a built environment characterised by a standard, unified template and the contiguous arrangement of its constructions now represents no more than 5 to 10% of the inhabited surface area. If we object that this 5 to 10% comprises urban centres, which are precisely what defines (subjectively) the city concept, it is easy to answer, on the one hand, that the Americans—who do after all have a certain number of urban settlements of appreciable size—regard their centres, or what replaces them, with fear or even abhorrence: for them the real city is actually the suburb.

On the other hand, a European informed that he or she would one day have to give up the concept of the city centre would regard such an assertion as positively perverse—this despite the fact that the urban sectors we regard as "centres", for historic reasons, have nearly all changed their function and therefore their meaning: they have been turned into business districts, divided up, multiplied. City centres are gradually being replaced by "central areas" hosting activities that relate to multiple territorial scales: communal, cantonal, national, continental, global.

Yet to address this issue of scale and to combat the amorphous character (I freely admit it!) of the European megalopolis-in-progress, what are we being offered? Almost exclusively solutions conceived by looking in the rear-view mirror. Homages to the small town, to medieval communities, even the dörfli (little village). Models such as this are not just inadequate—they are mystification, pure and simple. The medieval town, always presented as an ideal of harmony and managed development, was equally remarkable for its economic, sanitary and cultural backwardness—and that is without counting the fact that it embodied a strictly hierarchical, privilege-based society.

Rather than limiting ourselves to manically cataloguing losses of a historic identity, we need to be focusing just as much effort on discerning the new, emerging identity. This cannot happen without some painful revisions, starting no doubt with that of unfettered liberalism. And none of this will be possible while we fail to recognise that the cut-off point established by the Enlightenment had something irreversible about it—making obsolete the idea we continue to harbour of the city, along with many other received ideas.

The future megalopolis, identical with the territory it occupies, will contain a multitude of non-urban spaces which we will call nature. It will consist of multiple networks and will appear, to backward-looking eyes, like a kind of generalised non-place, because its interwoven hierarchies will perhaps be neither cumulative nor even apparent. To define it we could invert Pascal's famous saying: our descendants will live in "cities" whose circumference will be everywhere and whose centre will be nowhere … which does not necessarily mean that they will be worse than ours.

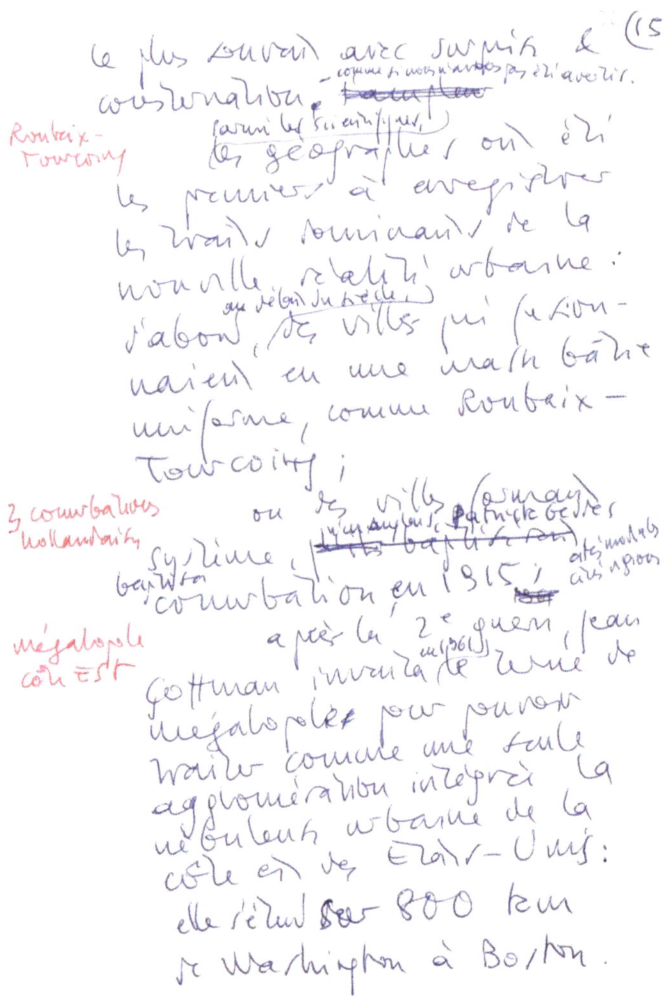

le plus souvent avec surprise & (15
consternation, comme si nous n'avions pas été avertis.
parmi les scientifiques,
les géographes ont été
les premiers à enregistrer
les traits dominants de la
nouvelle réalité urbaine :
d'abord, au début du siècle, des villes qui se fusionnaient en une masse bâtie
uniforme, comme Roubaix-Tourcoing ;
ou des villes formant système, qu'un Anglais, Patrick Geddes baptisa conurbation en 1915, cités modèles cités-région
après la 2e guerre, Jean Gottman inventa (en 1961) le terme de mégalopole pour pouvoir traiter comme une seule agglomération intégrée la nébuleuse urbaine de la côte est des Etats-Unis : elle s'étend sur 800 km de Washington à Boston.

Roubaix-Tourcoing

3 conurbations hollandaises

mégalopole côte Est

Fig. 24.1 «Towards the City Territory» notes for a lecture (page 15) "most often with surprise & dismay, as if we had not been warned. [Roubaix-Tourcoing] among scientists, geographers were the first to record the dominant features of the new urban reality: First, at the beginning of the century, cities that merged into a uniform built mass, such as Roubaix-Tourcoing; [3 Dutch conurbations] or cities forming a system that an Englishman Patrick Geddes named conurbation in 1915, model city city-region; [megalopolis east coast] after the Second World War, Jean Gottman invented the term megalopolis in 1961 to describe the urban nebula on the East coast of the United States as a single integrated agglomeration: it spreads over 800 km from Washington to Boston". Source: Corboz, A. (1989), «Vers la ville-territoire» and «De l'antagonisme ville-campagne au territoire urbanisé». Notes for two lectures in the French section of the ASPAN (Swiss Association for the national planning—*Association suisse pour l'aménagement national*) now renamed as EspaceSuisse. Saint-Gallen (07/06/1989) and Fribourg (14/10/1989). © Fondo André Corboz, Accademia di architettura Mendrisio, Università della Svizzera italiana

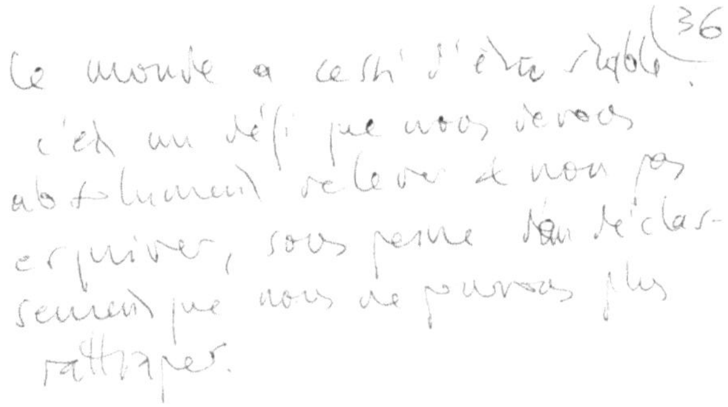
36
Le monde a cessé d'être stable.
c'est un défi que nous devons
absolument relever et non pas
esquiver, sous peine d'un déclas-
sement que nous ne pourrons plus
rattraper.

Fig. 24.2 «Towards the City Territory» notes for a lecture (page 36) "The world has stopped being stable: it is a challenge that we must absolutely take up and not duck, otherwise we will no longer be able to catch up".
Source: ibid. © Fondo André Corboz, Accademia di architettura Mendrisio, Università della Svizzera italiana

Fig. 24.3 View of Geneva from the Salève, 1997, picture taken by André Corboz. © Fondo André Corboz, Iconoteca dell'Accademia di architettura Mendrisio, Università della Svizzera italiana

Fig. 24.4 "Aerial view of Halen (Kirchlindach) in relation to Bern", second half of the 70s. © Fondo André Corboz, Iconoteca dell'Accademia di architettura Mendrisio, Università della Svizzera italiana

Chapter 25
Desakota (1991)

Terry McGee

Introduction

Qinyi Zhang
Lab-U (EPFL), Lausanne, Switzerland
qinyi.zhang@epfl.ch

Through the neologism "desakota", Terry McGee indicates a newly emerging settlement system, a phenomenon linked to a strong space-economy transition influencing specific Asian agricultural contexts, primarily engaged in wet-rice cultivation and producing territories with dynamic linkages between agricultural and non-agricultural activities. The author observes the gradual emergence of a new form of socio-spatial organization where "zones of intensive economic interaction between rural and urban domains" characterize increasingly large portions of metropolitan regions "stretching along corridors between major cities". Vast rice cultivated areas, due to their closeness to strongly growing urban centers and/or to important infrastructural systems, were featuring ever more "closely interlinked, co-penetrating rural/urban livelihoods, communication, transport and economic systems". While witnessing "desakota"'s different degrees of development, the author questioned the transitory nature of the phenomenon, its possible evolution and its fragility in relation to attraction forces coming from main urban cores and globalization.

To answer this question, McGee's further research explores an evolution of the desakota concept. If the first version[1] *emphasises the role of context, specificity,*

[1]McGee (1991, pp. 3–26).

T. McGee (✉)
UBC, Vancouver, Canada
e-mail: tmcgee@geog.ubc.ca

M. Barcelloni Corte and P. Viganò (eds.), *The Horizontal Metropolis*,
https://doi.org/10.1007/978-3-030-56398-1_25

and local activities, the later version[2] *attempts to integrate the increasing impact of the expansion of urban cores and the construction of new towns and industrial parks within the peri-urban and desakota area, to combine "above-ground knowledge" with "grounded knowledge".*[3] *However, the essence of the desakota concept—the mixture of agricultural and non-non-agricultural activities—remains while developing further into "the main zones of intense negotiation, contestation, and reconciliation of the forces coming from all social and political scales".*[4]

Today, the "city-centered" urbanization –with a poor awareness of the "grounded territorial knowledge"– has been predominant in the development of many rapidly urbanising regions, such as –for example– the Yangtze River Delta area, which today is confronting its economic, social, and environmental limits. In facing such a rapid transition exacerbated by global challenges such as climate change, the desakota concept becomes extremely useful (especially in its later form) to imagine and conceive, for these regions, a completely new urban paradigm.

25.1 The emergence of Desakota Regions in Asia (from: *The Extended Metropolis: Settlement Transition in Asia*)[5]

The emergence of Desakota Regions in Asia: Expanding a Hypothesis

The symbiosis of urban and rural in Megalopolis, creating new and interesting patterns of multiple-purpose land use over large areas, gives to the region a rather unique character. Like the downtown business districts with powerful skylines, this aspect of Megalopolis will probably be repeated in slightly different but not too dissimilar versions in many other regions of the rapidly-urbanizing world.

Jean Gottmann, Megalopolis: The Urbanized Northeastern Seaboard of the United States (New York: Kraus, 1961), p. 257.

Paddy has developed a strikingly similar landscape, broadly similar from the Ganga to the Yangtze... but no other way of life... has led to the evolution of a cultural system so stable and permanent as that associated with the great paddy-plains of Monsoon Asia.

O.H.K. Spate and A.I.A. Learmonth, India and Pakistan: A General and Regional Geography (London: Methuen, 1967), p. 202.

---

[2]McGee et al. (2007).

[3]McGee (2016).

[4]McGee et al. (2007).

[5]McGee, T. G. (1991). The emergence of Desakota regions in Asia: Expanding a hypothesis. In N. Ginsburg, B. Koppel, & T. G. McGee (Eds.), *The extended metropolis: Settlement transition in Asia*. Honolulu: University of Hawaii Press, pp. 3–26.

This chapter is a preliminary exploration of the emergence of what appear to be new regions of extended urban activity surrounding the core cities of many countries of Asia.[6] The ideas to be explored must be placed in the framework of the overall patterns of urbanization at a global and regional level, which are predicting a continuing increase in the proportion of the world's urban population. By the year 2020, the U.N. Centre for Human Settlements predicts that more than 57% of the world's population will be living in urban places.[7] This population will be unevenly urbanized, with levels of urbanization at almost 77% in developed countries and 53% in developing countries. Within the developing countries, the contrast will be even greater, with Latin America 83% urbanized, and Africa and Asia close to 50%. However, Asia's urban population will account for a very large portion of that of the developing countries. Bangladesh, India, China, Indonesia, and Pakistan together will contain 34% of the developing countries' urban population.

These United Nations predictions are largely based on assumptions concerning the growth of population in places defined as urban. The predictions are calculated using growth rates reflecting performance in previous decades. When projected forward, they appear to suggest a successful shift to urbanized societies and a repetition of patterns of the more developed countries. As Ginsburg has commented about urbanization in the United States, "This condition reflects the progression of the … space-economy to a state of what one might consider 'maturity,' that is, to a condition whereby areas possessed of substantial comparative advantage... would be drawn effectively, through improved transportation networks, into the national geographic structure."[8] The implication for the urban systems of the largest developing countries is that a continued growth will create cities of the immense size of about 16 million to 30 million. However, this may not be the only possible outcome for Asian urbanization.

The purpose of this introductory chapter is to draw upon earlier ideas put forward by McGee and Ginsburg to challenge this particular view of the urban transition.[9] The Ginsburg-McGee position essentially argued that in the Asian context the conventional view of the urban transition, which assumes that the widely accepted distinction between rural and urban will persist as the urbanization process advances, needs to be re-evaluated. Distinctive areas of agricultural and nonagricultural activity are emerging adjacent to and between urban cores, which are a direct response to pre-existing conditions, time-space collapse, economic change, technological developments, and labor force change occurring in a different manner and mix from the operation of these factors in the Western industrialized countries in the nineteenth and early twentieth centuries.

---

[6]Some of the ideas in this chapter were first put forward in an address to the International Conference on Asian Urbanization held at the University of Akron in April 1985. See McGee (1989). See also Chap. 3 in this volume.

[7]U.N. Centre for Human Settlements (1987).

[8]Ginsburg (1990).

[9]See McGee (1987). See also Ginsburg, "Urban Transition.".

To elaborate further, the conventional view of the urban transition is inadequate in three respects. First, it is too narrow in its view that the widely accepted spatial separation of rural and urban activities will persist as urbanization continues.

Second, it is inadequate in its assumption that the urbanization transition will be inevitable because of the operation of "agglomeration economies" and comparative advantages, which are said to facilitate the concentration of the population in linked urban places. The emergence of such a system was described by Jean Gottmann in 1961 as a "megalopolis" in which, when applied to the northeastern United States, the population was largely concentrated in the urban and suburban areas, but interspersed with areas of low population density used for agriculture and as leisure spaces by the population of the megapolitan areas.[10] In many parts of Asia, the spatial juxtaposition of many of the larger city cores within heavily populated regions of intensive, mostly wet-rice agriculture based on a mixture of "skill oriented" and "mechanical" technological inputs has created densities of population that are frequently much higher than in the suburban areas of the West.[11] This juxtaposition permits demographic densities similar to urban areas over extended zones of intensely cultivated rural areas located adjacent to urban cores. The considerable advances in transportation technology, particularly in relatively cheap intermediate transportation technology such as two-stroke motorbikes, greatly facilitate the circulation of commodities, people, and capital in such regions, creating, in turn, large mage-urban regions.

Third, the Western paradigm of the urban transition, which draws its rationale from the historical experience of urbanization as it has occurred in Western Europe and North America in the nineteenth and twentieth centuries, is clearly not neatly transferable to the developing countries' urbanization process. The uneven incorporation of these Asian countries into a world economic system from the fifteenth century onward created divergent patterns of urbanization, which reflect the different interactions between Asian countries and the world system.[12] For example, the British, French, and Dutch all developed the productivity of wet-rice agriculture in Southeast Asia.[13] In a similar manner, Japanese rule in Korea and Taiwan further accentuated the monocrop rice characteristics of parts of these countries as sources of supply for Japan's prewar empire. Geopolitical events determined that both these

[10]Gottmann (1961).

[11]This division between "skill-oriented" and "mechanical" technological inputs is used to buttress Bray's provocative arguments concerning the distinctive role of Asian wet-rice agriculture in the agrarian development processes in Asia. She argues persuasively that this agroeconomic system has created very different conditions from the "Western model." This position in certainly central to some of the arguments of this chapter, but its acceptance does not rule out the application of the "mechanical" technological input (as reflected by capital replacing labor in these regions), as can be seen in the growth of agribusiness in the chicken industry in areas such as the Central Plains of Thailand. See Francesca (1986).

[12]This argument is presented in much greater detail for Latin America in Morse (1974); and idem, "The Development of Urban Systems in the Americas in the Nineteenth Century," *Journal of Interamerican Studies* 17:1 (1975), pp. 4–26.

[13]See Baker (1981).

countries emerged into "fragile" independence with high rural densities and low levels of urbanization. On the contrary, British intervention in Malaysia created an urban system oriented to the production of export products on the west coast away from the heavily populated rice bowls of Kedah and Kelantan, limiting the possibilities of an emergent mega-urban region.

Because of these inadequacies in the conventional view, the concept of the "urban transition" needs to be positioned within a broader paradigm of the transition in the space-economy of countries. Such a paradigm would include (1) a heightened sensitivity to the historical elements of the urban and agrarian transition within specific countries; (2) an appreciation of the ecological, demographic, and economic foundations of the urban and agrarian transition; (3) an investigation of the institutional components, particularly the role of the state in the development process; (4) a careful evaluation of the transactional components within given countries including transport, commodity, and population flows; and (5) a broad understanding of the structural shifts in the labor force reflecting economic change. Essentially, such an approach is an attempt to investigate the manner in which particular sets of conditions in one place interact with broader processual change. It is not much concerned with the contrast between rural and urban as the space-economy changes but focuses instead on the interactions within the space-economy as they affect the emergence of particular regions of economic activity. This view has important implications for policy making in the Asian context, for it poses a challenge to sectoral approaches to development planning.

Definitions and Parameters

Since this assertion is quite challenging to those who have vested interests in the persistence of the urban-rural paradigm, it is necessary to spell out in some detail the definitional components of this broader view of the "space-economy transition."

Figure 1.1 presents a model of the spatial configuration of a hypothetical Asian country, which I will label Asiatica Euphoria for the purposes of this exercise.[14] In this example, five main regions of the spatial economy are identified as follows: (1) the major cities of the urban hierarchy, which are often dominated in the Asian context by one or two extremely large cities. (2) The peri-urban regions, which are those areas surrounding the cities within a daily commuting reach of the city core. In some parts of Asia, these regions can stretch for up to thirty kilometers away

[14]The diagram in Fig. 1.1 was constructed by Dr. M. Douglass of the Department of Urban and Regional Planning, University of Hawaii, and myself during the course of some extended evening discussions on the subject of the "urban transition." I am very grateful to him for his constant probing of my ill-formed ideas that formed the basis of earlier presentations of his model. See Douglass (1988).

from the city core. (3) The regions labeled *desakota*,[15] which are regions of an intense mixture of agricultural and nonagricultural activities that often stretch along corridors between large city cores. These regions were previously characterized by dense populations engaged in agriculture, generally but not exclusively dominated by wet-rice. (4) Densely populated rural regions, which occur in many Asian countries, particularly those practicing wet-rice agriculture. (5) Finally, the sparsely populated frontier regions found in many Asian countries that offer opportunities for land colonization schemes and various forms of agricultural development.

The model of the spatial economy is, of course, static and must change as the economy changes. The pace and characteristics of this settlement transition vary from country to country, reflecting the features of socioeconomic change at the macro level. The role that the growth of metropolitan cores and the *desakota* process play in this transformation is of major importance. The mega-urban regions that emerge often incorporate two large urban cores linked by effective transportation routes. These regions include the major cities, peri-urban zones, and an extensive zone of mixed rural-urban land use along such routes. Travel time between any two points in a region would probably be no more than three to four hours but in most cases is considerably less. Mixed economic activities may also occur in villages in these zones, which are less accessible and where economic linkages are more reliant on social networks.

It should be stressed that this model of the transition of the space-economy is not intended to be universally applicable, but to fit the situation where one or more urban cores are located in densely settled peasant rural areas.[16] There may also be cases where the urban cores are located in lightly populated regions of plantation agriculture as in the case of Kuala Lumpur in Malaysia. The contrast between the two agroeconomic systems as they are reflected in socioeconomic systems, export trade, and class relations is not a new theme. It has been used by Baldwin for a theoretical exposition of patterns of development in newly settled regions, by Dowd to explain the differences in the settlement patterns of the American west and south, and by Morse to explain the different urban systems that evolved in the "hacienda" and "plantation" regions of Latin America.[17] These writers are not suggesting that a particular urban system results from a pre-existing agroeconomic system, but rather

[15]The use of a coined Indonesian term taken from the two words *kota* (town) and *desa* (village) was adopted after discussions with Indonesian social scientists because of my belief that there was a need to look for terms and concepts in the languages of developing countries that reflect the empirical reality of their societies. Reliance solely on the language and concepts of Western social science, which have dominated the analyses of non-Western societies, can lead to a form of "knowledge imperialism." In this text I have used the term *desakota*, which can be used interchangeably with *kotadesa*.

[16]The term "peasant" in this context applies not only to those farmers who own their land but also to tenants operating small units of farmland.

[17]Baldwin (1956), Dowd (1956). See also Morse, "Trends and Patterns"; and idem, "Urban Systems in the Americas.".

that the existence of these agroeconomic systems provides the possibility for the emergence of certain urban systems and regions.

In the Asian context, the existence of high-density agricultural regions adjacent to large urban cores offers an opportunity for a particular form of mega-urban region to emerge. Their existence does not ensure the inevitability of the emergence of such regions. These will result from, for example, the policies of private and public sectors, the form of economic growth, and the position of the urban core relative to international connections. In the case of the Republic of Korea, with a precondition of high-density rice regions, on the one hand, the government adopted a strategy of concentrating on industrialization rather than agriculture, which led to slow growth in rural income and a release of surplus rural population into urban-based industrialization. Thus, South-Korea was characterized by a metropolis-dominated urban hierarchy. On the other hand, in a region of similar pre-existing rural densities such as Jogjakarta in Java, in a slow-growth situation there are only limited possibilities for drawing off surplus rural population to urban centers in other parts of the country, and the rural inhabitants engage in an intense mixture of nonagricultural and agricultural activity that permits survival but does not increase income.

There are at least three types of spatial economy transition occurring in Asia in regions that have the prerequisite of the historical evolution of high-density, mostly rice-growing agroeconomic niches.

First, there are those countries that have seen a decline in rural settlement, land use, and agricultural population as the population has moved to urban centers. In such countries, agricultural land use may remain important as a reflection of government land or agricultural protection policies. This pattern has been associated with overall increases in income and productivity in which rural populations fall well behind those of urban areas. South Korea and Japan are examples of such a spatial economy transition. These countries will be labeled *desakota* Type I or *konjuka*, the Japanese term for landscapes that have a mixture of small farm plots, residences, and industry. Such regions are characterized by rural landscapes in which most of the economically active work is in nonagricultural activities.[18]

Second, there are those regions in which, over varying periods of time, productivity gains in agriculture and industry, and secular shifts from agricultural to nonagricultural activities are focused particularly on the urban cores and adjacent regions. These changes are linked to rising household income, improved transportation linkages, and improved infrastructures. Examples occur in regions such as Nanjing-Shanghai-Hangzhou, the Central Plains of Thailand, the Taipei-Kaohsiung corridor, the Calcutta region, and Jabotabek in Java. These may be identified as *desakota* Type 2 and are characterized by rapid economic growth compared to other regions of the country.[19]

---

[18]Oshima (1986).

[19]For examples, see Chaps. 7, 8, 10 and 12 in this volume.

Third, there are those regions of high density in which economic growth is slow. Often such regions are located close to secondary urban centers that have slow economic growth and are characterized by continuing high population growth, surplus labor, and persistent low productivity in both agriculture and nonagriculture. Examples are the Jogjakarta region in Java, Kerala in South India, Bangladesh, and the Sichuan Basin in the interior of China. These regions will be labeled *desakota* Type 3 and are characterized by slow growth of income and involuntary economic activity.[20]

The Emergence of the Extended Metropolitan Region in Asia

Using data provided by a number of sources, Table 1.1 attempts to present some statistical information on these regions. Much of this material is taken from Vining, who documented the process of population redistribution toward what he labels the core areas of forty-four less developed countries. His definition of core areas as "the regions containing and surrounding the country's most important and dominant city (in a few cases, cities), which is generally but not always the capital city,"[21] is certainly not coterminous with the areas defined as *desakota* zones. Although the actual administrative definition of these cores, which he takes from the census, is broad enough to encompass sizable parts of them (for instance, Jabotabek, which includes areas with all the features of the *desakota* zones), generally, these core areas are too narrowly defined spatially to fit neatly with zones of *desokota*, which may extend over large areas between urban centers. Still, the data presented in Vinings's analysis give an overall picture of the emergence of *desakota* zones in Asia.

Following Vining's analysis, Table 1.1 indicates the growth of core population between 1960 and 1980 for eleven Asian countries. Data for Hong Kong and Singapore are not presented because as city-states they do not fit into a *desakota* definition. However, particularly in the case of Hong Kong, one may argue that a *desakota* zone is now emerging, stretching from Hong Kong through the Shenzhen Special Economic Zone and the Pearl River Delta to Guangzhou and Southwest to Macao.[22]

Generally, the table shows that core areas have continued to increase their share of total population most markedly in Korea and Taiwan, which have experienced rapid industrialization and increases in gross domestic product, and in the case of the Philippines, Malaysia, and Thailand, where industrial transformation has been

[20]Hugo (1975), Guiness (1986), Ali (1984).

[21]Vining (1986).

[22]Johnson (1986), Han (1988).

less marked. However, in countries that have large populations (the exception is Sri Lanka) and have experienced modest growth until recently, the growth of the cores is slower. It must be stressed that almost two-thirds of the core population in developing countries is located in Asia.

Figure 1.2 shows the location of these core regions in Asia as grouped into three main *desakota* types. First are those countries that have experienced a rapid transformation of the spatial economy in terms of rural-to-urban shift in population, although agricultural land use may remain quite persistent. Japan and South Korea are the most prominent examples.[23]

Second, there are those regions of countries that have experienced a rapid change in their economic features in the past thirty years. An example is the Taipei-Kaohsiung corridor of Taiwan, which has experienced a declining proportion of people in agriculture from 56 to 20% between 1956 and 1980, and a concurrent growth of industrialization. Speare et al. have estimated that the growth of small to medium-sized industries in rural areas slowed the growth rate of cities by 6% in the 1960s and 1970s.[24] At the same time, this region was characterized by a decline of staple crops as a proportion of the total agricultural value of production. Thus, over the past thirty years although the production of rice has increased considerably, the share of rice as a proportion of gross agricultural receipts has dropped from 50% in 1950 to 34% in 1980. At the same time, other agricultural products have increased from 20 to 36%, and vegetables and fruit from 7 to 20%. More recently there has been a rapid increase in fish farming (prawns), chicken rearing, and other forms of capital-intensive agribusiness. This shift has led to a significant change in the pattern of female employment, with a decline from 52.5% (1965) in primary industry to 16% in 1980, and an increase in secondary industry from 18.2 to 43.7%.[25] Similar patterns are being exhibited in the Bangkok-Central Plains region of Thailand and the four major coastal zones of China. These regions, with only 12.5% of China's population, accounted for 46.4% of the value of industrial production and 13% of the value of agricultural output in 1986.[26]

Third, there is a type of region that bears some spatial and economic resemblance to Type 2 but is characterized by changes that occur because of high population growth and slower economic growth. This situation results in the persistence of underemployment and self-employment in unpaid family work and enterprises. In such regions there may be a juxtaposition of elements of Type 1 and Type 2, producing a highly dualistic economic structure. Thus, technological inputs in agriculture may cause labor shedding and an increase in nonagricultural activities in the rural areas adjacent to urban cores. This phenomenon has been recorded in areas such as Kerala and Tamil Nadu. Although there often growth of small industry and other income opportunities, these regions are characterized by the persistence of

[23]Hayami (1976), Hasumi (1985), Nakai (1988), Lewis (1981).

[24]Speare et al. (1988).

[25]See Todd and Hsueh (1988), Williams (1988).

[26]See Table 5.9 in this volume.

low incomes, which reflect the slow structural transition in the allocation of labor. In some cases, regions continue to absorb population into agricultural areas (e.g., the Sichuan Basin) where nonagricultural employment has remained fairly static over the past forty years while the population has almost doubled in size.[27]

Conditions and Processes Underlying the Emergence of New Zones of Economic Interaction: *Desakota*

Given the diversity of these transitions, it may well be argued that the *desakota* have little in common. However, certain common conditions and processes occur in these regions.

First, virtually all these regions are characterized by densely populated, small-holder cultivation agriculture, commonly wet-rice, that involves careful water management and agronomic practices.[28] Densities frequently approach one thousand persons per square kilometer of cultivated land. Historically, Oshima has argued, the pronounced seasonality and the intense labor of planting and harvesting have had two results: increasing population density, which has led to increasingly smaller farm plots, and uneven seasonal demands for labor inputs. During the off-season, "This dense population must look for off-farm employment since most farms are too small to generate enough income to live on... Monsoon rice-farming never became separated from nonagriculture as it was in the West where cropping came to be combined with animal husbandry in large capitalistic farms."[29] He goes on to argue that in the prewar centuries "monsoon agriculture kept large masses of workers tied down in rural areas, rendering the large labor supply inflexible."[30] In a pre-Green Revolution situation, the capacity of the population of rice-growing areas to increase and still produce enough rice to avoid famine even with very limited infrastructural investment in the basic explanations that Geertz provides for the growth of Java's population from 5 to 25 million in ten decades of slow economic growth.[31] This argument can be applied to many of the other Asian rice bowls in India, China, Japan, Korea, and Southeast Asia. Bray reinforces this point with the following comment:

---

[27]I am Grateful to Rex Casinader and Wang Yaolin, Ph.D. candidates in the Department of Geography at the University of British Columbia, for information on developments in South India and Sichuan.

[28]An exception to this generalization appears to be the Beijing-Tianjin area, which is a region of intense mixture of crops. Since 1949 the southern part of the Shenyang-Dalian region has become an important area of rice production. The identification of "rice-growing" regions does not preclude the possibility of "mixed-crop" systems developing similar population densities. For example, see Hill (1982).

[29]See Oshima, "Transition from an Agricultural to an Industrial Economy," p. 784.

[30]Ibid., p. 785.

[31]Geertz (1963).

The organization of resources typical of a "skill-orientated" technology such as intensive rice farming dovetails very neatly with petty commodity production, which requires very little capital to set up a family enterprise and absorbs surplus labour without depriving the farm of workers at times of peak demand. It can be expanded, diversified or contracted according to market demands, but the combination with the rice farm guarantees the family subsistence. The products can be conveniently conveyed to local or national markets by merchants, who pay the villagers for their labour and often provide raw materials as well as information on the state of the market.[32]

Thus, it is possible to argue that the labor force of these rice-bowl areas was "culturally" prepared to commit its labor to various forms of "new" nonagricultural activity.

Second, in virtually all the rice-bowl areas there were large cities or clusters of cities such as Calcutta, Shanghai, Bangkok, and Guangzhou that provided both opportunities for seasonal labor and important markets for rural rice and other products. The linkages with these critics were important for the surrounding rural areas for cultural and economic reasons.

Third, these regions were frequently characterized by a well-developed infrastructure of roads and canals that allowed an intense movement of commodities and people. Indeed, one is constantly reminded of the importance of water systems for these regions as the fundamental sustainers of the ecological system. Colonial impact, whether by Britain, Japan, of Holland, did little to change these patterns. Indeed, by the provision of political stability and investment in infrastructure the colonial powers frequently enhanced the preconditions for growing populations.

Fourth, by the early 1950s all these regions were large, cheap labor reservoirs waiting to be tapped by state, international, and private capital investment. The manner in which this labor was incorporated into nonagricultural activity varied markedly from country to country and region to region. Most successful were Japan, Taiwan, and South Korea, where both industrialization and agricultural growth led the way in Asia. In these three countries, efforts were made to increase rural incomes through higher yields, guaranteed prices, diversification into nonrice crops, and increased opportunities for employment in rural industries. This effort was aided by physical infrastructure improvements such as rural roads, electrification, consolidation, and irrigation. Increased mechanization released labor, particularly females, to work in industry. Institutional changes, particularly land reform and the introduction of higher yielding rice varieties, were crucial in this process. In addition, the state invested in major transportation linkages such as freeways and electrified railways, which pulled these regions closer to the urban cores. Of course, these processes did not prevent the movement of labor out of agriculture into nonagriculture and urban centers, which accelerated in Japan in the 1960s, in Korea in the 1970s, and in Taiwan in the 1980s. But political and institutional imperatives,

[32]Bray, *Rice economies*, p. 135.

particularly the need for food self-sufficiency, encouraged the persistence of rice farming and created *konjuka* landscapes.

In other countries, these processes moved more slowly, partly because the institutional changes of the "Three Tigers" could not be implemented. The Philippines is an excellent example of initial success and then slowdown because of bad government. However, in the late 1970s other regions in Asia began to exhibit enough of the features responsible for the successes in Japan, Korea, and Taiwan to achieve considerable growth. For example, the four regions of Hong Kong-Guangzhou-Macao, Nanjing-Shanghai-Hangzhou, Beijing-Tianjin-Tangshan, and Shenyang-Dalian have exhibited all these features since 1978. Wang and Veeck have shown how remarkably these regions fit the model of *desakota*.[33] Other studies find the same processes occurring in the Central Plains of Thailand and event in the extended Calcutta Metropolitan Region, which has long had the reputation of being one of the main centers of world poverty.[34] Rice-bowl areas such as Kerala, Bangladesh, Jogjakarta, and the Sichuan Basin, less favorably positioned from the viewpoint of state and private investment, exhibit a persistence of low income and continuing pressure of population on available resources.

Fifth, all the *desakota* regions were characterized by highly integrated "trans-active" environments in terms of movement of people and commodities, for example. In many cases the dense network of rural canals and waterways was central to this integration in the pre-World-War II period. The onset of technological developments in intermediate transportation since the 1950s has greatly accelerated this process.

Finally, it is important to acknowledge the role of the expansion of the global economy and the international division of labor, which create a situation where national governments responsible for gigantic, cheap labor pools have adopted different policies with respect to permitting or encouraging their countries' labor to be tapped for national and international industrial growth. In this respect the location of *desakota* Type 1 and 2 regions adjacent to large cities and transportation points has been particularly important. From the viewpoint of many investors, investment in industry in these regions is cheaper for virtually all the factors of production, and they are able to avoid some of the diseconomies that exist in the large urban zones. Thus, these regions are important areas of subcontracting for a portion of the industrial production process.

To summarize, the regions designated as *desakota* have six main features. (1) They have been or are characterized by a large population engaged in small-holder cultivation of rice that in the pre-World War II period conducted considerable interaction through accessible transportation routes. (2) They are generally characterized by an increase in nonagricultural activities in areas that have

[33]Wang Yaoling's Ph.D thesis research on the Shenyang-Dalian region is being carried out in the Department of Geography at the University of British Columbia. See also Chap. 9 in this volume.

[34]See Douglass (1984); and idem, "Population Growth and policies in Mega-cities: Calcutta," *Population policy Paper No. 1* (United Nations, 1986).

previously been largely agricultural. These nonagricultural activities are very diverse, including trading, transportation, and industry. The increase in nonagricultural activity is characterized by a mixture of activities, often by members of the same household. Thus, one person may commute to the city to work as a clerk, another engage in farming, a third work in industry, and another find employment in retailing in the *desakota* zone. This mixture creates a situation in which the economic linkages within such a region may be as important as the dominance of the large cities in the megalopolis that draw the surrounding regions into their orbit. (3) The *desakota* zones are generally characterized by extreme fluidity and mobility of the population. The availability of relatively cheap transport such as two-stroke motorbikes, buses, and trucks has facilitated relatively quick movement over longer distances than could be covered previously. Thus, these zones are characterized not only by commuting to the larger urban centers but also by intense movement of people and goods within the zones. (4) The *desakota* zones are characterized by an intense mixture of land use with agriculture, cottage industry, industrial estates, suburban developments, and other uses existing side by side. Such a mix has both negative and positive effects. Agricultural products, particularly industrial crops, have a ready market, but the waste of industrial activity can pollute and destroy agricultural land. On the whole, these zones are much more intensely utilized than the American megalopolis, with regards to which Gottmann commented on the amount of woodland and recreational areas that exist. In the *desakota* zones of Asian countries, pressures of population place greater demands on the available space. (5) Another feature of the *desakota* zones is the increased participation of females in nonagricultural labor. In part, this feature is associated with a demand for female labor in industry, domestic service, and other activities, but it is also closely related to changing patterns of agricultural production in the *desakota* regions. Generally, agricultural production shows a shift from monocrop grain cultivation to increased diversity with production of livestock, vegetables, and fruit, sometimes for national and interregional consumption. (6) Finally, *desakota* zones are to some extent "invisible" or "grey" zones from the viewpoint of the state authorities. Urban regulations may not apply in these "rural areas," and it is difficult for the state to enforce them despite the rapidly changing economic structure of the regions. This feature is particularly encouraging to the "informal sector" and small-scale operators who find it difficult to conform to the labor or industrial legislation.

In essence, them, the central processes that shape these regions are the dynamic linkages between agriculture and nonagriculture, and investment seeking to utilize cheap labor and land within a distinctive agroecological setting. Ranis and Stewart have identified how expansion in agricultural output leads to an expansion in other activities and, conversely, how additional nonagricultural activity in the rural areas provides opportunities and incentives for raising agricultural productivity.[35]

In the *desakota* regions that show the greatest increase of both agricultural and nonagricultural income, there is a general rise in household income. Depending on

[35]Ranis and Stewart (1988).

the expenditure decisions of households (i.e., savings/expenditure ratios), there may be an increase in demand for local supplies, goods from urban centers, and imports. Ranis and Stewart show significant variations in different regions. Thus, in Taiwan an increase in agricultural income in rural areas was associated with increased nonagricultural production-related employment and high increases in all linkages. In the Philippines, slower rates of agricultural production associated with a large population increase led to an increase in low-income nonagricultural employment and limited linkages with larger markets. The Philippine situation is typical of the "involuntary" and "distress" features of surplus labor markets first discussed by W. Arthur Lewis.[36]

Questions Concerning the *Desakota* Regions in Asia

The region where the *desakota* processes are in full operation are clearly an important part of the "settlement transition" in Asia. For example, Zhou reported that the four key economic regions of China (Nanjing-Shanghai-Hangzhou, Hong Kong-Guanzhou-Macao, Beijing-Tianjin-Tangshan, and Shenyang-Dalian) with approximately 12% of the nation's population in 1986–87 were responsible for 47% of its industrial output.[37] These four key economic regions play a crucial role in the current phase of rapid economic growth in China. Similar arguments were presented by Liu and Tsai for Taiwan.[38] Thus, there is a great deal of support for the assertion that these regions are highly significant foci for the development process. However further data are needed to support the assertion, particularly with respect to the following questions:

In what manner are these new economic regions different from the "zones of urban influence" that are well established in the urban transition literature? In other words, if these zones are simply a greater areal extension of the "peri-urban" region of large cities that have been brought about by space-time collapse and transportation improvements, then what is different about them from the so-called peri-urban regions? In fact, the relationship between the urban cores and the adjacent regions is important to the sustainability of these regions, and therefore any study of these regions must investigate the urban core functions and the relationships with the surrounding regions.

Can economic data be collected to show processes operating that facilitate certain types of economic activities in these regions? For instance, are there diseconomies associated with large cities that facilitate the location of economic activities in such zones? Is the legislative environment in such zones more

[36]Lewis (1963). See also Geertz, *Agricultural Involution*; and P. Gourou, *L'Asie* (Paris: Librairie Hachette, 1953).

[37]See Chap. 6.

[38]See Chap. 11.

permissive (are taxes lower?) for certain types of activities? Are wage rates persistently lower? Are the social overhead costs lower in such regions? In other words, are there economic benefits that facilitate the growth of economic activities in these regions that may be a "growth generator" in the development process?

Are such regions likely to exhibit short-term persistence (up to fifty years), or are they simply a transitory phenomenon? Some evidence presented by Liu and Tsai suggests that in Taiwan the persistence of such a region, at least in the corridor between Taipei and Kaohsiung, is questionable. They show evidence that the population is concentrating in the northwest quadrant of Taiwan as the industrial and services characteristics of the economy change. This shift is associated with a decline in agricultural employment. Is there some point in the development process at which the agglomerative tendencies will take over? This type of question has important policy implications for countries such as China or India as their economies develop.

Issues of Policy Formation

Prevailing policy prescription for macrointervention in the spatial shift of population during development is broadly polarized between acceptance of big-city growth representing the "rational" development of economies of scale and agglomeration, and arguments in favor of the development of "small" and "intermediate" towns in the urban hierarchy, decentralization of industry, and frontier development.[39] Current developments raise the question of whether *desakota* regions represent a viable "middle" policy option and, if so, what kinds of policies need to be adopted. Whitney's chapter on Shanghai addresses the "problems" resulting from the growth of urban cores and the *desakota* regions adjacent to them, problems that emphasize environmental degradation, waste removal, and adequate delivery of social welfare.[40] Before the issue of some form of "middle planning option" is tackled, it will be necessary for the governments of the region to deal with the fundamental issue of the usefulness of a rural-urban distinction. In reviewing the debate on "urban bias," Harriss and Moore explain this point as follows:

There are two themes central and common to their ["urban bias" analysts] work. First, in attempting to explain national level patterns of economic resource allocation within a political economy framework, they use the concept of economy sectors, mainly the rural/agricultural versus urban/industrial categorization. Second, they suggest in varying degrees that the way in which sectoral conflicts influence the allocation of economic resources through state action has been the prime cause

[39]Yeung (1986).

[40]See Chap. 10.

of slow rates of economic growth (and in Lipton's work, of growth biased against the poor) in developing countries since the Second World War.[41]

But what happens if the rural-urban dichotomy ceases to exist, as in the case of *desakota* zones? Then the whole policy debate on urban and rural allocation of resources becomes fuzzy and meaningless unless one accepts Lipton's argument that "rural–urban is not a categorization of space alone. To see it like that is to underpin an incorrect, absolute distinction between geographical (residence, density), occupational, sectoral, and class categorizations of households."[42]

All societies have working spatial definitions of urban and rural areas. These definitions are highly variable from country to country and are often changed. However, most definitions have some common elements such as size and political definition. For instance, in some countries all gazetted towns exceeding a population of ten thousand are regarded as urban; everything else is rural. In fact, it is more important to know two important economic pieces of information: (1) What is the contribution of agricultural and nonagricultural activities to the GDP of a given spatial unit (nation, province, and so on)? and (2) What is the proportion of the working labor force employed in agricultural and non-agricultural work in a given spatial unit? If this information were available over given time periods, it would be possible to develop a more precise definition of urban and rural areas.

One could conceive a rather simple matrix (see p. 21) constructed at the level of small administrative units that would allow a fourfold spatial division of a country on a continuum from the most urban spatial unit to the most non-urban spatial unit. Assuming some ideal statistical base, this type of analysis would enable the estimation of the contribution of the urban spatial units to the GDP as compared to the non-urban spatial units. Temporal data if available would permit the assessment of the relative contribution of urban and non-urban areas to the GDP though time as well as the differences in the labor force over time. This kind of information would provide vital feedback to the government in assessing the spatial impact of its investment policies. Unfortunately, few developing societies possess data that can be analysed in this way, relying instead on macrodata that conceal these significant differences between urban and non-urban areas.

There appear to be six priorities for many Asian countries if they are to develop pragmatic strategies that attempt to recognize the importance of the *desakota* regions. First, the government will have to make some significant decisions with respect to agricultural policy. All indications are that in most Asian countries, the "agricultural issue" is of central importance to *desakota* regions. The problem revolves around the need for a sufficient supply of foodstuffs and the "cultural" demand for agricultural activity in Asian countries. The crucial issue is how long rice growing, for example, will persist in *desakota* regions as economic growth proceeds. In the cases of Japan, Korea, and Taiwan, it appears that policies will eventually reduce the role of rice in *desakota* areas. In other cases, labor is being

[41]Harriss and Moore (1984).

[42]Lipton (1984).

released from rice growing as a result of technological changes over which governments apparently have little control.

Second, Asian governments will seriously need to consider in what manner the release of labor from agricultural labor pools is going to occur. Policies that slow down geographical relocation and foster in situ development should be given priority. Given the already high levels of development in these regions, they should not be expensive. Of course, the problems of the more "traditional" regions of *desakota* such as Sichuan are more intractable and should be tackled by more conventional development inputs that emphasize population control, delivery of basic needs, and integrated rural development.

Third, Asian governments will have to recognize the reality of these zones of intense urban-rural interaction and direct much of their investment to these areas. This means making hard decisions against fostering small-town development and rural industrialization in less accessible areas without neglecting rural-integrated development schemes in such areas. Many governments should opt for policies of "modified regional growth pole" growth (i.e., emphasizing not just the urban pole but also the large mega-urban region of which it is part).

Fourth, Asian governments will need to monitor carefully the growth of economic activities in these zones for the obvious problems that will arise over conflict in incompatible land uses and environmental pollution, for example. This process will involve great care, for one of the major factors favouring the economic vitality of the *desakota* zones is the mixture of land uses.

Fifth, Asian governments will need to improve access in these zones of intense rural-urban interaction with improved roads and fast railway communication. In this respect, the building of the Shinkansen, the Seoul-Pusan Highway, and the Taipei-Kaohsiung Freeway have been crucial to the development of Japan, Korea, and Taiwan, respectively. This costly investment reaps developmental rewards.

Finally, Asian governments should develop new spatial systems of data collection similar to those of the "living perimeters" of Taiwan, which will enable them to monitor effectively the impact of investment decisions on labor force composition and income, among others, within the *desakota* zones.

Of course, the timing of government strategies and fiscal ability to implement them are not easy to gauge. However, the demographic and economic reality of the growth of the *desakota* means that planning decisions relating to them cannot be postponed.

Conclusion

The chapter that follow in this volume not only catalog the features of the emergence of new regions of economic interaction, but they also emphasize how the historical evolution of these regions has created a unique opportunity for Asian development. Whether this opportunity will be taken advantage of in all Asian countries remains to be seen, but certainly the challenge to urban versus rural growth, which seems so central to many Asian development strategies, can be resolved as least partially through the planned development of *desakota* regions.

References

Ali, A. (1984). Agricultural stagnation and population pressure: The case of Bangladesh. New Dehli: Vikas Publishing House.

Baker, C. (1981). Economic reorganization and the slump in South and Southeast Asia. *Comparative Studies in Social and Economic History, 23*(3), 325–349.

Baldwin, R. E. (1956). Pattern of development in newly settled regions. In *The Manchester School of economic and social studies* 24 (pp. 161–179)

Douglass, M. (1984). *Regional integration on the capitalist periphery: The central plains of Thailand.* The Hague: Institute of Social Studies.

Douglass, M. (1988). Urbanization and National Urban Development Strategies in Asia, Indonesia, Korea, and Thailand, Discussion Paper No. 8. Honolulu: Department of Urban and Regional Planning, University of Hawaii.

Dowd, D. (1956). A comparative analysis of economic development in the American West and South. *Journal of Economic History, 16*(7), 558–574.

Francesca, B. (1986). *The rice economies: Technology and development in Asian societies.* Oxford: Basil Blackwell.

Geertz, C. (1963). *Agricultural involution: The process of ecological change in Indonesia.* Berkeley: University of California Press.

Ginsburg, N. (1990). Extended metropolitan regions in Asia: A new spatial paradigm. In *The Urban Transition: Reflections on the American and Asian Experiences* (p. 21). Hong Kong: The Chinese University Press.

Gottmann, J. (1961). *Megalipolis: The urbanized Northeastern seaboard of the United States* (p. 257). New York: The Twentieth Century Fund, Kraus International Publications.

Guiness, P. (1986). *Harmony and hierarchy in a Javanese Kampong.* Oxford: Oxford University Press.

Han, B. (1988). *Industry booms on the Zuijiang River Delta* (pp. 18-23). *Beijing Review,* 22–28 Aug. 1988.

Harriss, J., & Moore, M. (1984). Editors' Introduction to special issue on development and the rural-urban divide. *the Journal of Development Studies, 20*(3), 1–4.

Hasumi, O. (1985). Rural society in post-war Japan, part 1 and 2, *The Japan Foundation Newsletter, 12*(5), 1–10, *12*(6), 1–7.

Hayami, Y. (1976). *A century of agricultural growth in Japan.* Minneapolis: University of Minnesota Press.

Hill, P. (1982). *Dry grain farming families: Hausaland (Nigeria) and Karnalaka (India) compared.* Cambridge: Cambridge University Press.

Hugo, G. (1975). *Population mobility in West Java, Indonesia.* Ph.D. Thesis, Australian National University, Canberra.

Johnson, G. E. (1986). 1997 and After: Will Hong Kong survive? A personal view. *Pacific Affairs, 59*(2), 237–245.

Lewis, W. A. (1963). Economic development with unlimited supplies of labour. In H. N. Agarwala & S. P. Singh (Eds.), *The economics of underdevelopment* (pp. 400–449). New York: Oxford University Press.

Lewis, J. (1981, June). The real security issue: Rice. *Far East Economic Review*, *19*, 70–71.

Lipton, M. (1984). Urban bias revisited. *The Journal of Development Studies, 20*(3), 139–166, specifically p. 155.

McGee, T. G. (1987). *Urbanisasi or Kotadesasi? The Emergence of New Regions of Economic Interaction in Asia, WP-87-8, WP-87-8*. Honolulu: EWCEAPI.
McGee, T. G., et al. (1989). *Urbanisasi* or *Kotadesasi*? Evolving Patterns of Urbanization in Asia. In F. J. Costa (Ed.), *Urbanization in Asia: Spatial dimensions and policy issues* (pp. 93–108). Honolulu: University of Hawaii Press.
McGee, T. G. (1991). The emergence of Desakota regions in Asia: Expanding a hypothesis. In N. Ginsburg, B. Koppel, & T. G. McGee (Eds.), *The extended metropolis: Settlement transition is Asia*. Honolulu: University of Hawaii Press.
McGee, T. G. (2016). Città Diffusa and Kotadesasasi: Comparing diffuse urbanization in Europe and Asia. A project in the making. In International Roundtable Conference: Territories of Metropolis: Compactness, Dispersion, Ecology: Comparative Perspectives Between Asia and Europe. April 5–7, 2016 Shanghai, China.
McGee, T. G., Lin, G., Marton, A., Wang, M., & Wu, J. (2007). *China's urban space: Development under market socialism*. New York: Routledge.
Morse, R. M. (1974). Trends and patterns of Latin American urbanization, 1750–1920. *Comparative Studies in Society and History, 16*, 416–447.
Nakai, N. (1988). Urbanization Promotion and Control in Metropolitan Japan. *Planning Perspectives, 3*, 783–810.
Oshima, H. T. (1986). The transition from and agricultural to an industrial economy in East Asia. *Economic Development and Cultural Change, 34*(4), 783–810.
Ranis, G., & Stewart, F. (1988). Rural linkages in the Philippines and Thailand. In F. Stewart (Ed.), *Macro-policies for appropriate technology in developing countries* (pp. 140–191). Boulder: Westview Press.
Speare, A., Liu, P., & Tsay, C.-L. (Eds.). (1988). *Urbanization and development: The rural-urban transition in Taiwan*. Boulder: Westview Press.
Todd, D., & Hsueh, Y.-C. (1988). Taiwan: Some spatial implication of rapid economic growth. *Geoforum, 19*(2), 133–145.
U.N. Centre for Human Settlements. (1987). *Global Report on Human Settlements*. Oxford: Oxford University Press.
Vining, D. R. (1986). Population redistribution towards core areas of less developed countries, 1950–1980. *International Regional Science Review, 10*(1), 4.
Williams, J. F. (1988). Urban and regional planning in Taiwan: The quest for balanced regional development. *Tijdschrift Woor Economische En Sociale Geografie, 79*(3), 175–187.
Yeung, Y.-M. (1986). Controlling metropolitan growth in Eastern Asia. *Geographical Review, 76*(2), 125–137.

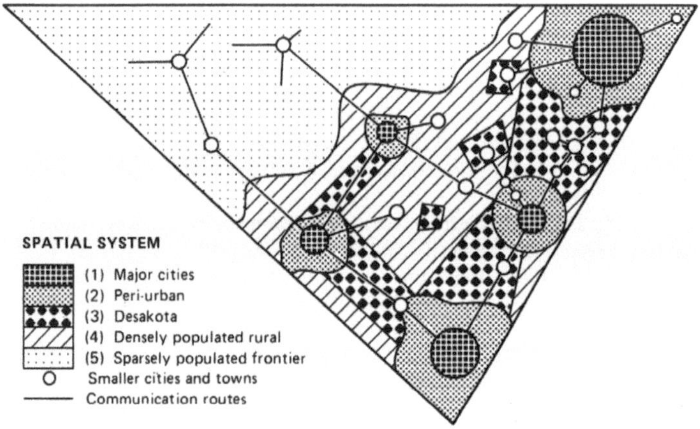

Fig. 25.1 "Spatial configuration of a hypothetical Asian country (ca. 2000)".
Source: McGee, T. G. (1991). "The emergence of Desakota regions in Asia: Expanding a hypothesis". In: N. Ginsburg, B. Koppel, & T. G. McGee (Eds.), *The extended metropolis: Settlement transition is Asia*. Honolulu: University of Hawaii Press, p. 6. © University of Hawai'i Press.
"This figure presents a model of the spatial configuration of a hypothetical Asian country, which I will label Asiatica Euphoria for the purposes of this exercise. In this example, five main regions of the spatial economy are identified as follows: (1) the major cities of the urban hierarchy, which are often dominated in the Asian context by one or two extremely large cities. (2) The peri-urban regions, which are those areas surrounding the cities within a daily commuting reach of the city core. In some parts of Asia, these regions can stretch for up to thirty kilometers away from the city core. (3) The regions labeled desakota, which are regions of an intense mixture of agricultural and nonagricultural activities that often stretch along corridors between large city cores. These regions were previously characterized by dense populations engaged in agriculture, generally but not exclusively dominated by wet-rice. (4) Densely populated rural regions, which occur in many Asian countries, particularly those practicing wet-rice agriculture. (5) Finally, the sparsely populated frontier regions found in many Asian countries that offer opportunities for land colonization schemes and various forms of agricultural development." (ibid., pp. 6–7)

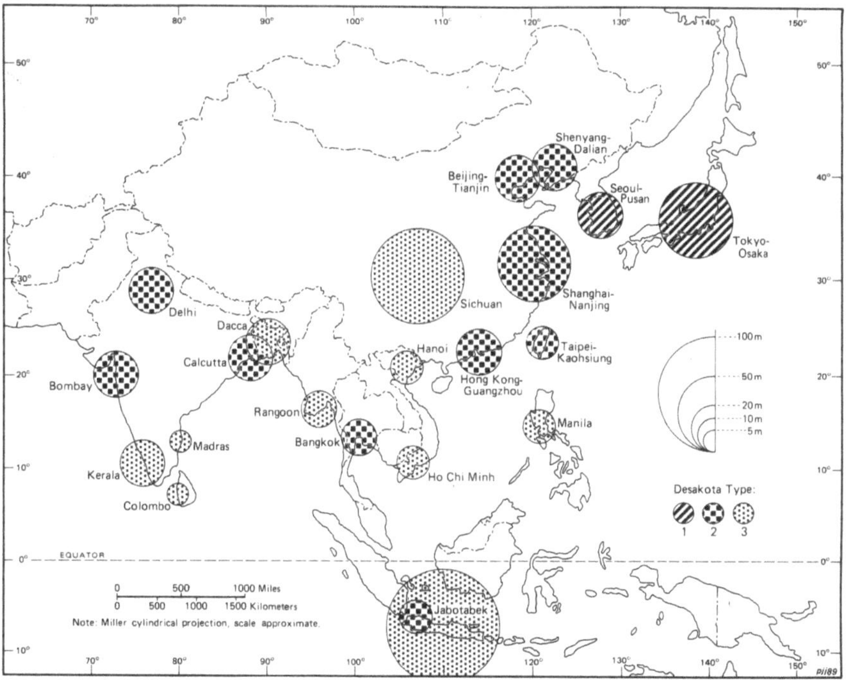

Fig. 25.2 "Growth of core areas in Asia".
Source: ibid., p. 13.
"This figure shows the location of core regions in Asia as grouped into three main desakota types.

- [Desakota Type 1]: countries that have experienced a rapid transformation of the spatial economy in terms of rural-to-urban shift in population, although agricultural land use may remain quite persistent. Japan and South Korea are the most prominent examples.
- [Desakota Type 2]: regions of countries that have experienced a rapid change in their economic features in the past thirty years. An example is the Taipei-Kaohsiung corridor of Taiwan, which has experienced a declining proportion of people in agriculture from 56 to 20% between 1956 and 1980, and a concurrent growth of industrialization. [...] Similar patterns are being exhibited in the Bangkok-Central Plains region of Thailand and the four major coastal zones of China. [...].
- [Desakota Type 3]: region that bears some spatial and economic resemblance to Type 2 but is characterized by changes that occur because of high population growth and slower economic growth. This situation results in the persistence of underemployment and self-employment in unpaid family work and enterprises. In such regions there may be a juxtaposition of elements of Type 1 and Type 2, producing a highly dualistic economic structure. [...] This phenomenon has been recorded in areas such as Kerala and Tamil Nadu [...]." (ibid., p. 12)

Chapter 26
The Transformations of the Urban Habitat (1993)

Bernardo Secchi

Introduction

Paola Viganò
EPFL, Lausanne, Switzerland
IUAV, Venice, Italy
paola.vigano@epfl.ch

The Wunderkammer *of the European territory offers itself to the gaze of the scholars, part of the research group coordinated by Bernardo Secchi at the beginning of the 90s on "The transformations of the urban habitat in Europe", who meet on given occasions at the IUAV of Venice to discuss fundamental questions posed by new forms of urban development, descriptive tools and cognitive strategies. The difficulties of naming the object of research hark back to the time of discovery of the phenomenon, so too the idea of the construction of an* Atlas *that might help to rearrange what is perceived as confused, when it comes to "having to simultaneously describe and interpret" what was not a city though which today is part of the same.*

The text here presented is therefore the opening of a research study that began to weave relationships and build a network of European researchers on the phenomenon of settlement dispersion, written by its coordinator who in nine points defines the "themes" and "problems" to be addressed.

Description, for Bernardo Secchi, is the first point that requires investigation: the great interpretative and over-interpretive richness of scholarly work risks veiling phenomena and processes, rather than revealing them. The reductionist move is no longer possible: "Our language and our discourses shy away from every 'reductionist' attempt". The difficult relationship with the modern project, its slogans, the tabula rasa*, opens up to situated approaches; the universal conceptions*

M. Barcelloni Corte and P. Viganò (eds.), *The Horizontal Metropolis*,
https://doi.org/10.1007/978-3-030-56398-1_26

and the equally general simplifications break and fracture against the things that change, that fall outside the main road of development, acquire complexity, and the same for the interpretations and theories concerning them. The differences between territories, researchers and strategies appear irreducible. This is a conviction that runs throughout Bernardo Secchi's entire thinking, that reads the process of progressive individualization of situations and people as a conquest and at the same time as a problem of contemporary society, which always needs to be taken into account: whether you design a new part of the city, or study a diffuse inhabited area, or design a public space. It is in the attention to the individual, in its different dimensions, that we find one of the most important and lasting interpretative keys of Bernardo Secchi's research on diffuse cities, city territories and horizontal metropolises. This has important consequences in the ways and tools of description he chooses, that do not stop at statistics, but that punctually and extensively resort to important field trips, to listening to places and people, to "technically relevant analyses" of the minimal and for some lesser details of urban space.

Fully aware of the era of transition, between the modern, post-modern era and beyond, to which the research belongs, Secchi defines the theoretical background, which is also open and uncertain, which assumes the exhaustion "of faith in development and in modernity" and the "great revolution the extent of which is still largely obscured by the banality of everyday life" (and here he cites Guido Martinotti, among the most influential geographers in Italy in those years, in his Metropoli: la nuova morfologia sociale della città[1]*). Even so Secchi shuns the grand scenarios, his hypotheses on these themes will always be cautious, careful, avoiding what appears at first sight evident, being at the same time reluctant to superimpose unlikely new overall frescoes over the great tales of the past. The need is for accurate research. And the research done on the territories of diffuse individualisation is the terrain of exploration and opportunity, starting from the space and the practices that they have been subjected to, to understand the underlying movements of society.*

Together with the growing process of individualization, heterogeneity marks the new territories and their constituent materials that are still to be investigated in the confused articulation of newly discovered sequences. They are not easy to decipher, in their programmatic ambiguity between "function" and "fiction". In these layers, imaginaries and ideologies accumulate that between 1993 and today have also been more clearly delineated from the political and cultural point of view: "Between affirmations of localism and denial of solidarity."

The research on the transformations of the urban habitat in Europe is placed within a set of consolidated traditions (as this Anthology also wants to sustain), which implies "a lot of re-reading". It is, at the same time, committed to interpreting the processes of restructuring the economy, its institutions, social relations and lifestyles, in relation to the transformations of space. The "flexible

[1]Martinotti, G. (1993). *Metropoli: la nuova morfologia sociale della città.* Bologna: il Mulino.

accumulation"[2] *is represented in* urban materials *that tell us of the new industrialization beyond and in substitution of urban ones. It sometimes reinforces more ancient settlement and production networks that have developed outside the main centres. The relationship with urbanism is complex, to be reinvented and above all marked by the cyclical movement determined by the modern urban project and the criticisms of the same. "An oscillation", writes Bernardo Secchi, "within which the modern project could and can still legitimately be considered incomplete".*

26.1 The Transformations of the Urban Habitat in Europe: Notes[3]

Those who think about the transformations of the European habitat, if only in terms of the city and the metropolis, cannot perhaps avoid certain "themes" and certain "problems":

1. The city and the European territory, and even more so their transformations in progress, seem to vigorously defy "description." For some years now we have seen a multiplication, within different ways of organizing the discourse, of attempts to convey the variety, ramification, complexity and chaotic character of the city and the European territory, and their current situations of ferment. At the same time, there has been a proliferation of words, categories, metaphors, forms of visual representation and organization of discourse through which to describe the city, the territory and their transformations. "The social sciences and prevalent viewpoints are trying to give a name and a conceptual definition ... to processes that are perhaps still too much in progress to be imprisoned inside consistent words and ideas" (G. Martinotti, *Metropoli: la nuova morfologia sociale della città*, il Mulino, Bologna, 1993). Every term, every representation has been intentionally laden with meanings partially compressed within the depth of what is not said. Our language and our discourse shun any "reductionist" temptations and this often makes conversation, exchange and discussion difficult.

2. It is always hard, in these cases, to say how much depends on the researchers and their imaginary, their strategies and cognitive techniques, and how much on the characteristics of the object of the research. Probably the answer is not equal for everyone and depends, for each, on the degree of moderation of their "realism" or "relativism," on the willingness to assign more or less importance to the gap that exists between "words and things."

---

[2]Vazquez-Barquero, A. (1992). Local development and flexible accumulation: Learning from history and policy. In G. Garofoli (Ed.), *Endogenous development and Southern Europe*. Aldershot: Avebury.

[3]Source: Secchi, B. (1993). Le trasformazioni dell'habitat urbano in Europa alcuni appunti. *Quaderno della ricerca sulle trasformazioni dell'habitat urbano in Europa*, 0, 7–15.

At first glance, however, the city and the European territory present themselves today as "cabinets of wonders," collections of different and apparently irreducible situations that can perhaps only be catalogued and documented in an "atlas."

The term "atlas" itself is highly evocative, and periodically returns in different forms in the history of European culture; perhaps in the periods of greater disorientation and uncertainty; perhaps when the need becomes more urgent for different disciplinary areas, geography and anatomy, botany and urbanism, to reorder their materials of observation.

The difficulties we have in giving a title to the research, "the diffused city," "the distributed city," "the low-density city" and, finally, "the transformations of the habitat," a formulation we still feel is weak; the shifts of the object of research from the "periphery" to "metropolitan sprawl," to "outer territories," only to then recognize the impossibility of talking about one without observing the others; the need to simultaneously describe and interpret the transformations of the city and the territory that was external to it, but is no longer extraneous; all these factors show how problematic it is to identify even a tenuous regulating and ordering device for the observations and situations that have stimulated our curiosity as themes of research.

Clearly, we live in a paradoxical situation: the one constructed by the acute awareness of a general transformation of the habitat of a large part of the European population, and by the clear perception of the impossibility of reducing this transformation within the geometry of the topographical map and narration, not to mention a rapid explanation.

3. Looking back on the well known essay by I. Hassan ("The Culture of Postmodernism," in *Theory, Culture and Society*, 2(3), 1985, p. 119–132), the forms of the new habitat, the city and the territory now appear to be "open," the result of "play," of "chance" if not of "anarchy," the outcome of "processes" in which we can recognize a form of "deconstructive participation" that is perhaps accompanied by the "dispersal" of subjects and their "rhetorics," the absence of recognizable "syntagmata," the invasion of "parataxis," of "metonymy," of "combination" that generates "counter-interpretation," where each person makes use of specific "idiolects" that provoke our "irony" but also generate "indeterminacy" and "uncertainty." Some believe they can glimpse the characteristics of the postmodern in all this, of the depletion—already from the last years of the period between the two World Wars—of the Enlightenment project, of faith in progress and modernity (Benjamin, Horkheimer, Adorno), and it is certain easy to perceive in the new European urban habitat the characteristics, for example, of "the crisis of modernity" underscored by David Harvey (*The Condition of Postmodernity*, Blackwell Pub., 1991), by F. Jameson (*Postmodernism, or the Cultural Logic of Late Capitalism*, Duke University Press, 1991), by many of the scholars brought together by C. Jencks in his recent anthology (The Post-Modern Reader, Academy Editions, London, 1992), or (even earlier perhaps) by M. Foucault.

The recent transformations of the European habitat have often been seen as indications of epochal ruptures. "The inhabitants of the contemporary urban world live in the midst of a great revolution whose spread is still to a great extent obscured by the banality of everyday life" (Martinotti). The discovery of a discontinuity in history, in past years, appeared with the characteristics of liberation and promise. Yet in times closer to our own the judgments have become more cautious and the desire to get away from the narrative of big images (of the "post-modern," the "post-industrial," the "global city") with precise research projects and technically pertinent analysis has become more urgent. The city and the territory, in their physical and visible constitution, perhaps construct a layer of important phenomena from which to observe other, perhaps more fundamental movements of the society.

4. What in evident ways grants these characteristics to the contemporary European city and territory is the heterogeneity of the constituent "materials"; the often sloppy and in some cases reckless ways in which "materials" with different roots and meanings are juxtaposed, contaminated and crossed by social practices that are apparently not consistent with them. This is an aspect that runs through the literature and visual areas of the last decades, and is now being investigated with lively anthropological interest.

Perhaps we know about the materials of the ancient city, and also those of the modern periphery, but a "technically pertinent" analysis of the materials of the diffused, low-density city or the new urban habitat has not yet been made. The image of their heterogeneity itself has perhaps prevented this, to date. Emphasis has often been placed on the ambiguity of the materials of the contemporary city; on the dual role, for example, of certain innovative materials; on their temporary, often ephemeral response to a "function" containing a high degree of "fiction" (of hyper-representation of the primary function, as in the shopping mall, but also its hypo-representation, as in the box-building suitable for all purposes): a point that certainly also has to do with the small house on a lot in the "diffused city" or the "green suburbs."

5. The first analyses of the constituent "materials" of the contemporary urban habitat in the various "case studies" examined thus far, in the Milan, Veneto, Liguria, Tuscany, Adriatic and Apulia areas, like the analysis of the areas of the Plateau Suisse, Flanders, the Dutch Randstadt, the metropolitan area of Porto or Thessaloniki, like the analyses that see the problem against a wider historical and theoretical background, demonstrate something that goes beyond these first impressions.

The study, for example, of certain innovative materials like the "market streets" has made it possible to associate the various inflections of their forms with strategies of enterprise that are sufficiently complex, though governed by high level of productive and communicative rationality.

This leads us to carefully observe a set of social and cultural practices, policies and modes of function of institutions, in the manner, for example, of the "regulation school" (A. Lipietz), a new version of institutional economics: policies, projects and practices that are constantly unstable, in pursuit of reciprocal regulation.

The study of the entire archive of "materials" of the contemporary habitat, especially of that less analyzed of the "diffused city," seems to demonstrate that the pieces of the "collage," apparently foreign to each other, establish reciprocal "bonds" (Dahrendorf) at a deeper level: in the imaginary and ideologies of certain groups ("minorities") or entire communities. The crisis of modernity is simultaneously crossed by images of the past and the future (Benjamin) revealed often in highly contradictory ways by artistic movements (the aesthetics of architecture, music, post-modern graphic design) and political movements (the new right). Just consider the relationships of denial that in one field are established between use of history and denial of continuity, and in the other between assertion of localism and negation of solidarity.

6. All this, at least in part, takes the study of transformations of the urban habitat in Europe back onto more familiar ground, making it possible to understand how many aspects of the contemporary habitat have attracted the attention of scholars in years and periods much earlier than our own. Adna Ferrin Weber, H.G. Wells, Frank Lloyd Wright, Lewis Mumford and Homer H. Hoyt, cited by Brian L. J. Berry and G. Martinotti, are perhaps the best known, though not the only ones.

Much of the literature of the start of the century, the years between the two World Wars, the 1960s, contains images, observations and thematic formulations of the field—if observed in today's perspective—that are now being "rediscovered." Each of the themes and issues raised by the transformations of the contemporary urban habitat is not new; the words may be new, perhaps, their constellations, their mutual relations. But even this is not certain. There is a big task of re-reading to be done.

Saying that the study of the transformations of the European habitat returns to the ground of more established research traditions does not mean that every answer is known; nor does it mean that the study of the transformations of the European habitat can use the same categories of the tradition. Many scholars have insisted, in recent years, on the need to change vantage points, systems of observation, naming, concepts, theories. The study of the transformations of the habitat also imposes reflection on our tools and our cognitive strategies.

7. The formation of the new habitat in the various European regions seems to be connected to vast processes of restructuring of the economy and its institutions, and therefore of social relations and lifestyles. Some of the regions taken into consideration and mentioned above, such as the Veneto area, for example, Flanders, the Adriatic area, the region of Porto, have been the protagonists over the two last decades of an intense process of growth that some observers have connected to a specific "path" of development ("industrialization without fractures," G. Fuà), to a specific "model" (the Veneto model) and to specific "social formations" (A. Bagnasco). In many other regions the formation of the new habitat seems to be associated with the constitution of "industrial districts" in the sense of Marshall, or of "areas of specialization": the area of compasses, ski boots, eyewear. The clarification of the spread of these processes, their nature, character and non-generic relations with the transformation of the habitat still remains an open program of research.

Some scholars have observed, for example, that in many European regions the process of restructuring of the 1960s and 1970s, often addressed in terms of "productive decentralization," "urban decline," "disurbanization," is associated with the "migration" of businesses, usually from old industrialized areas to new outlying areas. In the decade of the 1980s, on the other hand, the same process is associated with the birth of new businesses within a model of "flexible accumulation" (A. Vázquez Barquero, "Local Development and Flexible Accumulation: Learning from History and Policy," in: Gioacchino Garofoli (ed.), *Endogenous Development and Southern Europe*, Avebury, Aldershot, 1992). There is no doubt that this is accompanied by the appearance, inside the dispersion of these regions, of new "materials": the "home workshop," the multi-use "shed," the "industrial zone" later used for commercial, sporting and entertainment activities, etc.

8. Urban de-industrialization and industrialization of rural areas have prompted many scholars first to carefully observe, often with a taste for archaeology, the "abandoned areas," and then to reconsider the processes of productive, economic and social restructuring within a more extensive spatial and temporal perspective ("I territori abbandonati," *Rassegna* no. 42).

Extensive and in-depth analyses have demonstrated that in many rural areas industry has its own remote history, in some cases more remote than that of the industrialization of the country, the region, the main regional city. In many of these areas, as in the case of the Veneto, what seems like a dispersal of industry and, more in general, of settlements actually conceals an opposite phenomenon: of "densification" of an older and less evident network of settlement and production (Fuà, Vázquez Barquero).

In certain countries, like Spain and Italy, the process of industrial and urban concentration was launched when industrial production had already begun in small urban centers, while manufacturing activities were scattered in rural areas. This takes us back to what was written years ago, from two very different vantage points, by David S. Landes ("Technological Change and Industrial Development in Western Europe, 1750–1914," in *The Cambridge Economic History of Europe*, Cambridge University Press, 1965), and by Roberto Gabetti and Carlo Olmo, *Alle radici dell'architettura contemporanea*, Einaudi, Torino, 1989). Dispersion and densification, urbanization (H.E. Tisdale) and counter-urbanization (Brian L. J. Berry) are perhaps overly ambiguous terms and concepts.

Maybe it is not a matter of examining the formation of the new habitat starting with aggregated terms and concepts that attempt to describe the phenomenon in its principal, as yet unknown character, but instead of starting, with an "elementarist" attitude, from its physical and visible constitution, its constituent "materials," in order to observe the permanence and persistence, the disappearance and replacement on the part of innovative materials, the changing of function, role and meaning.

From this standpoint the history of the various areas examined may seem like the result of a process of "cumulative selection" (Alain Léveillé) during which the different constituent "materials" of the city and the territory are used, reused, modified, transformed, constructed, destroyed and replaced.

9. Within the processes of restructuring of the European society and economy and of the various regions, policies have particular importance, perhaps. Territorial policies, but also the economic, social and institutional policies whose results are associated with territorial transformations. The comparison among the different situations should shed a great deal of light in this sense. The abandoning of the meta-narratives that prevailed until the 1970s and the application in the following decade of the more open forms of the "new pragmatism" (H. Putnam), especially in its more reduced and instrumental forms ("truth is what it is useful for me to believe"), must perhaps be somehow connected to the fragmentary character of the transformations in progress, to the inability to legitimately trace them back to a single hypothesis of interpretation, to a single historical subject endowed with sufficient authority and power.

The transformations of the European habitat over the last two decades have formed the backdrop of a major transformation of urban planning policy, a transformation that has perhaps not yet been completed, which is not even accepted by many, and is interpreted by others in perhaps an overly self-serving manner as the "end of urbanism." Paradoxically, the new habitat, with its fragmentary, heterogeneous, chaotic character, has come into being "inside" urban plans constructed in the forms dictated by the great narratives of the past; they are still used as containers, while the new plans have perhaps not yet been able to find their own precise role in relation to the new situations and new phenomena.

But a deeper look at the long term recognizes that the history of urbanism in this century is ambiguously crossed by various interpretations of the "modern project" and an equal number of interpretations of the "critique of the modern project." The "post-modern" critique is but one episode inside this history. Both the modern project and its critique are reciprocally forced, throughout the century, to always provide new proofs of their legitimacy; both have never managed, throughout the century, to completely delegitimize the opposite camp, forcing European urbanism to constantly waver in a situation in which the modern project has been able—and is still able—to be legitimately considered unfulfilled.

The concrete experiences of planning and construction of the new city, like the policies of urban and territorial modification, are part of this ongoing process of "extreme efforts of the imagination" of the various urban and dis-urban utopias running through the century. The transformations of the European habitat in progress are also a result of those efforts.

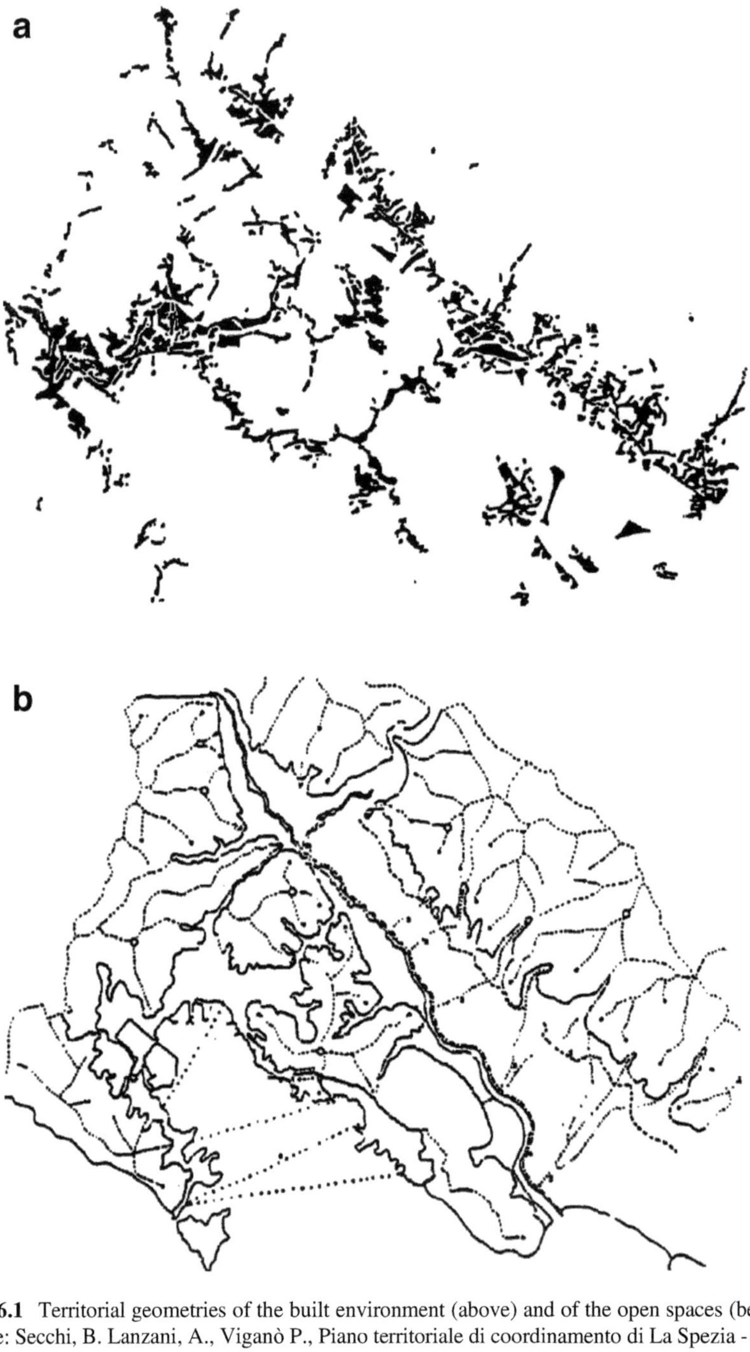

Fig. 26.1 Territorial geometries of the built environment (above) and of the open spaces (below). Source: Secchi, B. Lanzani, A., Viganò P., Piano territoriale di coordinamento di La Spezia - val di Magra. Published also in: Secchi, B. (1994). *Tre piani, La Spezia Ascoli Bergamo*, C. Bianchetti (Ed.). Milano: FrancoAngeli, p. 16

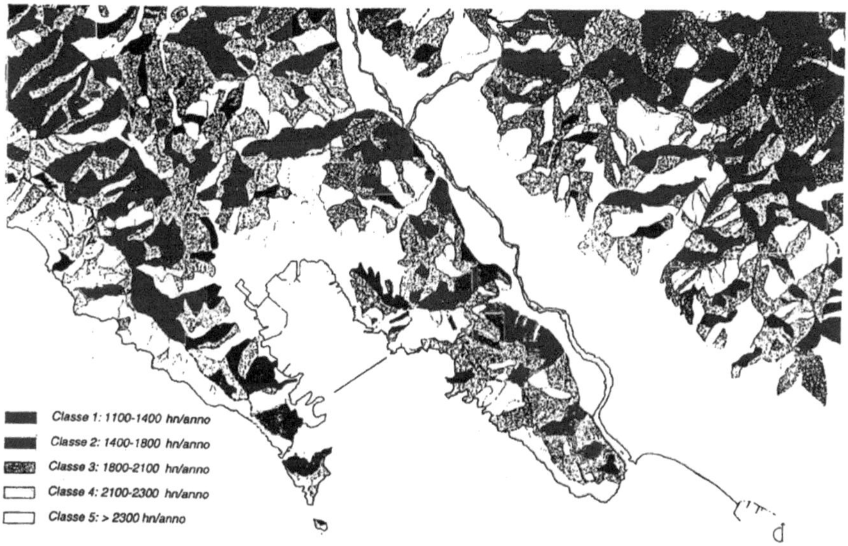

Fig. 26.2 Territorial sun exposure. This map was superposed to the settlements distribution, showing the correlation between well oriented parts and settlements dispersion.
Source: ibid., p. 25

Chapter 27
Agronica (1995)

Andrea Branzi

Introduction

Martina Barcelloni Corte
University of Liège, Liège, Belgium
mbarcellonicorte@uliege.be

In continuity with the author's longstanding research on forms of open, continuous and temporary space launched in the late sixties together with the Archizoom Associati (1969–72)[1] *and in accordance with his custom of using the urban project as social and cultural critique,*[2] *Agronica is one of the author's best-known theoretical projects.*

Commissioned by Philips Electronics and created in association with the Domus Academy, the project explores the idea of a "weak and diffuse" urban system of "continuous and extended mobility, fluidity and flux"; a "degree zero urbanization"[3] *enabled by neo-liberal economics and materialized in the horizontal spread of capital across a diffusely inhabited and extensively productive territory:*

[1]I refer to their research on "No Stop City". See Branzi, A. (2006). *No-stop city: Archizoom associati*. Orleans: Hyx.

[2]See Waldheim, C. (2010, November). Notes toward a history of agrarian urbanism. *Places Journal*. and Waldheim, C. (2018). Industrial economy and agrarian urbanism. In P. Viganò, P. Cavalieri, & M. Barcelloni Corte (Eds.), *The horizontal metropolis between urbanism and urbanization*. Springer.

[3]See Branzi, A. (2006). *Modernità debole e diffusa* (pp. 9, 10, 132). Milano: Skira.

A. Branzi (✉)
Polimi, Milan, Italy
e-mail: studioandreabranzi@gmail.com

M. Barcelloni Corte and P. Viganò (eds.), *The Horizontal Metropolis*,
https://doi.org/10.1007/978-3-030-56398-1_27

"a horizontal landscape without cathedrals crossable and reversable".[4] *Compared to other theoretical models of his, especially those emerged in the 60 s, Agronica is never disconnected from space, but constantly confronted with its limits (and potentials), with the very characters of the contemporary territorial condition. In the words of the author, the project embodies an "incomplete utopia", destined "not to produce global models of cities, but only conflictual parts of the same".*[5]

Through Agronica *Branzi explores the potential of a model of "weak urbanization" where the agricultural and urban spheres integrate and coexist thanks to a particular type of architecture transformed into "freely available and mobile" construction components, a system of objects dispersed in a semi-urbanized agricultural tissue and where new relationships—as those between agricultural and energy production—become possible. In the early seventies the author had already pointed out how the advent of new technologies and production logics were opening a new design season capable of questioning and upturning the very foundations of modern architecture. The risk was, in his view, that the design culture would not realize this mutation and persisted in living in reflection of the previous century's models/ideals.*

That portrayed by Branzi is a silent and profound mutation of the very concept of the city as an efficient and plannable organism where the market has denied the functional and programmatic logics elaborated by the industrial culture of the past and where a new season of experimentation and reformism is preparing to upset and overturn the inhabited space. The concepts of the "reversibility of uses", the inexpressiveness of artefacts, the absence of urban form's global significance, the total integration between the designed environment and the natural environment, the separation between typology and function, already present -in nuce- in other researches, take-on here increasingly clear and detailed characteristics.

Agriculture, portrayed as an evolved serial production model, as a system of environmental transformation able to "adapt to reversible programming" and "powered by weak, seasonal and eco-compatible genetic energies", is capable of scanning and ordering the inhabited space, of organizing its areas and rules of operation. It is a universe of complex natural technologies, as "wide and extensive as the territory itself" and able to provide a "self-regulated model of extreme management and production, sophistication based on governable natural energies".

An essential reference for a series of subsequent projects such as the Strijp Philips Masterplan of Eindoven, realized together with Lapo Lani in 2000, Agronica will maintain -strengthening it over time- its critical and projective relevance for following generations of architects and urbanists, not only for the visionary character of his proposals, but also for their ability to synthesize themes dear both to the

[4]See Branzi, A. (2003) «Weak and spread». Rotterdam, The Netherlands: Public lecture at the Berlage Institute.

[5]Branzi, A. Donegani, D. Petrillo, A. Raimondo, C. with Tamar, B. D. (1995). Metropoli simbiotiche: Agronica. In E. Manzini & M. Susani (Eds.), *The Solid Side*. Naarden: V+K (pp. 101–120). Reprinted courtesy of A Branzi.

nascent Landscape Urbanism as to the research that will soon focus on the analysis/study of extensive and "planetary" models of urbanization.

27.1 Symbiotic Metropolises: Agronica[6]

Agronica: an incomplete utopia

This research can be seen as a theoretical and design reflection in the context of the end of unity of the project. Just as positive utopias have existed (the Ville Radieuse), as well as critical utopias (No-Stop City), so there are also incomplete utopias (Agronica).

They are incomplete because they are destined not to produce global city models, but only (conflicting) parts of them. The city is a set of contrasting logics, and as such it cannot be resolved, but only managed inside a stratification of plans that are also in conflict and therefore exist in an unstable balance.

When we talk about the fall of the unity of the project we mean that urban design is divided according to opposing logics: projects of metropoles, of cities, small cities and villages. These realities are not destined to vanish in a process of inevitable transformation of the village into the metropolis; the village will remain a village, and the same is true of cities and small cities. So there are different projects for different urban logics.

But that is not enough: while in the village a unity still exists between technologies and urban image, this unity no longer exists in the metropolis. In the metropolis buildings, urban signs and interior design correspond to projects, technologies, separate logics that cannot be traced back to any unity.

Agronica is a partial utopia because it is positioned as "part" of what exists, alongside it and not as its replacement. This designs a part of the territory next to the metropolis, the existing city and village; alongside the traditional circulation system and the market of goods existing today.

Agronica develops an ulterior model of urbanization never previously explored, which consists of a weak system that guarantees the survival of the agricultural and natural landscape, with the presence of evolved but no longer totalizing urban services. It therefore tends to get beyond the old conflict between city and countryside through an innovative mediation, in the certainty that these two realities correspond to ossified concepts that are not capable of dividing the world. As an incomplete utopia, it is not based on the advent of new technologies (though they do exist), but on the desire to go beyond the boredom generated by the old question of the uniqueness of the identity of places and the unity of the project.

---

[6]Source: Branzi, A. Donegani, D. Petrillo, A. Raimondo, C. with Tamar, B. D. (1995). Metropoli simbiotiche: Agronica. In E. Manzini & M. Susani (Eds.), *The Solid Side*. Naarden: V+K (pp. 101–120). Reprinted courtesy of A Branzi.

A weak model of urbanization

In the previous phase of the industrial revolution a strong correlation was maintained between constructive solutions, territorial morphology and models of organization of activities: the factory, the office, networks of connection each had their own expressive form determined by the constraints of technical and functional organization.

Already for some time this correspondence has no longer been necessary. Different settlement realities reproduce codes that are simply citations of self-referential morphologies that have no real relationship with the activities that take place inside them.

The set of functions and operations performed in an office building, a university or an industrial workshop could be done in a group of farmhouses or an abandoned thermal power station.

Seven hypotheses thus arise.

- *Separation between technology and form*

A correspondence no longer exists between construction technologies and the form of (urban and useful) artifacts.

While there was a close relationship between mechanics and the objects derived from it, today there is no direct relationship between electronics and the form of the products it generates.

- *Separation between function and form*

Electronic instrumentation makes it possible to perform the same functions in any place. Places, as a whole, lose their functional identity. The function of places no longer corresponds to a visible stylistic code (house, office, factory), but to software that changes the use of places, i.e. a networked program that specializes activities in real time.

- *Weak urbanism*

Urban planning still pursues the goal of correspondence between physical and functional models. The word "plan" itself translates the illusion of being able to create a system of analogies that permits the transfer of the multiplicity of structural dimensions present in the metropolitan volume onto a single level.

The contemporary territory, on the other hand, is a sum of physical and virtual places that remain independent of each other and respond to different organizational logics, constantly interpenetrating and readapting.

All this makes the metropolis an environment with a great density of functions, but without syntax and structure. The terms and grammars still used today to describe it (center/suburb, housing/labor, houses/infrastructures, spaces of life/connection) are vestiges of previous realities (the military city, the industrial city, etc.).

What we consider to be the physical form of the city is actually composed of thickenings, seams, concretions of earlier fabrics that have been lacerated and dissolved.

In new edification that starts “from scratch,” without existing historical elements to which to refer, the present functional density could not correspond to any physical image, any model of reference.

This is the project diagram of the formless metropoles of the Third World.

A tracing of this new structure of the city can be found precisely in the traditional conception of the Japanese city, where the distinction between city and countryside has no precise definition.

In the ideogram for city, machi, both the rice field ta and the grouping of houses, cho, are represented. Likewise, in the term that comes closest to our concept of the countryside, inaka, the rice field ta appears again, with the house sha, this time subdivided into two ideograms. The ideogram for village, mura, contains the image of trees (ki) and an ancient unit of measure, sun, putting the accent on the village not as a grouping of buildings, but as a controlled territorial system. We might say, then, that Edo, the ancient Tokyo, was an example of weak urbanization, with the rice fields extending up to the walls of the castle of the shogun, studded by residences, markets, centers of crafts production, pleasure districts.

- *Habitation is a shopping bag*

We need to understand the dissolve between form and organization in a positive light.

The urbanized territory is no longer a geometric landscape made of valleys, constructive orders, recognizable physiognomies like those that stand out in the background of Renaissance paintings. The whole metropolitan reality has become an immense supermarket of existential offerings, where people go shopping and fills their bags with different selections of products.

The morphological order of the territory no longer represents its structure, the framework in which the different organizational processes of metropolitan life find their places, but a structure of physical levels of the offering, a system of perceptive stimuli like advertising, the faces and bodies of persons, the airborne pulsations of telephonic and televisual transmissions.

- *Virtual metropolis and natural metropolis*

The evolution of the metropolis has brought about the end of the mass society. Television broadcasting is the last of the frames: a system of information that still guarantees the universal reception of news and the coinciding of phenomena. With interactive television we will find ourselves definitively scattered in hypertexts in which encounters will be random. Living in these virtual spaces, however, opens up the possibility of pacifying the physical economics of relationships with nature. The settlement form will no longer have to be the expression of a social design, a level in which identity strategies and the regulations of models of cultural life are manifested (the form of labor, the structure of the family, etc.).

Its definition can be the result of the interaction of the human species with the environment, in which the constructive models that emerge will have a relationship not so much with social choices as with the technical and ecological balances the natural and artificial circuits will be able to establish, from time to time.

- *Weak urbanization*

In this overall reorganization, the historical terms of city and countryside lose their meaning. By now, settlement technologies are part of territorial metabolic systems in which the functions separate and scatter, only to regroup in combinations that would once have been unthinkable (coincidence of production and consumption of food products, residential spaces interspersed with agricultural or research activities, technical systems of movement at the service of transport and residence at the same time, etc.).

The metropolis no longer corresponds to the concentration of goods, technologies and information.

It corresponds to a model of weak urbanization that also coincides with the highest level of the available technologies, which is the "symbiotic" level.

- *The vineyard does not represent the wine*

The horizon of aesthetic quality changes. Agronica coincides with the end of the domestic composition as the glorification of a place, a microcosm in which to achieve and reproduce all the qualities offered by a productive system. Habitation becomes a dispersed form of utilization, a way of entering and becoming part of the surface of the territory.

Until today design culture has worked on the relationship between architecture and industry (modernity), in the certainty of a symbolic value of the latter; in the relationship between architecture and communication (post-modernity), in the certainty of a symbolic value of the latter…

Agro-electronics sets out to work in the relationship between architecture and agriculture, in the certainty of the absence of a symbolic value of the latter. The vineyard does not represent the wine.

Fig. 27.1 Article's first page original layout. Branzi's references are displayed in the different images, Parc de La Villette, Broadacre City and Non-stop City are positioned side to side.
Source: Branzi, A. Donegani, D. Petrillo, A. Raimondo, C. with Tamar, B. D. (1995). *Metropoli simbiotiche: Agronica.* In E. Manzini & M. Susani (Eds.). *The Solid Side.* Naarden: V+K. p. 102. Reprinted courtesy of A. Branzi.
"This research can be seen as a theoretical and design reflection in the context of the end of unity of the project. Just as positive utopias have existed (the Ville Radieuse), as well as critical utopias (No-Stop City), so there are also incomplete utopias (Agronica)." (ibid., p. 103)

Fig. 27.2 Article's original layout. Images of Agronica's model are juxtapposed to images of agricultural landscapes taken from the sky.
Source: ibid. p. 103. Reprinted courtesy of A. Branzi

Fig. 27.3 Images of student's projects from the course "Incubators": Bloom City Project—"cubic units system, located remotely and programmed to form an equipped virgin land. They can be arranged to accommodate various living functions and serve as store of removable parts, which organize in a flexible manner the entire territory. The result is a habitat that strongly relates to nature through different filters, and different activities depending on the seasons. Inside/outside relations are widened." Reprinted courtesy of A. Branzi.
Source: "Incubators", 1996, course held by Prof. A Branzi at the Domus Academy (Milan); assistants: Messori, F. Zini M.; student: Nulsuwan P., Thailand

Fig. 27.4 Images of student's projects from the course "Incubators" : Landscape at work Project. Source: "Incubators", 1996, course held by Prof. A Branzi at the Domus Academy (Milan); assistants: Messori, F. Zini M.; students: n.a. Reprinted courtesy of A. Branzi

Chapter 28
Zwischenstadt (In-Between City) (1997)

Thomas Sieverts

Introduction

Farzaneh Bahrami
Groninghen University, Groninghen, The Netherlands
f.bahrami@rug.nl

In 1997 Thomas Sieverts, German architect and urban planner, published Zwischenstadt (literally "in-between city"), on the state of the contemporary city "in-between place and world, space and time, city and country". Grounded in the observation that the open countryside was subject to a profound transition, from being the background of the city to being a figure in se, Sieverts described not only a formal but also a cultural shift in the German territory, as well as throughout the world. He defined Zwischenstadt as an "international phenomenon", taking it from the early twentieth century prophecies on the probable diffusion of cities,[1] *to more contemporary discourses on the urbanised countryside in the European context.*[2] *Focusing on the apparent insoluble contradiction of the loss of the compact city in*

[1]Wells, H. G. (1902). *Anticipations of the reaction of mechanical and scientific progress upon human life and thought (8th ed.).* London: Chapman & Hall. (Chap. 5 in this book).

[2]See, for example, Neutelings, W. I. (1995). Erkundung des Wunderlandes. Eine Fahrt durch die Peripherie der Niederlande' [The exploration of the wonderland. A journey through the periphery of the Netherlands]. *Westfälischer Kunstverein Münster (Hg.): Die verstädterte Landschaft, München.* Nicolin, P. (1992). 'Notizen zu: Peripherie - Metropole - Loslösung' [Notes on Periphery - Metropolis - Detachment]. *archithese*, 6, pp. 57–59.

T. Sieverts (✉)
University of Darmstadt, Darmstadt, Germany
e-mail: tom.sieverts@googlemail.com

M. Barcelloni Corte and P. Viganò (eds.), *The Horizontal Metropolis*,
https://doi.org/10.1007/978-3-030-56398-1_28

the open countryside, Sieverts proposed to generate an up-to-date interpretation of the periphery in which a special role was assigned to the periphery as "the mediator between place and *world", as "a place in which the exchange between the city and the rest of the world is played out". Building upon and learning from the already emerged discussions on the city-territory and the renewed status of the periphery within the European debate, Sieverts was able to widen the field of investigation—beyond his case studies—outlining a broader theory on the contemporary city. Deconstructing the myth of the "Old City", he calls for a revision of specific key concepts such as urban-ness, centrality, and ecology (understood as the most compatible integration of the city into the natural cycle), and puts forward a more articulated, fine-grained interpretation of the new built-up area, describing a new form of city. Sieverts, hence, recognized in what was commonly described as lack of urbanity, "a decentralized cultural diversity", a "unique archipelago" within which new social, economic, environmental, and structural features gradually emerge.*

His attentive observations and experience as one of the directors of the Internationale Bauausstellung Emscher Park (International Building Exhibition Emscher Park)[3] *became a substantial asset for the development of Zwischenstadt. While observing the mutual penetration of the built fabric and landscape, and the gradual disappearance of traditional hierarchical patterns, Sieverts made a critical analysis of the socio-spatial characteristics of this new kind of emerging urbanity. Foregrounding the potentials of such territories, he made a plea for consideration of the new political opportunities that Zwischenstadt could offer. In an attempt to propose future perspectives for regional planning, the book discussed extensively the challenges of these territories and their risk of disintegrating in political, social and cultural terms, and specifically identified sustainability as a major pressing concern for the future.*

Through the work on the Zwischenstadt, a double observation was made: the dissolution of the preindustrial compact city on the one hand, and the discrepancies between this new reality and the collective representations (of what the city is), on the other. To overcome this gap and to be able to regenerate the identity of space and society within this new urban condition and as an urgent task on the agenda of planning and design, Sieverts underlined the need for the intelligibility and legibility (and thus imageability) of the Zwischenstadt, as the most critical conditions for future development.

[3]A major urban renewal project in the previously industrial heart of Germany in the Ruhr area. See Shaw, R. (2002). The international building exhibition (IBA) Emscher Park, Germany: A model for sustainable restructuring?. *European Planning Studies, 10*(1), 77–97.

28.1 In Between City. Between Place and World, Space and Time, City and Country[4]

Foreword

This book deals with the dissolution of the compact historical European city and with the treatment of a completely different and new form of the city, which is spreading across the world: the urbanised landscape or the landscaped city. For the sake of simplicity I call this *Zwischenstadt,* meaning the type of built-up area that is between the old historical city centres and the open countryside, between the place as a living space and the non-places of movement, between small local economic cycles and the dependency on the world market.

In order to avoid one misunderstanding in advance, let me make it clear that this book is not a plea in favour of dispersed settlement, of urban sprawl. I am just as keen on the form and nature of the historical European city and the historical cultural landscape as my colleagues who still today heroically try to defend it as a model for the future against all the devilish forces of dissolution. Who would not be pleased if cities could be kept compact, villages intact and the natural landscape unscathed? Perhaps it is even justified to continue to call for the old conventions and objectives in the hope that at least something of them remains valid, although, the reality is getting further and further away from these ideals. By calling into question the validity of conventions which the guild of planners has adopted as its broadly agreed principles of planning—such as the principle of 'decentralised concentration', of 'regulating intervention and compensation', of 'land saving through high building densities', or even the ordering system of 'centralised locations' and 'small grain mixed use'—we find ourselves occupying an uncomfortable position between all camps. The idealists among the planners accuse the critics of a betrayal of sublime principles—and that is painful; investors and technocrats of all types could use the criticism to justify their limited actions—and that would be dangerous; and the politicians could then feel themselves challenged to pay still less attention to spatial planning than they have done so far, instead of seeing the major new political design opportunities which the *Zwischenstadt* offers—and that would be disastrous!

The dilemma seems unavoidable, because the 'Golden Age', of an enlightened, well-funded and well-equipped European system of public intervention with a 'golden rein'—where town planning, despite much uneasiness, could feel at home and where a ceasefire, if not exactly a peace, prevailed between municipal corporations and capital—is over. Sometimes, among my guild of architects and urban

---

[4]Excerpts from: Sieverts, T. (1997). *Zwischenstadt, zwischen Ort und Welt Raum und Zeit Stadt und Land.* Berlin: Vieweg. Source used in this publication: Sieverts, T. (2003). *Cities Without Cities, An interpretation of the Zwischenstadt.* London: Spon Press and New York: Routledge, pp. xi-xvi and 1-12. Reproduced by permission of Taylor & Francis Books UK.

designers, I feel like a guild master at the beginning of the nineteenth century, previously able to maintain the old rules and practices to some extent against mercantilism—although not without losing some of his beautiful, colourful plumage—but who now has to protect himself against the rise of capitalism and so suffers one defeat after another, because he misjudges the forces of the age. Sometimes I have a frustrating feeling that, with our efforts to save the archetype of the historical European city as a model for the future, we urban designers today are only delivering the public fig leaf for developments which run in a quite different direction and in a completely different environment. With an ecological approach, which predominantly concentrates on the symptoms only, not much is going to be achieved. Sometimes I get the impression that our guild has forgotten how to think politically and to draw distinctions between social changes, movements and forces - which we, whether we like it or not, have to recognise as preconditions for planning because we can change hardly anything about them—and such developments as we can influence and structure. However, only when we can do that, will we be able to get involved with any prospect of having an effect.

The fundamental changes that we will have to accept as preconditions are well known, for example:

1. the worldwide distribution of labour in the economy and the change that this means for the position of the city in the structure of the world economy;
2. the dissolution of the cultural binding forces of the city and the radical pluralisation of city culture which is connected with it;
3. the fact that the natural world has now been almost completely penetrated by artefacts and the contrast between city and nature has therefore dissolved.

In combination, these changes and forces are leading to a profound transformation of the city which could destroy the European city but which—and this is the core thesis of this book—is also opening up the possibility of new design perspectives. In the context of a globalisation which can no longer be held back, thinking through these changes confronts us with the task of finding and developing, in the course of its transformation, new forms of the European city. In these the historical city is protected and sublimated as a special part of the city if only because, once it is destroyed, it cannot be reconstructed. To be sure, in the interest of maintaining its essential characteristics, the historical core will have to give up specific central tasks and thus become one part of the city among others—albeit of a unique kind. In this way, and by responding to place-specific conditions, Europe could make a unique contribution to achieving a European characterisation of the global city model. Although this would have links with the great traditions of the European city, it would not be in a defensive, retrospectively orientated manner, but in an extensive perception of the immense new possibilities which globalisation offers in addition to all its negative aspects. I am convinced that this task opens up a field for design which is as productive as it is rewarding for politicians and town planners, even though this might seem almost utopian today.

In Europe, town planning and urban design were children of the crises of liberalism in the age of the Industrial Revolution. Initially they arose to repair the worst side effects of city expansion: epidemics of cholera and typhus, outbreaks of fire and desperate living conditions. Modern town planning and urban development as approaches to the taming and cultivation of capitalism only came into effect much later, but even in the early utopian and reforming conceptions they were laid out in perspective. As far as I am concerned, even today urban design still belongs to this tradition of European town planning, as a constant challenge to protect the endangered weaker elements of the city—human beings, nature and culture—and develop them in their own values with their own inherent characteristics. The conditions for efficacy in urban design construed in this way have altered sharply again since the epochal change in world history in 1989, a turning point to a new expansion of global market forces which transcend nation-states and in turn are having highly deleterious ecological and social side-effects. Confronted with this, should we not still continue to believe that society will be able to mobilise counterforces, to discover new opportunities and to answer the questions which today seem to be put from an almost utopian perspective? Do the debates about communitarianism, about the necessary new reorientation to social well-being—and also the discussions about a new balance of competition and cooperation, of planning and self-regulation, of state and intermediate institutions—not reveal approaches to a new understanding of town planning? These debates are very much in a European tradition. The present book is intended to make a contribution to this debate. It is a plea to perceive the opportunities of the *Zwischenstadt.*

The subject is not a new one; seen in the historical perspective, urban development is usually concerned with recurrent themes in a long tradition of ideas and concepts. With my questions, I connect my subject with a European debate about modernity from immediately after the First World War which, here in Berlin, where this book was written, is connected with Bruno Taut and Hans Scharoun. 'In his work *The Dissolution of Cities,* Taut formulated the prospects of human globalisation in a pacifist euphoria with dreamy visions of bands of settlements covering the world' (Durth).[5] Hans Scharoun's concept of the urban landscape, as, for

---

[5]'The concept of the dissolution of the compact city have a longer history of ideas'. In *Tomorrow: A Peaceful Path to Real Reform* (London: 1898). Later published as *Garden Cities of Tomorrow*, Ebenezer Howard had already regarded the 'small garden city' as a 'working model'. It would be followed, after experience had been collected, by a 'cluster of garden cities' in order finally to carry out 'a complete reconstruction of London'.

Bruno Taut sketches with expressionistic pathos the 'dismantling of the stone city' and writes in the sketch: 'Let them collapse, the horrid constructions ...', so that in floral forms of settlements visions of new symbiosis of man and nature, built structure and landscape can be designed (in Bruno Taut, *Die Auflösung der Städte oder Die Erde eine gut Wohnung* [The Dissolution of the Cities or the Earth, a Good Home], Hagen, 1920, p. 1).

From the criticism of the large city in the nineteenth century, through the garden city and life reform movement, through the manifestos and programmes for the new start in 1919, various paths lead to the concept of the 'urban landscape', which was formulated around 1940 and becomes the guiding principles for the reconstruction after 1945. Generations of planners yearned to dissolve example, in his plan for Berlin after the Second World War, also falls into this tradition. There are many other examples from the history of ideas of urban development that could be mentioned which, rooted in a criticism of the densely composed city, demand the dissolution of the city. These range from the early socialist Robert Owen with his communes, in which industry and agriculture were to be combined, through Frank Lloyd Wright's archetypal American Broadacre City, in which land ownership forms the basis of social collaboration, up to the half-European half-American Ludwig Hilberseimer with his bands of settlements cutting through the US prairie. However, historical perspectives are not the subject of this book (individual, detailed references can be found in the notes). The emphasis is on the present and on the argument about a restorative conception of the city, which has been predominant for a generation and not only in Berlin. No attempt will be made to base the force of the argument on the history of ideas, and the line of argument will frequently appeal less to seminal texts than to their specialist interpreters. The book should be regarded not as an enlightened analysis without anger and passion but as a polemic and a challenge to action. The five chapters each consider the *Zwischenstadt* from a different perspective. Description, analysis and proposals for action interlock, in order to render the relation to action of the line of thought palpable. This means that questions about images and about the readability of the *Zwischenstadt* are constantly taken up.

The argument presented by this book is structured as follows. The first chapter focuses on the urban environments of the *Zwischenstadt.* A brief description of the urbanised landscape is given and the myth of the Old City is subjected to a critical analysis of five key concepts, the meaning and use of which are at present prominently oriented towards such a myth.

In the second chapter, building up from the basis of this criticism, a new perspective for the urbanised landscape is developed. This perspective is subjected to a political critique with reference to the theoretical stance of Alain Touraine. This determines a political position from which seven theses on the need for the intelligibility and legibility of the *Zwischenstadt* are developed.

In the third chapter, the urbanised landscape as a space for day-to-day life is considered and criticised from the political position that has now been developed. A perspective is drafted for a living space characterised by the principle of a 'cycle economy', in which awareness for sustainability becomes paramount.

The subject of the fourth chapter is a cultural and design-oriented interpretation of the *Zwischenstadt* in its diversity as against the familiar historical city and the importance of cultural policy for its design.

---

the 'urban settlement mass' which was to be 'segmented and dispersed' in order to grow together with the landscape. At the start of the 1970s (heritage year 1975), the return of the yearning for the Old City becomes the counter-movement for the next generation against comprehensive redevelopment and the Modern Movement. (On this in detail, see Durth and Gutshow: *Träume in Trümmern: Stadtplannung 1940–1950* [Dreams in the Dust: Urban Planning 1940–1950], paperback edition, 1933, p. 214, chapter: Letbilder für Stadtplanung [Guidelines for town planning].

In the fifth and final chapter an attempt is made to sketch a perspective for a new form of regional planning and regional policy.

This book was written in the academic year 1995/96 in the stimulating environment of the Institute for Advanced Studies in Berlin, which I would here like to thank for a unique quality of life. Petra Sonnenberg transferred the various forms of my hand-written manuscript into computer form and read through the corrections with patience and care. Ulrich Conrads, as a reader and friend, prepared the manuscript for printing. Important discussion partners for the work as co-fellows were Susanne Hauser and Michael Mönninger, the latter precisely also with his differences of opinion. Werner Durth, who was to be with me in the Institute but was unfortunately unable to take up the invitation, as a friend of many years, helped me with references to the literature on the history of ideas and with well-grounded criticism; what a pity that we were not able to discuss these matters in the Institute's environment! With Sibylle Ebe I examined the manuscript for intelligibility and applicability to practice and in walks through the *Zwischenstadt* checked it for clarity. Hille von Seggern, with whom I have also had many years of stimulating discussions, contributed to clarifying my ideas with questions on my basic understanding of planning, as also did Ursula Stein. Hanns Adrian should really be mentioned as the co-author of two chapters, provided that he should not as result have to bear responsibility for the remainder of the book. All these friends and colleagues I would like to thank warmly for their constructive criticism. However, without my five years of experience as one of the Directors of the 'Internationale Bauausstellung Emscher Park' (International Building Exhibition Emscher Park) in the Ruhr area, without many thought-provoking discussions with my colleagues there—above all with the animator and driving force of the 'Emscher Park'—this work could not have been written. Consequently, the book is also a way of saying thank you to Karl Ganser.

The living space of the majority of mankind (Chapter I)

An anonymous space with no visual quality

Since the railway, car and electronics have exploded the spatial limits imposed by the muscular capabilities of human beings and animals, the city has been extending almost without restraint into the countryside. Its expansion and the degree of its dispersal follow the relevant traffic and communications technologies: the railway results in a star-shaped, linear expansion, the car fills in the surfaces and electronics leads to 'borderless' expansions. However, this development is based not only on technical inventions but also on far-reaching historical causes. The forces which produced the compact city and kept it together for 150–200 generations—by which I mean the priest-kings and religious associations, temples and churches, walls and markets, feudalism and the guilds—had definitively come to an end before these technical inventions.

It is possible that the compact city is itself only an interlude in the development of human civilisation. According to interpretations of the theory of evolution, human beings are primates, social animals living in loose groups who prefer life in the wide-open savannah and at the edge of light forests. The compact, walled city would therefore be an historically imposed form which would dissolve 'naturally' the moment the causal social forces are taken away.

Against this it might be argued from the perspective of cultural history that the cultural development of mankind in the past 5,000 years was inseparably connected with the development of the compact city. As result, it is part of the essence of human beings as cultural beings and with its dissolution the cultural development of humanity might also be jeopardised.

The dispute as to whether the disappearance of walls and moats around 1800 was a liberation or a loss of uniqueness and safety is as old as the event itself. Many citizens felt that through the opening up of the city to the open countryside they were being exposed to anxiety, whereas others greeted the liberation from narrowness and constraint. Goethe was one of those who welcomed the opening of the city:

> "Even relatively large cities are now pulling down their walls, the moats even of princely castles are being filled in, cities are only large blotches, and if you see such a thing on your travels you think general peace is confirmed and the Golden Age is at the door. Nobody feels comfortable in a garden which does not look like open countryside; there should be no reminders of art or constraint, we want to be completely free to breathe the air without restriction."[6]

Whatever the reaction of citizens at the time may have been, it must be recorded that throughout the whole world the 'city' of the modern age extends into its environment and thus creates the peculiar forms of an urbanised landscape or a landscaped city.

Following a venerable tradition we still call distinct regions of settlement 'cities'. Or we describe them with such abstract concepts as 'city agglomerations', 'areas of concentration', 'urbanised landscapes', etc., because we note how inappropriate the concept of 'city' is when applied to these fields of settlement as they evoke completely different associations. For want of a better term, we shall call these structures which consist of 'fields' of various uses, construction forms and topographies *Zwischenstädte.* They take up large areas, and they have both urban and rural characteristics. The *Zwischenstadt* stands between the individual, special place as a geographical and historical event and ubiquitous developments of the global division of labour; between the space as an immediate living area and the abstract traversing of distance which is only measured in the consumption of time; between

---

[6]See Martin Warnke, "Nature nach dem Fall der Mauern" [Nature after the fall of the walls], quotation from Goethe's "Wahlverwandtschaften" [Elective Affinities] (1807) (in *Dialektik* 2/ 1994, zur Ästhetik des Territoriums [On the aesthetic of the territory].

the mythical Old City which is still very effective, and the Old Cultural Landscape which remains anchored deep in our dreams.

The *Zwischenstadt* as an international phenomenon

This *Zwischenstadt,* which is neither city nor landscape, but which has characteristics of both, neither has a suitable name nor is it concrete. Despite the fact that it has no name, it can be found all over the world. Intermediate cities with 20–30 million inhabitants have emerged in Asia and South America. For all the massive differences, depending on the state of economic development, culture and topography, they have the shared feature that they have almost nothing to do with the relevant local pre-industrial urban traditions. Across all cultures of the entire world, they share specific common characteristics: a structure of completely different urban environments which at first sight is diffuse and disorganised with individual islands of geometrically structured patterns, a structure without a clear centre, but therefore with many more or less sharply functionally specialised areas, networks and nodes.

We find *Zwischenstädte* of this kind particularly conspicuous in areas in which cities grow beyond their own extensions into a city cluster, but most clearly where the historical, traditional city-composing forces never really took effect, such as in the Ruhr area, or also in the metropolitan areas of the Third World. In the *Zwischenstadt* the ratio of open landscape and built-up areas has frequently been reversed; the landscape has changed from being an all-inclusive 'background' to being a contained 'figure'. Conversely, the settlement surface has increased in size and openness and has acquired something of the character of a surrounding landscape. This *Zwischenstadt* is a field of living which, depending on one's interest and perspective, can be, interpreted either as city or as country. Although the causes of this diffuse form may differ, they share the feature that the historical city-forming forces and the limits imposed by them had reached their end.

The *Zwischenstadt* as a result of innumerable individual rational decisions

As a whole, the diffuse city gives an 'unplanned' impression, but it has arisen out of innumerable individual, and—considered on their own—rational decisions. A typical example from an old industrial region might run as follows: a road exists, a factory is built, either because agricultural products are to be processed or because mineral resources are available, the processing of which might supply a growing market. The factory attracts worker settlements to it, and gardens are allocated to the workforce to enhance its self-sufficiency. The population next needs schools and shops. The growing employment and consumption market attracts further institutions and, as the social richness increases, a basis is formed for new specialisation and further division of labour. New traffic routes and public establishments become necessary, and in this way the evolution of the city continues on the principle of 'settlement creates settlement', without following any pattern planned in advance.

Another example, but coming from the Third World, may be that: an old city functions as a centre of attraction for city migrants who leave their villages for the widest possible range of reasons—usually there are several—ranging from overpopulation or lack of basic foodstuffs, caused by unemployment, to the promise of emancipation offered by migration to urban centres. Such migrants look for a place of settlement in which they have access to the blessings of the city but can still operate a modest semi-urban agricultural economy. The consequence of these decisions, each of which is logical in its own terms, is a less structured, more open settlement between city and country, which develops further with its own workplaces and facilities into a more or less independent *Zwischenstadt.*

Structurally comparable results are produced by the behaviour of house buyers in our cities. They are looking for properties which they can still afford, from which the core of the city can still easily be reached and at the same time access is open to the country. The consequence of such an accumulation of decisions, each cogent in itself, is the 'settled' landscape, which initially is almost exclusively residential and, after a period of intensification and consolidation, attracts workplaces and consumer provision. Only then does it develop into a *Zwischenstadt* which frees itself from its original dependency, supplies itself and enters into a relationship of mutual exchanges with the original city.

Statistical analysis in Germany shows that it is not so much the small towns, the central places prescribed for this purpose by town planning, which attract new inhabitants, but rural communities on the periphery of the 'central places'. From a forecast of the Federal Research Institute for Regional and National Development it is apparent that the borders of today's settlements, according to a conservative estimate, will expand by a further 10% by 2010 (against a core growth of only about 2–4%). 'It is becoming increasingly obvious that the residential environment is becoming more and more decisive for where we live and no longer, as in the past, the proximity of the workplace.'[7]

In the USA a similar development is taking place on a much larger scale; the triggers in many cases are motorway exits, major shopping centres and, for several years now, major office complexes on motorway intersections. These are at the same time the consequences and causes of settlement activity. Also these very widespread *Zwischenstädte* have long since separated themselves from the original city, but here the dependency relationship has often been reversed—the impoverished core city now looks for its work places in the surrounding *Zwischenstadt.*[8]

[7]See Rainer Mackensen: 'Ist Stadtentwicklung planbar?' [Can urban development be planned?], lecture on the retirement of Reinhart Breit, Technical University of Berlin, 1996 (unpublished lecture manuscript). Andreas Kagermeier: "Jenseits von Suburbia – Tendenzen der Stadtentwicklung in der Region München aus verkehrsgeographisher Sicht" [Beyond suburbia—trends in the evolution of settlements in the region of Munich from the perspective of traffic geography] (in *Mitteilungen der Geographischen Gesellschaft in München* [Proceedings of the Geographical Society in Munich], 97th volume, Munich, 1994, edited by Reinhard Poesler and Konrad Rögner). Bundesforschungsanstalt für Landeskunde und Raumforschung [Federal Research Studies and Planning Research Institute], *Nachhaltige Stadtentwicklung: Herausforderung an einem resourcenschonenden und umweltverträglichen Städtebau* [Sustainable Urban Development: A Challenge to Resource Conservation and Environment-friendly Town Planning], Bonn-Bad Godesberg, February 1996).

[8]See Fishman (1987, 1991), Witold (1995), Louw (1985).

Even in cases in which the planning of major interconnected urban expansions offer possibilities for strong concentration, these newly planned urban configurations are rather uniform, only marginally centralised and differentiated, because the configurations of day-to-day living are relatively diffuse and changing.[9]

In the sequence of the development of the *Zwischenstadt,* internationally comparable stages are apparent.

> After a phase of rapid urbanisation, accelerated by migration between country and city, there is usually, a slower phase, in which the surplus birth rate is normally the main cause of growth. In still later phases, as is apparent in Western Europe, the annual growth of the city falls to below 1%, and the rate of immigration increases again. This also conceals the consequences of excess ageing in the cities and the emigration of families with small children and the prosperous to the attractive surrounding small towns and villages.'[10]

Over time, the original freedom of the choice of location becomes increasingly restricted. The area fills up, new developments have to respond to a context which is becoming more and more built-up and constricted. At some stage, the available land is fully developed, the *Zwischenstadt* has 'grown-up', and further development can and may only be implemented through intensification, reutilisation and the regeneration of disused sites. Old urban fabric and uses become superfluous, they are converted, reused, rebuilt and finally removed. All this taken together produces a carpet of settlement which appears to be without any plan but has the nature of a palimpsest in which old, superfluous and deleted text and images glimmer through the new text.

The unique character of the *Zwischenstadt*

Urban development in different parts of the world is subjected to very different forces. Whereas urban expansion in the Ruhr, Wallonia and the English Midlands is directed by heavy industry, in countries such as India, Africa and South America it is subject to the pressures of overpopulated rural areas. In the USA and Western Europe trends follow the investment of wealth in private residential areas or the compulsion towards affordable housing on inexpensive land. Despite the massive differences in the forces behind urban development, the result in each case is the diffuse form of the *Zwischenstadt,* which separates itself from the core city—if one still exists—and achieves a unique form of independence.

These characteristics link the area of Greater Tokyo with the Ruhr area, São Paulo with Boswash—the area between Boston and Washington in the USA which has merged into an integrated city environment—and Mexico City with Bombay. Even the area of Greater Stuttgart or the Rhine-Main area could be characterised in this way. Major differences are based on the varying densities of residential and car use. The *Zwischenstädte* in the Third World are on the whole still denser and more compact than the more dispersed forms in the industrialised world. When compared

---

[9]Examples of contemporary plans for major city expansions in Germany and Austria can be found in Gudzent and Sieverts (1995). Also DISP, July 1995, pp. 43–52.

[10]Wärneryd (1995).

with the giant *Zwischenstädte* in Asia and South America, comprising more than 20 million people, perhaps we should regard the increase in areas of concentration in European Countries as one single *Zwischenstadt*.[11]

What Karl Ganser says about the Emscher region applies to all cities across the whole world as soon as they have grown beyond the size of small major cities or, indeed, have formed in a period in which historical city-forming forces were no longer effective.

As, for example, in the Ruhr area:

> This 800 sq. km large settlement area ... is essentially fully developed. It is a *Zwischenstadt,* which does not correspond to our ordinary image of the city and our yearning for an intact landscape. With the weak growth potential of the time ahead of us, this settlement structure can no longer be reconfigured. We must assume that it is a given and develop its hidden qualities.'[12]

In all intermediate cities, characteristic patterns of interpenetration of open spaces and built form have become apparent. A generally common characteristic of the *Zwischenstadt* is the continued search for the implementation of the 'Tucholsky principle', the simultaneous yearning for both the urban and rural ideal which he describes in one of his famous poems entitled 'The Ideal'. The Tucholsky principle is the search for reconciliation between the contrasts of participation in human society, the life of the city, and participation in nature. In other words, it is the yearning for a combination of pastoral romanticism and the comforts of the city.[13]

---

[11]See Eckard Ribbeck, 'Von der Peripherie zum Zentrum? – Verstädterung in Asien, Afrika, Lateinamerika' [From the periphery to the centre? – Urbanisation in Asia, Afrika and Latin America] (*Deutsches Architktenblatt* DATT 12/95, p. 2330). Jürgen Oesterreich, *Elendsquartiere und Wachstumspole* [Slums and Growth Centres] (Deutscher Gemeindeverlag and Kohlhammer Verlag, Cologne and Stuttgart, 1980; Espacially p. 95: Überleben in ökonomischen Nischen [Survival in economic niches], Section: Hoher Grad der Raumnutzung [High degree of space utilization].

[12]Ministerium für Stadtentwicklung, Kultur und Sport des Landes Nordrhein-Westfalen [The Ministry of Urban Development, Culture and Sports of the Land of North Rhine Westphalia], *Memorandum II zu Inhalt une Organisation der Internationalen Bauausstellung Emscher Park* [Memorandum II on the Content and Organisation of the International Building Exhibition Emscher Park, p. 9 (Düsseldorf, 1995).

[13]Kurt Tucholsky weote the poem "The Ideal" in 1927. The first verse goes as follows:

> 'That's what you want, isn't it?:
> a villa in the countryside with a large terrace;
> at the front the Blatic Sea, at the back Friedrichstrasse [Berlin's main street];
> with a beautiful view of mondaine countryside,
> from the bathroom you can see the Alps –
> but in the evening you don't have far to the cinema.
> The whole thing simple and modest.'
> See Rowe (1991). Klaus Humpert, Sibylle Becker une Klaus Brenner, 'Entwicklung grossstädtischer Agglomerationen' [Development of urban agglomerations], In *Prozess une Form 'natürlicher Konstruktionen'* [Process and forms of 'natural constructions'], edited by Klaus Teichmann and Joachim Wilke, Ernst und Sohn, Berlin, 1996, p. 182).

This search for a resolution to the underlying paradox of a combination between urban centrality and contact with the countryside was also shared by the early socialists Owen and Fourier, by the reformers Cerdá and Howard and by the visionary architects Frank Lloyd Wright, Le Corbusier and Hilberseimer.[14] It is also still being pursued by innumerable home and land owners in day-to-day practice and is leading to a maximisation of the interface between built-up area and the countryside. These boundaries are of particular ecological interest and fit in well with the 'primate nature' of man. In the last few years they have excited the interest of 'fractal researchers' who have attempted to represent these growth processes in mathematical terms, with extremely interesting results.[15] They show, that at the macro level of the city region, growth processes and the distribution of the different sizes of settlements occur in major agglomerations throughout the world quite similarly and similar patterns of developments appear at different levels of scale. These patterns form virtually independent of the political, cultural and socio-economic backgrounds.

In Germany, these settlement patterns have been characterised by the particular form of local authoritie's self-administration—with heavy competition between communes.

> The developments have been driven by immigration and segregation against which town planning is powerless. At the same time, these are the problem areas which result from the decisions of the inhabitants which affect them most directly. This means that definitions of the problems of urban development would more sensibly be concentrated on not denying these trends but rather by accepting them as critical conditions for the solution of the problem.
>
> In addition, the developments are based on the fact that the local authorities each have to pursue their own interests. Their competition forms the basis for the tendency to exploit the advantages of the peripheral location against the settlement centres and to plan for residential and commercial areas which only drive forward the process of settlement expansion with all its consequences. Against this, city development planning is powerles— at least as long as it conceives itself as 'local authority development planning'. It will not be able do anything else until such time as the local authorities in the agglomeration areas merge into political units (and not only sectoral specialist associations). For this purpose, however, they must largely give up or restrict their political existence; but this cannot be expected from them.'[16]

However, planning across local authority boundaries is restricted due to local sovereignity and public expenditure limits. Simultaneously, with the restriction of public funds for publicly subsidised housing, infrastructure and public facilities, urban and regional planning is losing one of its last instruments for an effective, active influence on the evolution of urban areas.

---

[14]See Wright (1963), de Bruyn (1983).

[15]See Pierre Frankenhauser, 'Fraktales Stadtwachstum' [Fractal city growth] (*Arch + Heft* 109/110, December 1991). Klaus Humert, Sibylle Becker and Klaus Brenner, 'Entwicklung grossstädtischer Aggolmerationen' (see note 9).

[16]See Mackensen (note 3).

Zwischenstadt and landscape

Whereas obvious, though sometimes politically and culturally modified international comparisons can be observed at the macro level of the *Zwischenstadt,* there are clear differences at the micro level of the three-dimensional urban structure caused by cultural and socio-economic factors. These can be classified among the theoretical models for the 1920s and 1930s as being between the *Ville Radieuse* of Le Corbusier (or the 'Big City Architecture' of the early Hilberseimer) on the one hand, and Wright's Broadacre City (or Hilberseimer's later diagrams for US cities) on the other. The grain and density of development of the individual urban areas and the degree of penetration with open spaces and landscapes determine the specific character of each *Zwischenstadt.* Thus, whereas large Asian cities are tending more and more strongly towards Hilberseimer's type of 'Big City Architecture' and the US major cities are polarising between 'Big City Architecture' and Broadacre City, Europe's main cities mostly constitute mixed forms in the manner of the *Ville Radieuse* and Broadacre City. The *Zwischenstadt* can develop any diversity of settlement and built form, so long as, as a whole, they are intelligible in their settlement network and, above all, remain embedded as an 'archipelago' in the 'sea' of an interconnected landscape. In this way the landscape becomes the glue of the *Zwischenstadt.* In assessing and evaluating the character of this sort of landscape, perspectives on the *Zwischenstadt* differ: if we take the large 'consumption of landscape' or 'transformation of landscape' as an argument against the *Zwischenstadt,* one must bear in mind the fact that not only in the Garden City, but particularly in the compact city, the open space is always regarded as an indispensable complementary component, even in cases in which its character has changed from being the necessary source of food into an area for ecological balance or a place for recreation.

Even cities with a large number of inhabitants, such as Shanghai or Calcutta have, until recently—thanks to very high residential and occupation densities—remained so restricted in their extent that every day they could be supplied from the country with handcarts and wagons laden with fresh goods. The city area of Mexico City (approximately 20 million inhabitants) is the same as that of Berlin (3.5 million inhabitants).[17] In almost all parts of the Berlin of the nineteenth century, landscape was present in the form of small allotments only a walking or cycling distance from the densely inhabited residential areas. Once we take these into consideration, the densities of different cities and neighbourhoods which, at first glance, seemed so exceptionally different, become relative. Even the 'Märkische Viertel' in Berlin, by our standards a very dense, high-rise peripheral estate of the 1960s, has a rather moderate overall density if we allow for the recreational areas and open spaces which are intrinsically associated with it.

[17] See Ribbeck (note 7).

In-built redundancy versus capacity to adjust and the conservation of resources?

Special conditions of our mature industrial societies apply to the *Zwischenstadt,* on the one hand, low densities of activity and, on the other, freedom in the location of functions. The density of occupation is low and the building mass is comparatively large. Viewed in rational economic terms, we already have too much built fabric, which is used only for relatively short periods of the day and of the year. The question therefore arises whether, in the long-term, we will be able to meet the running costs, the costs of energy consumption, ongoing maintenance, repairs and renovation or whether we will have to find more economic forms of the *Zwischenstadt* because of this excess load on the economy. With the demise of location-dependent heavy industry and the increase of services, dependence on specific locations has relaxed significantly, thanks above all to the availability of good roads and communication networks. Therefore the location pattern of building fabric, of functions and of open space can, in principle, be selected freely. Thus, functions and structures can develop in response to the different natural and cultural characteristics of the region and in response to different socio-economic conditions (prices and charges) and socio-cultural challenges (lifestyle, purchasing power). The latter are changing profoundly.

Reduced working times, blocks of several days of free time during the working week, abbreviated working lifetimes in the form of 'early retirement' or 'training years' and working from home, are affecting the patterns of travel times between living 'in the country' and 'working in the city'. This will promote a further spatial separation of work and living places rather than a mixture of uses, except for specific services in the medium-sized economy.

These developments are causing a further expansion of, and the segregation of land uses in, the *Zwischenstadt.* This could have the effect that the resulting urban structures due to their major transport costs, consumption of resources and mono-functional, unilaterally determined land uses—can no longer adapt to a profound structural change caused by an ecological crisis.

Increased freedom for the location of functions in the city would, with the support of microelectronics, permit mixed and intensified urban structures in which the use of all the possibilities of telecommunications would reduce the transport volume and engender a more settled way of life. The *Zwischenstadt* would indeed make it possible to develop such urban structures, but this would presuppose a different lifestyle. The question remains open as to whether and how we would be able to prepare our cities for such a structural change and the drastically reduced consumption of resources which would result.

For whatever reasons, in the future about half of the population of the world will live in *Zwischenstädten.* These structures are often so large, in some cases with 10–30 million people, that their inhabitants have no opportunity whatever to escape them in their day-to-day living. As a result, all the living requirements, including the production of food in an urban farming economy, must exist within these *Zwischenstädten.* The urban areas in the *Zwischenstadt* must also meet all the requirements of an ecological balance, since there will be no other areas to offset any imbalance.

The fate of the *Zwischenstadt* is shared by all humanity, and in this regard no country has an 'inborn' advantage. Every culture can learn from the others. The distinctions into the First, Second and Third World become increasingly harmful, and it is we Europeans who must give up the arrogant role of the teacher and enter into a dialogue in which we are prepared to learn from other cultures.[18] This applies in particular to those economic and ecological conditions of scarcity to which Europe will have to adjust either voluntarily or amid catastrophes brought on by the necessity of adjusting to the rightful amount of global resources it is entitled to. The globalisation of the economy is changing the world into a system of communicating tubes. Exporting jobs and capital is inevitably connected with the importation of poverty, which will force us to alter our lifestyle drastically. In the foreseeable future, the borders between poor and rich will no longer be between north and south, but across all the cities of the world, and Europe will not be exempt.

Despite the worldwide distribution of the *Zwischenstadt,* when confronted with it as a political and design-task, all cultures find it a problem which so far appears to be insoluble and lack any strategy. There are several reasons for this:

- The *Zwischenstadt* does not have an independent identity either in the imagination of its occupants or as a subject for politics.
- The shaping of the *Zwischenstadt* can no longer be achieved by the traditional resources of town planning, urban design and architecture. New ways must be explored, which are as yet unclear.
- Last but not least, the fascination of the myth of the Old City clouds our view of the reality of the periphery.

The obstruction of our view by the myth of the Old City

In view of the massive and constantly accelerating expansion of the cities' built-up areas, there is a massive campaign in support of the traditional densely packed European city. This city, with its mixture of uses, its plot structure and its public spaces delineated by the walls of buildings is propagated as the immediately available and exclusive model for current urban design and development.[19]

---

[18]See Rainer Leppenies, 'Das Ende der Überheblichkeit' [The end of presumption] (in *Die Zeit* 48, 24 November 1995, p. 62).

[19]See, for examples, Feldkeller (1995). Hoftmann-Axthelm (1995). Fritz Neumeyer, 'Im Zauberland der Peripherie: Das Verschwinden der Stadt in der Landschaft' [In the enchanted land of the periphery: the disappearance of the city into the landscape] (in *Die verstädterte Landschaft* [The Urbanised Landscape], p. 31, edited by the Westphalia Arts Association, Münster, Aries Verlag, Munich, 1995). Günther Moeves, "Die Stadt, die Arbeit und die Entropie" [The city, work and entropy] (*Jahrbuch fürArchitektur,* pp. 29–45, Frankfurt, 1995). Henning Kahmann, 'Was geändert warden muss, damit sich wirklich etwas ändert' [What has to be changed to achieve any real change] (Techninal University Brunswick, Institute of Urban and Landscape Palnning, unpublished manuscript). Andreas Brandt and Rudolf Böttcher, *Bauten und Projekte* [Buildings and projects] (Ernst und Sohn, Berlin, 1995), especially p. 46: Projekt für eine Stadt mit 200.000 Einwohnern in Karow und Brandenburg bei Berlin [Project for a city with 200,000 inhabitants in Karow and Blankenburg near Berlin].

There are good reasons supporting this view. In energy terms, the densely packed city is very favourable because of its relatively small surface area given its large construction volume. It optimises the utilisation of the built-up area, particularly so when uses with different daylight needs and different sensitivities can be mixed vertically on a single plot of land. Density and mixture also contribute to short travel distances. This mixture plus density leads in turn to an animation of the public space and to a richness of experience, especially for children. Last but not least, this diversity leads to an unambiguous, contrast-rich delineation between city and country.

Despite all these plausible advantages, today we can only reproduce this type of city or district in exceptional circumstances, so profoundly have the social, commercial, cultural and political parameters changed.

All attempts to translate the image and structural type of the historical European city more or less directly into a general model for the future are, in my opinion, destined to fail. Indeed, I will go further still. The task is not only to acknowledge this fact but with an appropriate element of regret, to bid farewell to this well-loved image, despite its inexhaustible cultural diversity and the resurgence in city tourism. However, we should be aware of the fact that the love of the Old City is a relatively new phenomenon which dates back only one generation.

References

de Bruyn, G. (1963). *Die Diktatur der Philantropen* [The Dictatorship of Philanthropists]. Berlin, Frankfurt, Vienna: Bauwetlfundamente.

Feldkeller, A. (1995). *Die zweckentfremdete Stadt: Wider die Zerstörung des öffentlcihen Raums* [The City Alienated from its Purpose: Against the Destruction of Public Space]. Frankfurt and New York.

Fishman, R. (1987). *Bourgeois utopias: The rise and fall of suburbia*. New York: Basic Books.

Fishman, R. (1991, December). 'Die befreite Metropolis: Amerikas Neue Stadt' [The liberated metropolis: America's new city]. (*Arch + Heft 109/110*).

Gudzent, K., & Sieverts, T. (1995). *'Platz für neue Wohnungen, städtebauliche Herausforderungen bei Stadterweiterungen der Zukunft'* [Place for new homes: town planning challenges in city expansions of the future]. Düsseldorf: Information Forum for Town Planning Development Measures in North Rhine Westphalia.

Hoftmann-Axthelm, D. (1995). *Die Dritte Stadt* [The Third City]. Frankfurt: Suhrkamp.

Louw, R. (1985). *America II: The book that captures Americans in creating the future*. New York: Penguins Books.

Rowe, P. G. (1991). *Making a middle landscape*. Cambridge, MA: MIT Press.

Sieverts, T. (1997). *Zwischenstadt, zwischen Ort und Welt Raum und Zeit Stadt und Land*. Berlin: Vieweg. Source used in this publication: Sieverts, T. (2003). *Cities Without Cities, An interpretation of the Zwischenstadt*. London: Spon Press and New York: Routledge, pp. xi–xvi and 1–12.

Wärneryd, O. (1995). *Urban corridors in an urbanized Archipelago*. Lund: University of Lund.

Witold, R. (1995). *City life: Urban expectations in a new world*. New York: Scribner.

Wright, F. L. (1963). *When the democracy build*. Chicago: University of Chicago Press.

1850

1950

1995

Fig. 28.1 "Evolution of the region of Stuttgart between 1850 and 1995".
Source: compiled by Prof. Dr. Roland Wick, Stuttgart/Darmstadt and part of a joint lecture manuscript by T. Sieverts and R. Wick, published in: Sieverts, T. (2003). *Cities Without Cities, An interpretation of the Zwischenstadt*. London: Spon Press, pp. 133-134. Reprinted courtesy of T. Sieverts.
"The space of Stuttgart shows the development of a *Zwischenstadt* particularly clearly. The various settlement areas (...) have merged with the multivarious landscape into a regional city whose individual locations can be reached from each other by car or train in a maximum of half an hour." (ibid., p. 133)

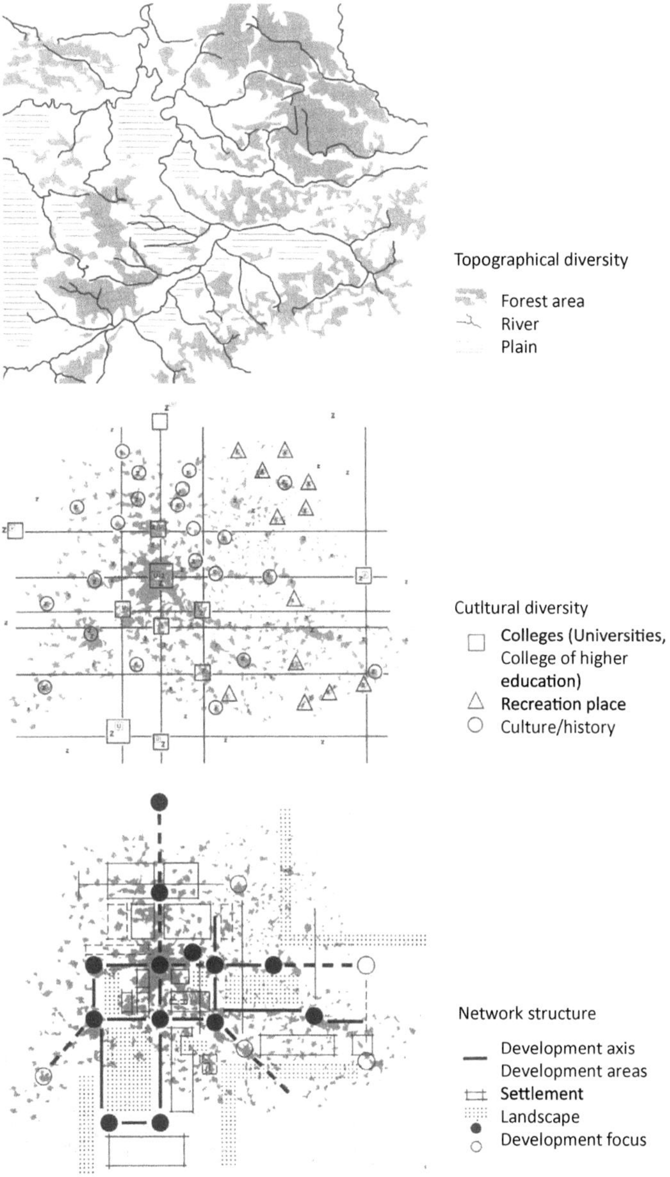

Fig. 28.2 Rationalities of the *Zwischenstadt* of the region of Stuttgart.
Source: ibid., pp. 135–137. Reprinted courtesy of T. Sieverts.
"The regional city shows a distinct network structure and a decentralised cultural diversity in association with the attractively structured landscape. This has enabled the space of Stuttgart to evolve into a multiple-centred metropolitan region, with eleven regional city components each of 150.000–200,000 inhabitants, each with their centres and with different 'features, aptitudes and qualities' which are mutually complementary." (ibid., p. 133)

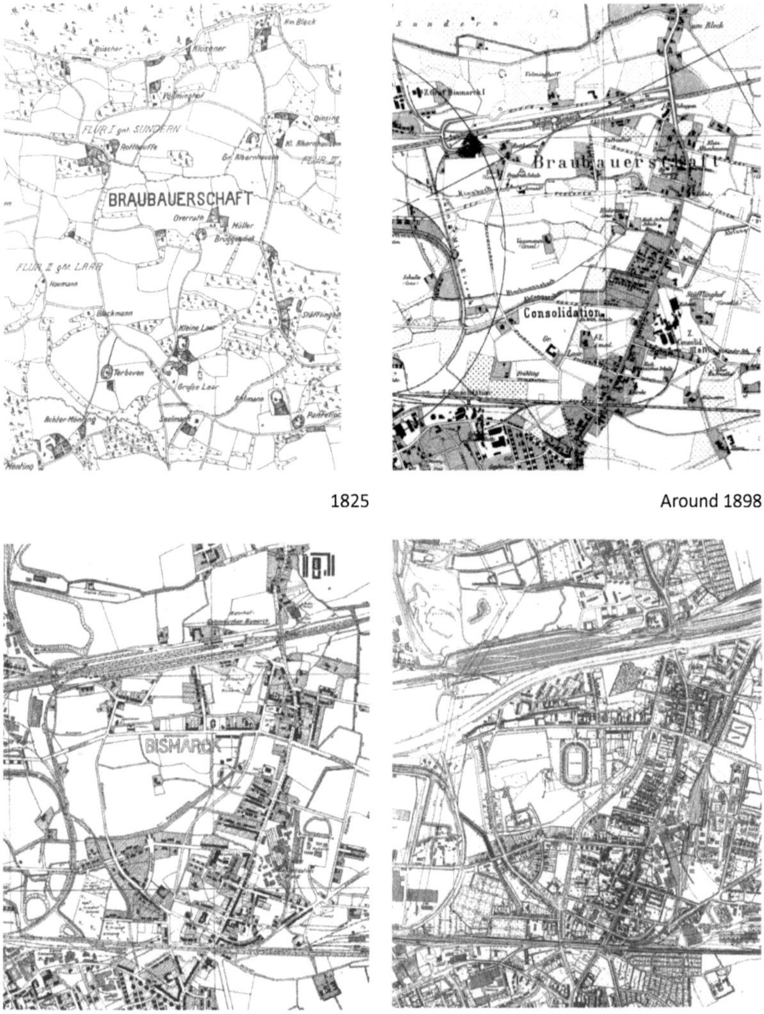

Fig. 28.3 Evolution of Gelsenkirchen-Bismarck between 1825 and 1961.
Source : 1825: Original land register of 1825 (reworked in 1925); around 1898: map of the District of Gelsenkirchen; 1961: Land Registry Office of Gelsenkirchen.
Published in: Sieverts, T. (2003). *Cities Without Cities, An interpretation of the Zwischenstadt. London*: Spon Press, pp. 63-66. Stadt Gelsenkirchen; Data licence Germany – attribution – Version 2.0; https://www.gelsenkirchen.de/de/_meta/Buergerservice/Onlinedienste/Geoportal/index.aspx. © Stadt Gelsenkirchen

Fig. 28.4 "Aerial photograph of Gelsenkirchen-Bismark and its environment, June 1996". Source: City of Gelsenkirchen, Office for Urban and Economic Development, published in: ibid., p. 68. Stadt Gelsenkirchen; Data licence Germany—attribution—Version 2.0; https://www.gelsenkirchen.de/de/_meta/Buergerservice/Onlinedienste/Geoportal/index.aspx. © Stadt Gelsenkirchen.
"The historic maps show the process of development of an industrially characterised *Zwischenstadt*. Until late into the nineteenth century, there was a poor agrarian economy with individual, evenly distributed farmhouses but without real villages. In the last third of the nineteenth century, major railway systems developed and workers' colonies arose along the old roads as a result of coal mining. Before 1990, the first schools and churches, graveyards and religious social facilities followed. (...) For several years, industrial development has been in recession, mining has ceased and the disused sites are offering new opportunities for development." (ibid., p. 62)

Part III
A Contemporary Reconsideration

Chapter 29
A Contemporary Reconsideration: Introduction

Martina Barcelloni Corte

On completion of the first two parts of the book, which proposed a double selection of authors which, during two centuries, have anticipated—through radically new gazes—the profound transformation of cities and territories around the world; the third part of the book draws on a series of six interviews conducted to some of the authors presented in the second part. The aim was here to test the Horizontal Metropolis' hypothesis through the voices of its protagonists, while setting new insights on the most recent debate on the contemporary city.

What interested us while conceiving the different interviews was, not only the scholars' individual positions and the critical distance they could assume from their own work after three to four decades, but also the possibility of building—through the layering of their different stances—an overall figure, a palimpsest able to articulate a wider evaluation of our hypothesis.

The six authors: Jean-Michel Roux, John Friedmann, Francesco Indovina, Terry McGee, Andrea Branzi and Thomas Sieverts, were duly questioned on their work, they were asked to draft a contemporary appraisal of their original hypotheses, a "personal" assessment in the light of the city's most recent transformations and in relation to a wider hypothesis: the one on the Horizontal Metropolis.

Surprisingly enough, the different positions, all related by a common framework, built a composite though concerted analysis in which the individual positions, although diverse and guided by different entry points and contexts, seemed to mutually reinforce and complete each other.

The territorial conditions in which the different authors were confronted at the beginning of their work were very different among each other but had in common the fact of dealing with unprecedented characters: they were characterized by "radically new urban qualities" (Smets), defined by "new practices and needs" (Indovina) and offering to the imagination "alternative forms of development"

M. Barcelloni Corte (✉)
ULiège, Liège, Belgium
e-mail: mbarcellonicorte@uliege.be

M. Barcelloni Corte and P. Viganò (eds.), *The Horizontal Metropolis*,
https://doi.org/10.1007/978-3-030-56398-1_29

(Friedmann) and design; they tendentially inspired projects that no longer required a "fixed structure but rather a great transformability of architectural aggregations, all provisional" (Branzi). The scholars' common—yet unusual—awareness concerning territories geographically and culturally diverse, to some extent reflected the different work's publication dates, which, despite the fact that most of the authors had only come into contact with each other very recently, in many cases were surprisingly close (Indovina, McGee, Branzi and Sieverts, for example, all published their work between 1990 and 1997).

Although in different ways and in different moments in time, all authors expressed a strong critique to consolidated urban models, often "unable of interpreting and coming to terms with the changes underway" (Indovina). The applicability of the traditional dichotomy between rural and urban and the resulting policies designed to keep them separate were among them commonly seen as a manifestation of a "culturally artificial division" (McGee) which had to necessarily be overridden and which, already at the time, had no place in reality. The overcoming of the rural–urban opposition matched the overcoming of obsolete building and territorial typologies and the introduction of more flexible types (Branzi), able to suggest original and innovative modes of territorial management (Indovina) or the possibility of new uses and practices (McGee).

Albeit with different degrees of intensity, all authors interestingly report a reaction of strong rejection to their texts from the general public. While John Friedman's *Agropolitan Development* model was named "Mr. Friedman's development nightmare" by recognized scholars, Thomas Sieverts was "publicly ridiculed" and called the "traitor of the original European city"; he was "heavily attacked and accused of being an advocate of urban sprawl and of the destruction of the landscape". Bauer and Roux were also strongly criticized for their study *La Rurbanisation* that, according to the critics, was related to an "excessively empirical" and "immoral subject", a form of "insubordination to national policies". The emphasis with which the different authors recount the misadventures of the past, clearly echoes their late success, related to a recent wave of deep reinterpretation and renewed interest in their work.

The positive approach to a phenomenon commonly interpreted as being negative pools all the interviewed authors which, in contrast with the trends of the time, understood the phenomenon of dispersed urbanization more as an opportunity to be grasped than a problem to be solved: a landscape with evident negative externalities but characterized by "economic vitality and flexibility" (Sieverts), by "relevant environmental qualities" (Indovina) and able to "reduce collective costs (Friedmann) and "increase self-determination" (Smets).

The call for a new era of "better planning" (Indovina), capable of deploying "new knowledge and interdisciplinary teams" (McGee), and to counteract the individual and partial responses of self-organized processes, is a general one. In most cases, even if through different terms and different ways of problematizing the matter, it becomes clear how the possibility of conceiving a project for these territories is directly related their capacity of being "imagined", in their ability of producing a sense of belonging in their inhabitants (Sieverts).

The maps and images supporting each interview, are conceived with the intent of immerging the reader in the territories originally analysed by each author; the effort is made to portray and convey—albeit very broadly—the different territorial materials, their structuring and prevalent elements together with their evolution in time. The space is pictured from above—through a set of drone sequences—and from below—through a collection of eye level images—portraying the landscape in its material terms, through the interweave of its vegetal and mineral elements, its mix of functions and their re-organization in time. The hybrid composition of detached houses, religious buildings and spaces, "big boxes", factories and industrial platforms, small and medium-sized centres, strongly modified and adapted to accommodate the needs of rapidly changing economies and lifestyles are visible through both sets. In these images, as in the maps it becomes visible how agricultural, industrial and residential space, along with spaces for work, trade, services and leisure time, reveal continuous variations, gradually transformed according to repeatedly different territorial speeds and logics.

While certain authors commented on the difficulty of actualizing their model, to some extent excessively rigid and normative (Friedmann), and others considered their intuitions current to the point of feeling the necessity of updating and reinforcing them with a set of new elements (Mc Gee, Branzi, Indovina), the hypothesis according to which the described modelizations—despite notable differences—may represent different facets of a common dynamic, and are therefore not isolated episodes but comparable attempts to recognize a new urban principle within the profound mutation of the contemporary city, seem to find in these interviews fertile ground.[1]

[1]Part of this text has been published also in: Barcelloni Corte, M. (2019). In the course of recognition: Notes on a new urban imaginary. In C. Cavalieri & P. Vigano (Eds.), *The horizontal matropolis. A radical project*. Zurich: Park Books.

Chapter 30
J. M. Roux on *La Rurbanisation*

Jean Michel Roux

J. M. Roux (✉)
Grenoble, France
e-mail: roux.transversal@wanadoo.fr

M. Barcelloni Corte and P. Viganò (eds.), *The Horizontal Metropolis*,
https://doi.org/10.1007/978-3-030-56398-1_30

30.1 *La Rurbanisation*: an appraisal

Today I still believe that, globally, the "dispersed", "scattered", or "diffused" city represents the normal form of expansion and transformation of contemporary urban centers. Whatever its denomination: suburbanization, peri-urbanization, rurbanization... Every point in a territory is linked to others and physically and virtually accessible via the means of transportation and telecommunication that we have at our disposal, across the historical urban extensions.

Against this background historical poles still exist. But most of them are fragile (even in some metropolitan aeras), because they are transforming under the pressure of factors that are neither synchronous nor mutually coherent; they are always on the verge of crisis, although at times they have possibilities for recovery and [positive] transformation. I believe that today's urban policymakers should address these existing hubs, at least for reasons of history and heritage. But those policies cannot forget to plan and organize their suburban territory.

Given these generalities, there are still national and regional particularities that we must take into consideration. I will therefore talk about the French case, in very genera terms, since this has been the original subject of my work on suburbanization, in the 1970s, later completed by investigations abroad.

After 1945, France had to rebuilt destroyed towns, like others western European countries, but with important specificities. For more than one century, it had experienced little population growth and moderate industrialization. So much of its land remained rural, with strong remembrance of the historical peasant society: in 1946 (census), one third of the national labour force was still employed in agriculture. The ensuing period witnessed the disappearance of this peasant society as much as the rapid increase of urbanization.

So, beyond the Reconstruction, there was a push towards modernization and industrialization, coupled with an exodus from the farmlands, and an unexpected demographic growth. France found itself facing a very severe housing crisis, with a lack of infrastructures.

Rapid solutions had to be found. In fact, there were decided belatedly, at the end of the 1950s. Powerful policymakers of the National State opted for highly regimented planning schemes: "Grands Ensembles", "Cités", mainly for public housing. They were developed to a large extent, in keeping with the principles of "Modern" urbanism (building blocks and towers, large neighborhoods without plots), breaking traditional urban network.

This pattern continued into the mid-1970s, then they were stopped by decision of the national government itself. Not far from 5 million dwelling units had been built, which are today places of crisis, in spite of planned demolitions. These forms of urban developments, indeed, would soon be challenged by a number of factors. First of all, many inhabitants didn't appreciate them, thus a trend towards abandonment occurred. Secondly, mortgages were introduced in those years to facilitate home ownership, allowing many people to buy their homes, and especially single-family houses (mainly detached, usually called "pavillon" with some condescension). Finally, the growing use of the car made peri-urban developments a concrete possibility.

We must keep these dynamics in mind if we are to understand that suburban, peri-urban, or rurban phenomena in France constituted a still unforgiven rebellion against national urban policies (and may be still a rebellion, regarding the recent movement of the "Gilet Jaunes"/ *Yellow jackets, born in the suburbia). It is thus subject, more than anywhere else, to deprecation and unrelenting criticism by administrators, planners, architects, academics... So, the suburban phenomenon remains an unplanned fabric of fragmented and decentralized small-scale initiatives, by small builders, in small municipalities, lacking any kind of organization. Unplanned but not unruly: the zoning ordinances (Plan Local d'Urbanisme) allow space for dwelling units.*

Regarding this background, I began working with Gérard Bauer on what we would later call Rurbanisation, *a book we published in 1976. It explained the phenomenon in a very empirical way, based on our surveys, conducted in some regions where urban scattering was taking hold. It was a rather basic book where we described how this particular space was produced, why people chose to live there, and how they lived.*

We essentially affirmed that the phenomenon was going to last and that it had to be organized. We drew measures to improve it, particularly by producing projects and planning the mobility. Note that few people in France were informed about European or American suburban developments.

The book received mixed reviews. While it had an unexpected impact, it was also strongly criticized. Because we were unfairly talking about a subject on which we should have kept silent. Also because we had chosen an empirical method, with figures and surveys, at a time when academic urban research assessed the city and its territory as a product of society, mostly concluding that the family house was a myth and worse: a way to enslave the lower middle class.

After this first publication, we continued our fourhanded work, until the end of the 1980s (Gérard Bauer died in 2013). Regarding the previous remarks, we have never been in a position to publish additional or updated editions. However, we followed two principal lines of inquiry.

One consisted of illustrating, proposing, promoting, and defending ambitious—in terms of both content and scope—urban development models inspired by the garden-city concept, of which many examples and debates exist in Western Europe and the United States. We have published books[1] *and studies on the subject and assembled a photo library. But we faced mainly (1) a lack of developers for large-scale suburban projects, which required long-term returns and sustainable*

[1]Gérard Bauer: Un urbanisme pour les maisons 10/18, Paris, 1979.

Bauer et al. *Banlieues de charme, ou l'art des quartiers jardins* (Pandora éd.). Aix-en-Provence. 1980 *investment capacity (2) political and financial weakness of small peri-urban municipalities.*

The other line of action after the publication of La Rurbanisation *involved critical analysis about centralized regional planning,*[2] *and work on decentralized local development, on the creation of financial tools for small businesses, and also on real estate and land market issues which is still my main field of study.*

What can I say today about the phenomenon and the issues associated with it? Of course, there are a lot of different subjects to address. I will provide a few general ideas, still about the French case.

First of all, we need to consider a new form of urban development, while growing in a rather non-organized manner. Successively, we have the historical center of the city, then the first ring of dense suburbs, representing growth in the years up to the 1960s, and then a very large and recent rurbanized *outer ring, where approximately ¼ of national population live: about 100/150,000 new houses/year, half century, plus previously existing housing. There should be residential mobility among the different rings, but there is little, fiscal and market issues being significant impediments.*

These peripheral zones have been developed at low cost, with poor planning, architecture and landscape (but rather comfortable and inexpensive homes). They suffer a lack of facilities, regarding especially mass transportation and urban amenities. For example, to put it simply: in almost all large French conurbations, families with children are suburban in an increasing number, while the major educational and cultural institutions are still, and may be more and more, located in the center of historical cities.

So living arrangements change significantly as one moves from the historic centers to the outskirts. Nevertheless, recent surveys show that the outskirts do adapt, with regards to transportation, energy use, leasure, commercial outlets, or social relationships. Their flexibility leaves them sensitive to oil prices for instance (see political news) but makes them probably sustainable.

Land consumption is the object of heated conflict in France (and anywhere else in the world). "Consumption", in the sense that a garden of a garden is artificialized, and a field treated with glyphosate is natural. The most often quoted statistics (Teruti Lucas) is also the most alarming (with unexplained differences from one year to the next), and frequently used to conclude that detached housing is the principal culprit.

I would like to point out that we have other data that support different conclusions. We cannot say that one position is more reliable than the other but simply that they do not share the same bases. This is not a simple issue and we must define exactly what we mean by land artificialization (building, impermeabilized ground, wasteland, gardens, etc.) with a scale of observation. Cutting to the point, we may

---

[2]Roux, J.-M. (1980). *Territoire sans lieux, la banalisation planifiée des régions*. Paris: Dunod. 1980) and later Roux, J.-M. (2006). *Des villes sans politique* (Gulf Stream ed.), Nantes.

therefore safely state that we cannot attribute the lion's share of blame to detached housing and there is no need for debate on this, because we have the data.

Since the turn of the century, we are witnessing a two-faced crisis in housing costs. On one side, we note sharp increasing prices in many metropolitan centers, mainly in Paris.

As far I can see, these increases (over mediatized) do not derive from a scarcity of lodgings but instead are the result of speculative processes. It began in the upscale neighborhoods, and then slowly spread across the suburban peripheries. We may say that housing is not expensive because it is scarce, but that it is scarce because it is expensive. From this point of view, the slogans about 500,000 new homes per year in France, or 70,000 new homes in Paris Region (Ile-de-France) are quite arbitrary, at least in quantitative terms, but most importantly, they have no impact on the real problem.

*On the other side, most small and middle range town centers (even some large cities) suffer a decline of prices, which discourages investment. Population decreases, with high percentage of vacant housing and commercial space. Therefore, French government launched an ambitious plan of renovation (*Action Coeur de Ville*) in 2017. We will see…*

In these two process, the suburban prices (land and buildings) grew (or declined) more slowly than in center, or eventually be stable. This buffer effect may be cause and effect of the sustainability of suburban rings.

The last point is perhaps the most important: we must be more respectful of the suburban people. There has been a debate in France, about hypothetical relationships between suburbia and the Front National *(now* Rassemblement National*). A mass of data—in my opinion often distorted but meriting discussion—sought to prove that owning a home in a peri-urban area, made one more likely to vote far right. I still believe that voting choices are the outcome of social and economic conditions, more than distance from the city center.*

But I would like to point out that the Gilets Jaunes *movement, emerged in 2018 against an increase of gas tax, was broadly a suburban phenomenon, with demonstrations on roundabouts and riots in downtowns. Political parties and unions are still not be able to frame it. That indicates a serious problem of local democracy, with a steady decline in voter turnout in local elections. Our territorial institutions have changed little since the French Revolution, even with the so-called* "Décentralization". *There need, in my opinion, an extremely radical reform. But this renewal meets the political class, comfortably ensconced in the existing perimeters and legal competences.*

The most striking case is the Paris area, where a multitude of institutions is no longer compatible with the metropolitan development. This produces glaring inequalities between the Ville de Paris *(2 million inhabitants, 110 km^2), and the rest of the metropolitan area (at least* Ile-de-France, *12 million inhabitants, 12,000 km^2), and a special disregard for the suburban areas.*

However, more than ten years of negotiations lead to visions that will never be carried out (International workshop of Greater Paris), and a rump Métropole du Grand Paris, *with poor competence, without direct elections, excluding the suburban areas. It remains a huge investment (still left largely unknow) in rail system (*Grand Paris Express*), forgetting most of suburbia), lead by the National State. We may add a "Playmobil politics", including for example the 2024 Olympic Games, or debatable projects for some well-known waste or unbuilt lands (*Triangle de Gonesse, *for example).*

Our first studies were funded by the Ministry of Agriculture. Villages and farmland were just beginning to be urbanized. During the first step of this "rurbanisation", *some large "horizontaly" planned neighborhoods were achieved, some of them consisting of several thousand family houses. They have mostly aged well. Some became chic in the western suburbs of Paris (Maule or St Nom la Bretèche) or on the Riviera. But in a general sense, these ambitious projects have now disappeared. There are even few significant allotments.*

In the same way, the architecture of these houses has simplified over time. This has both economic and regulatory reasons. Particularly, the planners who issued building permits in the 1970s strove to eliminate everything that might be considered a pastiche, and ultimately, everything beyond the basic cube. This resulted in a production of inexpensive cookie-cutter houses, with repetitive designs, however relatively well built. The new houses today are less expensive per living m^2 *than collective housing in similar perimeters, even though they face a higher cost/ living* m^2 *for the land.*

Today, those builders are still mainly small companies: someone buys a parcel of land and choose a builder. But this light production method exploits all opportunities offered by land regulations, with the agreement of the landowners. Let us looking at those regulations, in general: large buildings areas, strict zoning, lack of master plans (see German Bebauungsplan*). We remain in situation of large urban scattering, with poor landscape and public facilities. So, we need to advocate for more planification, with more investments for developments, and more brainpower. We must also maintain and improve this huge housing stock.*

We often hear that the peri-urbanization is a system of relegation: people leave the dense urban areas because they do not have enough money to stay there. This is not true, with notable exceptions, mainly Paris, and exceptions within exceptions. In general, people know quite well what they are doing. It is often reported that this suburban system will decline because people will not be able to pay off their mortgages, nothing like this has happened. There was no subprime mortgage crisis or anything like that. Banks were quite cautious and people had to tighten their belts to be able to afford a house.

This system does not age badly, but that does not make it a "place to live". It's still pretty basic, there is no organization, the public outside space is neglected. There is considerable criticism as to the outcome, but it should be directed towards the lack of a planning framework, the lack of care or attention given to these urban forms.

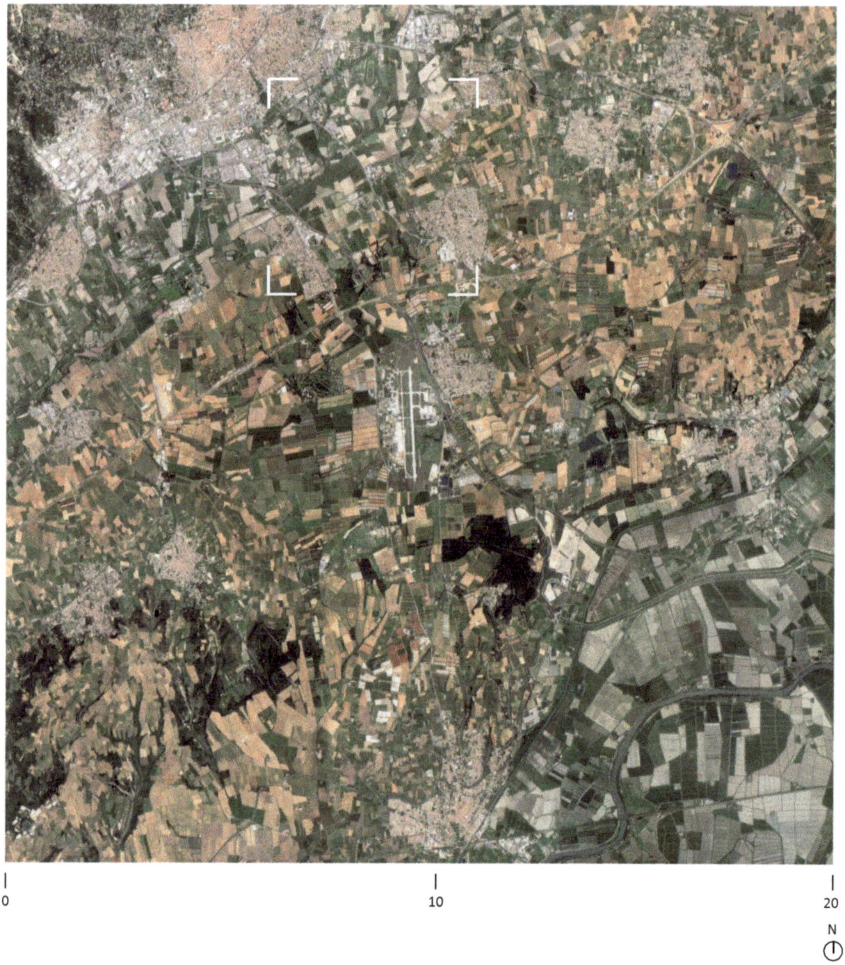

Fig. 30.1 Satellite image (20 × 20 km) of the Nîmes Metropolitan Area (south), including Saint-Gilles, Bellegarde and Nîmes on the top left.
Together with the areas around Rouen, Orléans, Paris and Lyon, the Nîmes Metropolitan Area was one of the territories selected by G. Bauer and J.-M. Roux during their surveys to describe the phenomenon of *Rurbanisation* in France.
Source: Google Earth, 2018

Fig. 30.2 Built fabric of the southern Nîmes Metropolitan Area (20 × 20 km), including Saint-Gilles, Bellegarde and Nîmes on the top left, 2018.
Credits: IGN France, map elaborated by C. Küpfer

Fig. 30.3 Satellite image (5 × 5 km) of the Metropolitan Area of Nîmes (Southeast), including Bouillargues, Caissargues and Rodilhan, 2018.
Source: Google Earth

Fig. 30.4 Evolution of the built fabric, Southeast Metropolitan Area of Nîmes (5 x 5 km), including Bouillargues, Caissargues and Rodilhan, 2018.
In black the village's ancient structure, in grey the transformation observed by the authors and described in the book La *Rurbanisation*, in red the transformation occurred after the publication of the book.
Credits: IGN France, map eaborated by C. Küpfer.
<1962: black.
1963–1988: grey.
1989–2018: red

Fig. 30.5 Aerial sequence (drone) of Boulillargues.
Credits: E. Barcelloni Corte, 2020

Fig. 30.6 Mix-use area between Manduel and Redessan, 4 km from Bouillargues. Credits: E. Barcelloni Corte, 2020

Chapter 31
J. Friedmann on Agropolitan Development

John Friedmann

M. Barcelloni Corte and P. Viganò (eds.), *The Horizontal Metropolis*,
https://doi.org/10.1007/978-3-030-56398-1_31

31.1 Agropolitan Development: an appraisal

Agropolitan Development *is a neologism, a new word, a new construct, composed of two elements: the agrarian element and the "polis" element where "polis" stands for a democratic and self-governing city.*

The "Agropolis", an "agricultural city" or "city in the fields", is how I originally introduced my concept in the 1970s, notion that I have subsequently developed in a series of articles with particular reference to rural areas in Asia.

It struck me, at that time, that the rural densities in the coastal areas of China, in many parts of Indonesia (particularly in Java) and in other parts of Asia, including India, had such high densities that exceeded those of a suburban area in the United States.

This idea gradually led me to question myself about the necessity of bringing the city to the countryside instead of waiting for people to move in the big cities and live in self-constructed housing projects, the slums.

This is how this notion of "Agropolitan Development" began to emerge, which some people called "in situ (in place) urbanization", proposing it as a solution to a problem.

The problem -at the time- was that there was a surplus population in the rural areas for which there was no space in the country's "economic life". On one side - in fact- the rural areas, on a per-capita *basis, had, in many cases, shrunk to less than half an acre per household (a multi-person household) and on the other, in the urban economy the only employment that people could find, for the most part, was in the so-called informal sector, informal activities much of which was not producing a lot, so that the productivity of labor was a declining one.*

At that time, the surplus population estimates were very imprecise but still one could argue that 40% or more of the population was surplus to production in the rural areas and not greatly more productive in the urban areas.

In the 1970s I started to think about these issues because of my personal interests. I completed my PhD in 1955 with a dissertation called "The spatial structure of economic development in the Tennessee Valley" which was an attempt to study the economic development of the large-scale River Basin development of the Tennessee Valley. This is where I began to formulate two main ideas: the first was related to the fact that economic development is actually urban based and the second that it is concentrated in certain key locations which later-on came to be called "growth poles". This was a notion that was introduced by François Perroux and some of his colleagues in France who formulated this idea of an economic development model which is based on concentrated development in key urban areas. This was the beginning of my thinking about urban-based development and, in fact, until the mid-seventies I hadn't given a great deal of thought to the question of rural development. Nonetheless, there were already several comments in the literature, especially coming out of Sussex University, that were arguing the case against what they called "urban bias" in development studies. That made me -at least conceptually- aware that this "urban-centered development" which I had been pursuing in my own writings and in my applied work, was due for a revision.

Subsequently, in a series of articles, I started to question myself about the possibility of making the rural population more productive in such a way that they could experience a sense of empowerment in the course of producing their own lives. Some of the essays that dealt with the "Agropolitan Development" concept and related issues were subsequently brought together in a book called "Life space and economic space: essays in third-world planning".

The notion of "life space and economic space" as a dualistic but also dialectical model became very useful and I kept using these terms throughout, together with another set which was argued in a book called "territory and function: the evolution of Regional Planning" which I wrote together with one of my students, Clyde Weaver in the 1980s. I didn't meant this dualism to be understood as an opposition but rather as a dialectical relationship between territory and function and between life space and economic space, as in the yin-yang philosophy of ancient China. My own preference for dialectics really stems from that time and part of the world rather than from the Hegelian dialectics introduced by Karl Marx, of a progressive upward movement of dialectical opposites.

Anyway, what I was arguing for at that time was essentially a form of bottom-up development, an endogenous or self-reliant form of development.

The context for this reasoning was the widespread felt need, particularly on the part of people working for the United Nations and for the International Labor Organizations, of an alternative form of development.

Of course, there was no agreement on what a viable alternative form of development might be in third world countries, in fact while I was attempting to formulate the "agropolitan alternative", the mainstream was represented by the more powerful advocates of an urban-based development with global reach and related to the reduction of the role of the national state in development processes.

I have to say that people got violently emotional about my model. From one end of the world to the other they strongly rejected the idea of shifting the balance from the city to the rural areas of Asia and develop here an urban based form of industrialization that would allow some accumulation of capital to occur.

Some people, particularly Indian economists, thought that I was trying to bring back the Gandhian model which -of course- was uniformly rejected by Nehru's Indian government. Then there where the famous Professors of economics from Yale and Harvard that ganged up on me and decided to "have a go" at what they called "Mr. Friedman's development nightmare". My development nightmare was - in reality- the large-scale unemployment or very low productive employment of these regions.

Many of the discussions on urbanization, at that time, where underlined by the Rostowian's model ideological tract. W.W. Rostow was arguing for a very rapid transition—over two or three decades from a state of under-development to a state of mass consumption that would "magically" lift the masses of the population out of poverty and turn them into urban consumers. That, to me, was not only unreal but something that I couldn't even accept on a philosophical level, since it aimed to turn people—who were productive in the creation of their own livelihood-suddenly into a mass of passive consumers whose only understanding of happiness was to go to the supermarket. It concerned a model of development where the people would become happy and look for recreation, for spending money in the context of the artificial environment of the "supermarket for little happiness".

Elements of a successful "Agropolitan" model can be found in China today and in earlier times as well. The idea of a more balanced rural-urban relationship was originally announced in one of Mao Zedong's essays on "the ten relationships between economics and society". In typical Chinese fashion having both elements and a dialectical relationship between them, he argued that the time has come to shift towards and emphasize the rural, while continuing to work in the cities. Of course the long-term outcome of such a model might be a kind of urban-based society but in this case, the Chinese model was assisted by what are widely considered to be successful albeit unusual policies: one was the one child per family policy which definitely lowered the birthrate and the second was to control people's movements through a system of registration cards called "Hukou". This prevented for a long time the mass exodus of the population into Chinese cities and was definitely helpful in supporting a rapid urbanization process which very quickly spilled out over the borders of large cities into the rural areas.

We cannot say that they were producing an "Agropolitan Development" but after the reform era that was introduced by Deng Xiaoping, during the 1980s and 1990s, they did succeed in stimulating industrial production in the rural areas, many of which were located close to large cities along the coastal regions of China. By the late 1990s about a third of the industrial production of the country had its origin in the rural areas. The rural industrialization was taking place together with the introduction of other elements of the urban: electric power, paved roads etc.

In China there was virtually no self-government so, to that extent, it wasn't an Agropolitan development model, the "polis" part is not so prominent in the Chinese case but certainly the "in situ urbanization" could be talked about and people started to connect up with what used to be isolated rural areas, aprocess of gradual urbanization of the countryside and successful industrialization took place.

In relation to the contemporary relevance of the Agropolitan Development model, I would say that I just learned that Mr. Modi has become a great admirer of Mahatma Gandhi and he's digging into the past history of India to help consolidate his own power. Nonetheless I have to say that it doesn't seem to be all that much emphasis on that concept today, the government rather continues to pursue major policies of accelerated urbanization throughout the country, along specific corridors that they have identified.

The point for me is that there are already people living in these corridors or future urban corridors, there are hundreds of millions of people living in these areas, they may be called rural but they are not all engaged in agriculture. They are engaged in non-farming activities, in small cities and towns. Here the country people which have been having a bottom-up industrialization just like in China in the 1990s, are now in India in the twenty-first century and will not let the government run over them and turn their modest little towns and villages into some magnificent contemporary era, conceptualizations of what the urban should look like, meaning high towers and all the rest that goes along with that.

In this frame, I think that "Agropolitan Development" as a normative concept must still be useful as a way of capturing what has been actually happening in part of Asia. But again it is a question of who is the steering wheel and I think, in spite of Mr. Gandhi, that the Prime Minister of India today is making political use of Gandhi more as a kind of a way of strengthening his ideological base than as a way of actually embracing the Gandian ideal. The spinning wheel is still out and it is not really about spinning wheels anymore but about other forms of urbanization from below.

Chapter 32
M. Smets on the Radiant Periphery

Marcel Smets

M. Smets (✉)
KULeuven, Leuven, Belgium
e-mail: marcel.smets@kuleuven.be

M. Barcelloni Corte and P. Viganò (eds.), *The Horizontal Metropolis*,
https://doi.org/10.1007/978-3-030-56398-1_32

32.1 The Radiant Periphery: an appraisal

As an urbanist, but I would even say as an inhabitant of an area, you are really intrigued by what is happening around you, you try to understand it. When I started my career, I noticed that the environment in which I lived and which was built around me, Belgium and -more specifically- Flanders, Belgium's most densely built part, was very much in contradiction with what was supposed to be the city and the good living at that moment in urban theory, when the ideal of the urbanists was the "densely inhabited city". The architects looked at the pavillionnaire, *at the* sub-urbia, *as something despicable.*

When, subsequently, I worked on my doctorate on the "garden suburbs", within Unwin's theory, I realized that there had already been the idea of building upon a qualitative suburban development. I started to look with new eyes to my own environment, I started to think that if all people appreciated that condition, they couldn't be all wrong, there was something behind it. This brought me to work for a long time of my life on what I later called "the nebulous town of Flanders".

I started to ask to myself why Belgium was so different from other countries as France or Spain -for example- where there is a much more clear division between country and city. In Belgium -on the contrary-the territory was a patchwork of many different things, very particular and difficult to analyse.

The beginning of it goes back to the Middle Ages when the Delta of the Scheldt, the Meuse and the Rhine Rivers, an extremely rich and fertile territory, very well suited for agriculture, was one of the densest inhabited areas in Europe. Already in the fifteenth century there was a development of many cities, which were all- more or less-located from 20 to 25 kms one from another. This was the distance that people could cover in one day if they lived in the countryside and needed to go shopping and return before night.

This gave an automatic distribution of cities relatively close to one another at critical distances, in a very densely inhabited area. Furthermore, in between these cities, there were small villages, hamlets, and a particular organization of agri-culture, which was based on the fact that people had to cover the distance from the farm to their land and back, and had to be secured from the Vikings or other dangers. The settlements were thus, already at an early stage, very close to one another in a kind of "patchwork" of city and villages.

In the second stage, in the early nineteenth century, Belgium was one of the first countries in Europe -after Britain- to industrialize. This industrialization was -again- very different from the one occurred in France or in Germany, where the countryside was completely empty and people lived together in big cities that gradually become very difficult to run. In Belgium, on the contrary, as we had so many cities and we had already certain types of manufactures, realized in the seventeenth and eighteenth centuries, the mechanization was more a question of how to transform manufactures into industries. From this moment -in fact- also the industrialization followed a very dispersed pattern.

There were major cities, like Gent and Liege, that grew from 1 to 3 in population in less than 60 or 70 years and medium cities, like Malines or Aalst, that grew from 25 to 50 or 60, but they were all dispersed and with the same kind of industrialization.

This was made possible because the Belgian state, which became independent in 1830, decided to build a very dense railway network (Belgium has the densest railways network of Europe) which brought the coal and the iron through the centres of labour to the cities. This meant that the industrialization didn't take place where the ores were, since the cores were transported to the places where people inhabited, where they could work in factories, a dispersed pattern of factories.

These dynamics produced an increase of population that led to land speculation. Belgium, again, didn't have large development companies and banks like in Germany where big new urban developments were built to be rented or sold, Belgium had a particular system of small allotments. When needed, a piece of agricultural land was gradually transformed into allotment by individual, small scale, middle-class investors who erected one to four houses on these plots and rented or sold them accordingly. This lead to a gradual extension of the cities which also allowed for social distinction within the cities: by changing the building typologies, the richer people had larger plots and larger houses, the middle classes had terraced housing on six-meter plots and the poorer workmen, very small lots and four-meter houses. But the system remained the same, creating unlike Paris or Berlin -for example- a much lower density.

This lower density was made possible because the public authority was creating the infrastructure necessary to make these plots accessible.

On a socio-economic level, the Belgian state was willing to do these investments because they suited everybody, there was a secured return with no risk. People were putting their savings into the construction of a house which provided, through the rent, a small interest on the savings; there weren't great difficulties and everybody did it in the same way.

On a political level, it suited both, the catholic, right-wing and conservative people and the left-wing, more progressive ones. For the catholic this kind of development represented more control and order, a community based and self-disciplined society; for the progressive it represented a more "hygenic" condition with a better price-quality situation than in German big cities, for example.

To obtain this result, the Belgian State took a sequence of important acts which redefined the country's development.

In the mid nineteenth century a law was created to avoid migrations; a person moving from one city to another couldn't benefit from any allowance of charity, before the sixth year of residence.

Subsequently, in 1898, the first law on affordable housing was introduced, which extended to the working class the conditions for loans and mortgages guaranteed - until that moment- to the middle-class. Thanks to this law, people from the working class could obtain a loan and become owners of their house if they could prove their working status; in this way, many of them become part of the middle class.

The third important act avoided migrations through the guarantee of low fares for daily train tickets for workers. Workmen had special fares which allowed them to commute easily from the countryside to the cities daily. Before the First World War, this right was even extended to the "vicino railways", *steam tramways organized to move people from the most profound places of the country to the railway stations from which they could then reach their work.*

Belgium become a very particular territorial condition where the whole countryside was made accessible to the cities and vice versa.

Through these measures the belgian landscape began to change, people who gained money working in the city returned to the countryside and build city homes while people that left the city to live in the countryside, began to build country homes; all of them, together, composed a mix of city houses, country homes, new developments, etc. It corresponded to a very individualistically oriented society, closely knit together in a local society of *mutual comprehension. Belgium -as a whole- became a particular kind of Garden City.*

This is what -in my article- I have called La Banlieue Radieuse, *"the shining periphery" with a clear allusion to the idea of Le Corbusier who spoke about* La Cité Radieuse. *The fundamental difference between the two is that in* La Cité Radieuse, *the architect determined how people should be happy and how they should live, whereas in* La Banlieue Radieuse, *people chose for themselves how to be happy.*

After the Second World War, with the massive introduction of the car in the 1950s, the dwelling ideal continued to inspire while being enlarged in scale and dimension since, at that time, all places became very easily accessible. This dynamic produced -to a certain extent- a loss of the local "togetherness", a territorial condition having more to do with what Melvin Webber called the non place *urban realm, which guaranteed complete accessibility of people and activities through a perpetual connection to the network.*

Again, the Belgian state did a great effort: the motorway construction led to the densest motorway network in Europe and unlike what happened in France, Italy or Spain, where motorway exits are placed thirty kilometres apart, in Belgium we had exits every two or three kilometers. Every small city was connected to the motorways and people owning a car could -more or less- live and work anywhere.

Whereas originally, during the nineteenth century, the centralities were represented by the cities, bigger and smaller, in this phase places as hospitals, schools, universities or industrial estates just at the exit of the motorway or close to it, become places of attraction, centralities. The territory gradually became like a mosaic in which all places overlap and become fragments in themselves, where you could find a university and 500 m further a park within an estate and a piece of agriculture. All of that is contributing, in my opinion, to something that is basically qualitative; we could imagine that the agriculture becomes a park where people could go for walks and recreation.

The fact that you can ultimately go everywhere is quite intriguing. Yet, of course, there is a limit to this kind of carpet metropolis *or mosaic and today we are reaching this limit. It has to do with the dependence of the car and with the fact that you cannot keep building motorways which get quickly saturated. They get so saturated that the distance you have to cover in order to go from one place to another is no longer foreseeable. You get to a situation in which you are insecure about planning your own life and this kind of mosaic city doesn't allow you anymore to proceed as you would want to. This is exactly what we witness today.*

We are -in fact- experiencing re-concentration movements and the emergence of new dynamics as -for example- the densification around areas that are more accessible by other means, as public transport. In small and medium sized cities, new developments are built and occupied by people who used to live in villas which -in turn- get to be subdivided producing denser populated areas. Because of this increase of density, it becomes possible to have more amenities, even restaurants, originally reachable by car, now follow their customers in these denser area where you can walk and find schools restaurants shops etc.

During the last 20 years, cities have also been densifying areas around railway stations with housing, offices and services.

To conclude, and here I refer to the latest research that I have done with my doctoral students, we should imagine a system of tramways connecting the country's denser points, as the junction that is linking the pearls on a necklace.

In this way, we could keep the idea of shining periphery and make it work again. Perhaps this is what we should be thinking about and instead of dreaming of living in a dense city as Paris (don't forget that Parisians, if they can afford it, have a second house in the countryside) create an environment that suits all purposes at the same time.

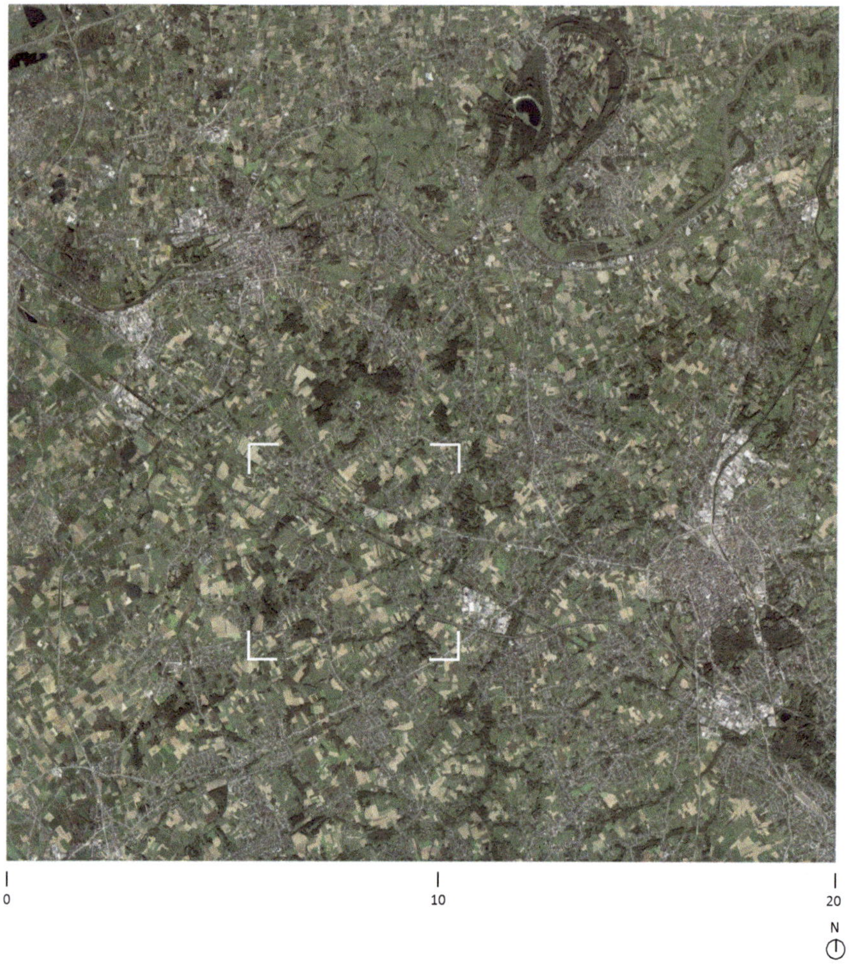

Fig. 32.1 Satellite image (20 × 20 km) of the area between Bruxelles and Ghent including Alost, Wetteren and Hautem-Saint-Liévin in the Flanders Region.
Source: Google Earth, 2018

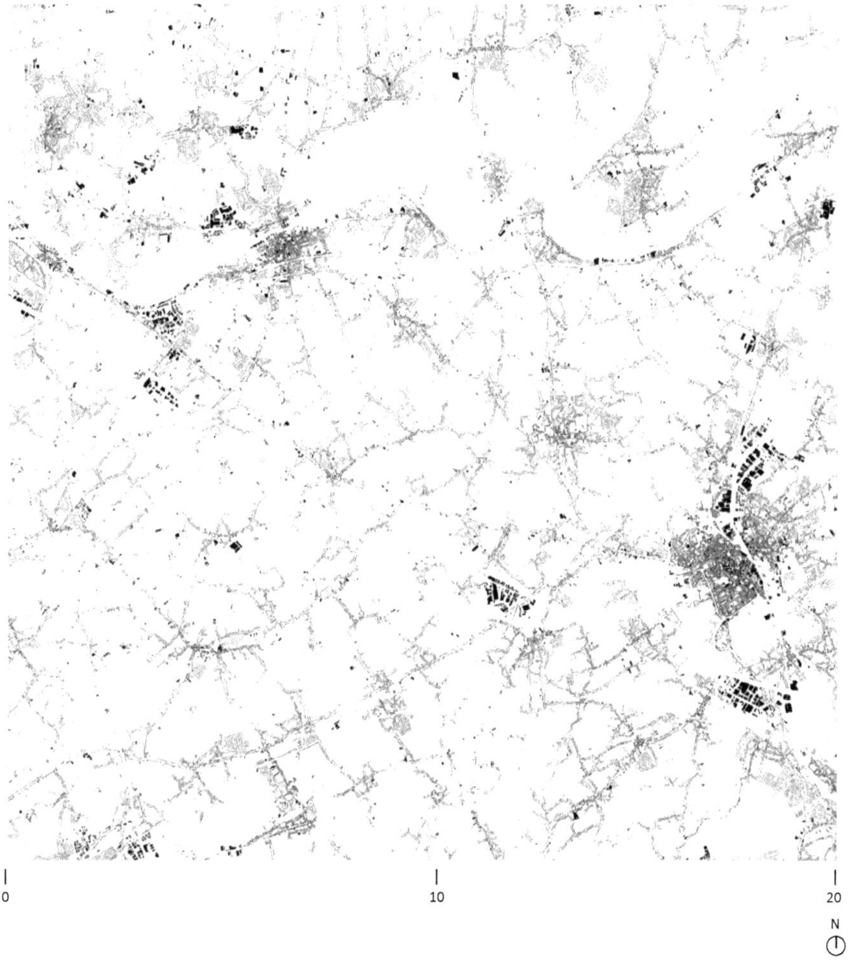

Fig. 32.2 Built fabric (20 x 20 km) of the area between Bruxelles and Ghent in the Flanders Region, 2016.
Credits: SPF/AGDP, map elaborated by G. Vanneste.
This map shows the built fabric only. Located in the Flemish diamond, this densely built area, characterized by a fragmented pattern of land and by heterogeneous uses and functions, is supported by a highly infrastructured mesh, made of streets, railways, paths and highways (not visible in the map). The map clearly shows ribbon developments where detached single family houses are juxtaposed one to one other, along main streets to form a characteristic pattern. Next to the houses, one will find small enterprises, warehouses, neighborhood facilities, public or private plots of open land

Fig. 32.3 Satellite image (5 × 5 km) of the area between Bruxelles and Ghent including Oordegem, Vlekkem and Vlierzele (Flanders Region).
Source: Google Earth, 2018

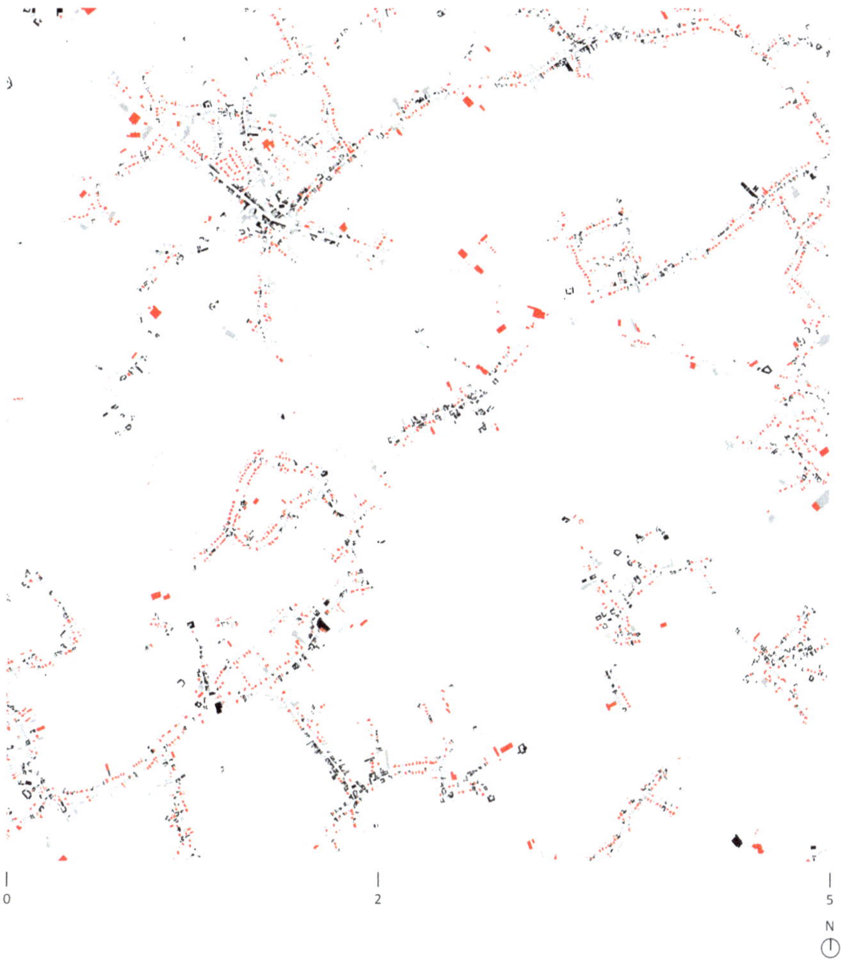

Fig. 32.4 Evolution of the built fabric, area between Bruxelles and Ghent (Flanders Region, 2016). In black the ancient structure, in grey the transformation observed by Marcel Smets, in red the transformation occurred after the publication of his article.
Credits: SPF/AGDP, map elaborated by G. Vanneste. < 1951 : black 1951–1980 : grey 1981–2016 : red

Fig. 32.5 Aerial sequence (drone) of Destelbergen (Gent Metropolitan Area).
Credits: Gogolplex; Antoine Chambre, Aurélien Garcia, Philémon Siesling, 2018. Video projected for the exhibition "The Horizontal Metropolis: A Radical Project" at BOZAR—Center for Fine Arts Bruxelles

Fig. 32.6 Mixuse and Ribbon development at Destelbergen (Gent Metropolitan Area). Credits: G. Vanneste, 2016

Chapter 33
F. Indovina on *La Città Diffusa*

Francesco Indovina

F. Indovina (✉)
IUAV University, Venice, Italy
e-mail: indovina@community.iuav.it

M. Barcelloni Corte and P. Viganò (eds.), *The Horizontal Metropolis*,
https://doi.org/10.1007/978-3-030-56398-1_33

33.1 *La Città Diffusa*: An appraisal

Traditionally it is customary to look at the expansion of the city via its outskirts or periphery, the space across which the city expands, widens, becomes ever larger. The periphery can take on very different forms, in general it is said that "the city spreads like an oil slick".

The new form of expansion that we—as a research group—began to study at the end of the 80 s seemed to us of great interest, precisely because it undermined the very concept of the periphery as such and proposed a new reading of the urban condition.

We had noticed that the city was expanding outwardly, into its surrounding territory, and that there were many reasons for this expansion.

The first was the rejection of agricultural work and the abandonment of agriculture due to its drudgery, "the land is a long way from the hands" the farmers used to say. The farmers' children, while remaining close to their family of origin, were gradually changing their lifestyle, choosing jobs different to those of their parents.

The lifestyle of the inhabitants of the compact city was also changing, it was becoming tougher and evermore conditioned by a less favourable, evermore congested, dangerous and expensive environment. So too their desires and ways of living, were influenced—on the one hand—by new technologies and—on the other —by the strong increase in mobility.

This meant that the population left the traditional, compact city and relocated to the countryside, where new building complexes, houses but also popular housing schemes were built. Hence the countryside became urbanized -because of the population leaving the cities but also due to the persistence of the "natives", those who already lived in these places, who had abandoned agriculture but not the field, whose children had found work in nearby industrial zones. The area we were dealing with was—in particular—that of Porto Marghera, a large industrial zone that had been an important center of attraction for this population.

And what does the young farmer do once he or she has abandoned farming, once they have secured a safe, permanent income? They think of building a family and above all a home. This led to the urbanization of an agricultural territory that was no longer agricultural, that had lost much of its connotation and that had become, to some extent, urban. It remained, however, the result of self-government, determined by individual decisions.

The inhabitants of the city therefore chose to live in the countryside where they could afford a new outdoor lifestyle, own a garden. At a certain point, however, even their children grew up and a new demand for services emerged, the population increased and with it a private interest to construct new centers in this territory, service centers, which can be defined as "trivial", such as for example—shopping centers or supermarkets. Hence this territory, dispersed with houses, provided itself with services. At the same time, many companies, especially small and medium-sized ones, left the city and relocated to the outlying areas, finding convenience both in the low cost of the land and lesser control.

This is how this territory was created, a highly composite amalgama*, which we have called "diffuse city", very much a contradiction in terms. When we conceive of the city—in fact—we do not think of it as being spread out, we think of something compact, dense, intense, ongoing and continuous.*

We have decided to call it "city" because the use that the inhabitants made of this territory was an "urban" one, the same as that made by the citizens of the compact city.

In this territory though, the two constituent elements of the city, historically considered inseparable, these being its morphology *(roads, buildings, universities, prisons, churches, the actual, physical territory comprised by the same) and the* urban condition*, that develops thanks to social relations, have become split.*

If a young person—for example—a resident of a compact city wants to go to the cinema, they read the newspaper, choose where to go, arrange to meet their friends and go to the cinema.

In the diffuse city, if a young person wants to go to the cinema they check their news source for films available within the widely diffuse area, they agree with their friends, who probably live far off, to meet in what is called a "non place", a motorway exit—for example—where they then meet up and go to the cinema, and the same goes for buying a pair of shoes or a wedding dress; everything happens like in the city, only the size and space of this city is much more extensive.

We have to think of the diffuse city as a concentrated city that has spread, which has "spilled" out onto the territory, preserving everything or almost everything.

In actual fact there are elements that—for a certain period—remain within the dense fabric, these are the elements of government, economy and communication or those of the excellence of research, of the university. But at some point they too find it convenient to expand across the territory and do so; the big banks and universities move because their customers are now scattered across the territory.

The diffuse city is enhanced with important elements that make it a territorial metropolis of millions of people, a non-concentrated metropolis.

It is said that there is a typology of typically European metropolis that contrasts with the metropolises of the rest of the world, that reach dimensions of 10, 15 up to 34 million inhabitants. In Europe, except in two cases (Paris and London) the metropolises are almost all modest sized "territorial metropolises", which have the advantage of size but don't have the disadvantages of the concentration of size.

Though let us not forget that these urban episodes are based on the existence of an ancient and extremely rich urban network, a structure on which this metropolis rests.

Despite the great potential of this new form of city, its problems cannot be ignored, we are still talking about self-organized processes, processes that usually give partial, individual answers and that have little to do with collective and general interests. A city of this type hence would have a huge need for a far greater and better planning, it would need a rational and integrated governance structure.

The concept of "diffuse city" was conceived by our research group while we were studying the industrial areas of the Central Veneto, that is the area between the cities of Padua, Venice and Treviso. In those years the faculty of urban planning was immersed in this territory, in the villa at Preganziol, and it is from here that a series of research opportunities started out, mainly oriented towards the analysis of the industrial zones, of the medium-small districts where washing machines or footwear were produced, but also in the larger industrial areas such as that of Marghera; this is how we necessarily came across the urbanized countryside of the Veneto and began to be struck by its strangeness.

Once we discovered this phenomenon we built an extremely composite research group that dealt mainly with this topic. Things were not easy at the time because the concept of the diffuse city did not arouse much enthusiasm among my fellow urban planners, who generally misjudged it. They loved and love the compact city the way many of us do, as I do too. The problem however is that a demand exists which the population has responded to which needs to be regulated by urban planning. This is what we have attempted to do with the subsequent works covering the European metropolis, concerning the metropolitanization of the territory.

Fig. 33.1 Satellite image (20 × 20 km) of a portion of the PaTreVe (Padova, Treviso, Venezia) Metropolitan Area including Camposampiero and Campodarsego.
Source: Google Earth, 2018

Fig. 33.2 Built fabric (only) of a portion of the PaTreVe (Padova, Treviso, Venezia) Metropolitan Area including Camposampiero and Campodarsego (20 × 20 km). The organisation of the diffused city following the roman *centuratio* drainage system is clearly visible. Source: IGM, 2018

Fig. 33.3 Satellite image of an area (5 × 5 km) north to Padua, including Camposampiero, Campodarsego and Curtarolo.
Source: Google Earth, 2018

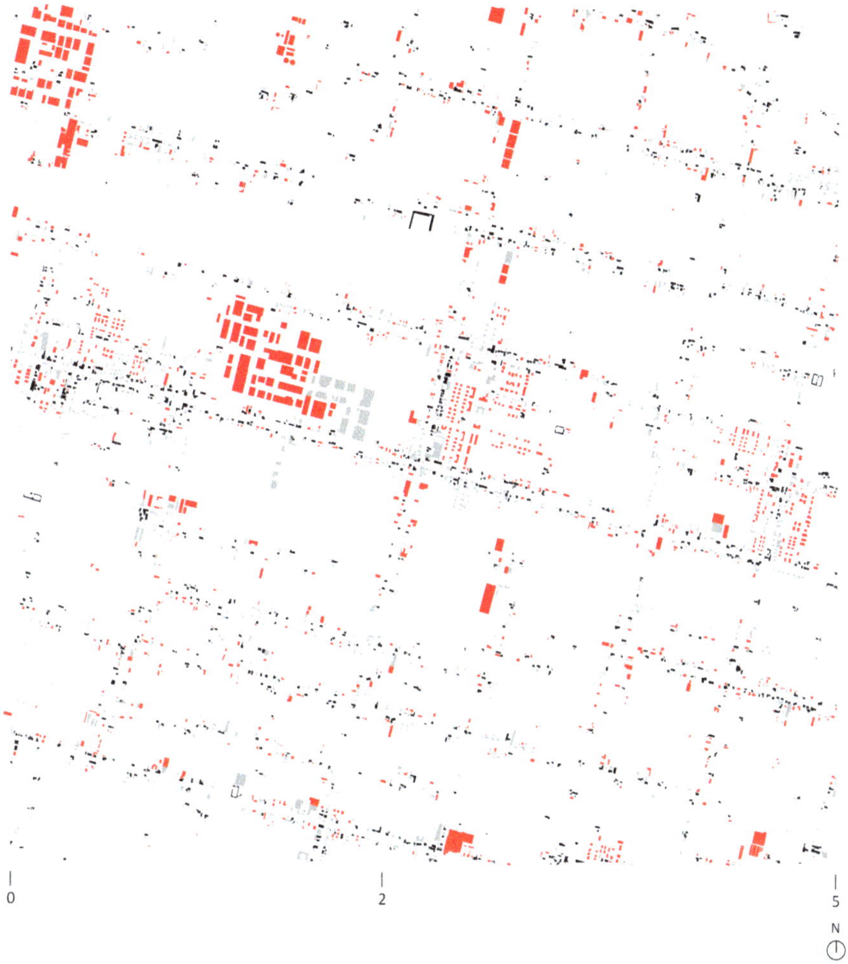

Fig. 33.4 Evolution of the built fabric.
In black the ancient structure, in grey the transformation observed by Francesco Indovina, in red the transformation occurred after the publication of the "Diffuse City".
Productive activities, services and housing are mixed and visibly distributed along the roman *centuriatio* mesh.
Credits: IGM/Google Earth, map elaborated by C. Küpfer.
<1966: black.
1967–1987: grey.
1988–2018: red

Fig. 33.5 Aerial sequence (drone) south of Camposampiero (PaTreVe Metropolitan Area). Credits: Visual Working, Paolo Brianzoni, Eugenio Somaini, 2015. Video projected for the exhibition "The Horizontal Metropolis: A Radical Project" in occasion of: XV Mostra Internazionale di Architettura di Venezia, Venice, Italy (2016); La Biennale di Architettura di Lugano, Istituto internazionale di architettura, Villa Saroli (2016); and at Archizoom, EPFL (2015)

Fig. 33.6 Industrial platforms agriculturalr and residential space in the Camposampierese Area (PaTreVe Metropolitan Area).
Credits: M. Barcelloni Corte, 2020

Chapter 34
T. McGee on *Desakota*

Terry McGee

T. McGee (✉)
UBC, Vancouver, Canada
e-mail: tmcgee@geog.ubc.ca

M. Barcelloni Corte and P. Viganò (eds.), *The Horizontal Metropolis*,
https://doi.org/10.1007/978-3-030-56398-1_34

34.1 *Desakota*: an appraisal

It is now almost 35 years since I began to develop the concept of Desakota *together with a group of PhD students, many from Asian countries.*

Desakota is a coined word from the Indonesian language (Bahasa) where desa *means village and* kota*, city. We applied this word to regions of mixed agriculture and non-agricultural activities that -at that time, in the eighties- were growing up at a very rapid rate, around the cores of many of the major urban areas in Asia.*

The idea of the concept emerged through a set of intellectual concerns, which have affected me and have been part of my thinking for more than 50 years.

The first was the concept of "rural–urban difference".

In the early sixties I was working in Malaysia and -more specifically- on Malay migrants moving to Kuala Lumpur City for work. I was engaged in anthropological fieldwork and living in the city for a year with a Malay family which had relatives coming from the countryside and staying with them (living in their house) to work in the city. We were clearly observing a fusion of rural and urban transfers and lifestyles, this led me to believe that there was a need for a radical rethinking of the concept of "rural–urban difference" that became the subject of my first published paper in 1961.

In the same years I was also concerned about the meaning of "rural" and "urban" in Malaysia. Initially I tried to understand and interpret these conditions through the concepts of "rurality" and "urbanity", dominated -at the time- by the writings of researchers from the developed world. I gradually felt that their interpretation was based upon a particular and empirical condition that characterized Europe and the United States in the late 19th and early twentieth century. I realized that there was a concern with the applicability of these concepts in the contextual settings of Asian urbanization. These theories had -in fact- emerged from the observation of very different contexts from those of Asia and of Asian countries as India, Southeast Asia, China, Taiwan, Japan and others. Interestingly, in those countries, local people were already naming these particular hybrid conditions in their own language, the Japanese -for example- began to call it Konjuka *which literally means "mixed use". I realized that, to interpret a process of this kind it was not possible to analyse it from the outside, it was necessary to experience it within the society, to observe the way people were relating to it and to its effects, to truly understand and interpret it properly.*

I was then concerned to rethink the "spatiality" of these processes through ideas of what I started to call "centrality" and "peripherality", today, a better term seems to be "horizontality".

As mentioned above, the term Desakota grew out of a period of fieldwork I was engaged in, in Java in the 1980s. Java is one of the most intensely populated islands in the world. At that time, it had a population of roughly 120 million, on a land area comparable to that of England. The level of urbanization measured statistically the 28% and it was characterized -throughout the whole island- by a mixture of agricultural and non-agricultural activities, by a high growth of population due to migration and population movement, rapidly changing structure of labour force from agriculture to non-agriculture.

The invasion of Desakota activities by industry, commercial and residential land uses provoked many environmental challenges to the local ecosystem, often characterized by irrigation, a highly volatile and fragile system wonderfully engineered and in place since thousands of years, but which under the pressure of urban activities became threatened. In some ways this mixed landscape cleared the way for further invasion since previous rice-land no longer had water so it had to be sold or redeveloped for non-agricultural activities.

This dynamic presented also major problems of governance and management which the separate administrations of rural and urban had no ability to handle: the rural administration -for example- had plenty of agronomists but no urbanists and -thus- had no true understanding of what was happening. It was also a very enriching process for the developers who were driving this dynamic through the corruption of local politicians.

The distinctive features of this process in Java led us to think that similar processes were occurring also in other areas of Asia, in fact we realized that it was the case in China in the Yangtze River delta, in Guangzhou, in the Pearl River Delta, in India, in the region outside Calcutta along the Ganges and further South, and in many Southeast Asian countries as well, Bangkok, in the Chao Phraya delta and in the Philippines, around Manila.

The idea of Desakota provoked much debate. It was developed in the context of the re-emergence of neoliberalism, a doctrine which reflected the collapse of socialist and communist societies and was spreading throughout. This particular formulation of Desakota was seen as a threat to the ideas of neoliberalism because it suggested an alternative form for the use of urban space.

Since then, we have published a lot of work on this topic, on Bangkok, Kerala, on many parts of China, Shanghai, Guangzhou, etc. Today we work on the idea of "rethinking Desakota": these intellectual paradigms interpret specific conditions and the intellectual currents of their time which are subject to continuous change. Processes change as people's interaction with the same processes.

Thus, the original idea of Desakota, even if still valid, has changed in one important way: Desakota has begun to spread outwards from the city cores along corridors, often major arterial routes, along rail routes, leapfrogging through -what I call- "polyadic centres", secondary urban centres, around which further patterns of mix-use, which are reproducing the process of the first Desakota (the form emerged during the 80 s), have developed.

This means that Desakota has become a phenomenon, a form of highly hybridized landscapes that leapfrog into large international space, it is in fact possible -for example- to find Desakota in places like Lhasa, away from the urbanized, industrialized, coastal regions of China.

I call this phenomenon the "new Desakota" which, for me, poses new policy challenges (at the macro and micro level) and "challenges for thinking", it creates new challenges for the people concerned to train the practitioners which will be responsible for managing, planning, designing, building, the new urban shape, the urban form of the future.

It has been estimated (UN) that over the next 40 years, around 60% of population growth in Asia will occur in areas outside the urban cores, this makes about 40% of the overall urban increase in the next 40 years. This means that the city of the future will develop in these places, it is -thus- extremely important to get right the intellectual ideas about what's happening and begin to think about the implications for policy.

To begin, we need to revise the visioning of the spatial structure of national entities. We need to introduce ideas of "national hybridity" of economic activity, we need to accept that we are dealing with a transactional environment that is very different to the one that characterized these countries in the eighties. Today Java -for example- has one of the major number of "tweeters" than in any country in the world, yet it has a highly unequal income and is developing a transactional environment which is very similar to that of the developed countries.

These spaces need also to deal with new forms of governments: much of desakota activities -for example- are concentrating along corridors spreading outwards from major urban cores, thus, in order to deal with policy implications of these new urban spaces, we need to think about new kinds of "corridor governments" which will cut through previous administrative divisions, particularly for macro land use plans.

These new forms of government have also to be integrated with existing ecosystems and environmental dynamics, the process of urbanization described above are -in fact- mightily destructive of ecosystems. When this kind of urbanization occurs in the context of climate change or resource depletion, it becomes extremely challenging to maintain the viability and liveability of these regions. Today, the physical indications of such mutations are increasingly evident in Asia through growing episodes of flooding, seasonal fluctuations and uneasy delivery of water systems; the new forms of government have the challenge of creating equitable and just societies also in these terms.

Decentralization has become a buzzword for neo-liberalization where local governments are being asked to take on fiscal and administrative responsibilities that they didn't have before. This is fine if governments develop a fully decentralized government system, but this is not the case in countries as Indonesia -for example- or China.

Despite all the optimism generated from the neoliberal paradigm, it is still far behind that of Europe or other developed countries; resource challenges related to environment, energy and food thus remain.

To conclude, there are the challenges posed to practitioners and policymakers, planners, architects, urban managers and institutions that train them or create the ideas that drive them. This means we need to rethink what kind of planners we need for these regions. My idea is that we need teams, planners working together with agronomists, physical scientists, engaged soil scientists, hydrologists, social scientists etc. The challenge of Desakota is that it does create the need for an interdisciplinary framework to develop viable policy solutions.

Thus, what we really need to do is to develop a broad, new urban agenda for the new Desakota for the next forty years.

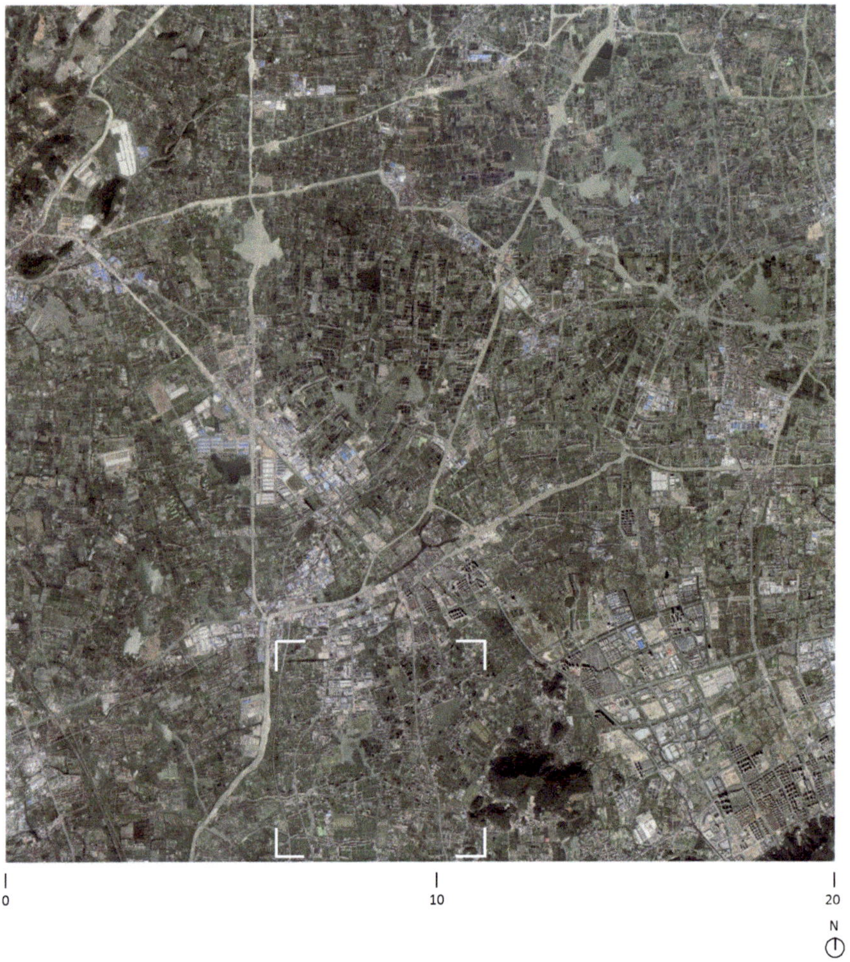

Fig. 34.1 Satellite image (20 × 20 km) of the area between Hangzhou and Huzhou, including the northern part of Hangzhou Metropolitan area. Hagnjiahu Plain is the Southern part of the Yangtze River Delta.
Source: Google Earth, 2018

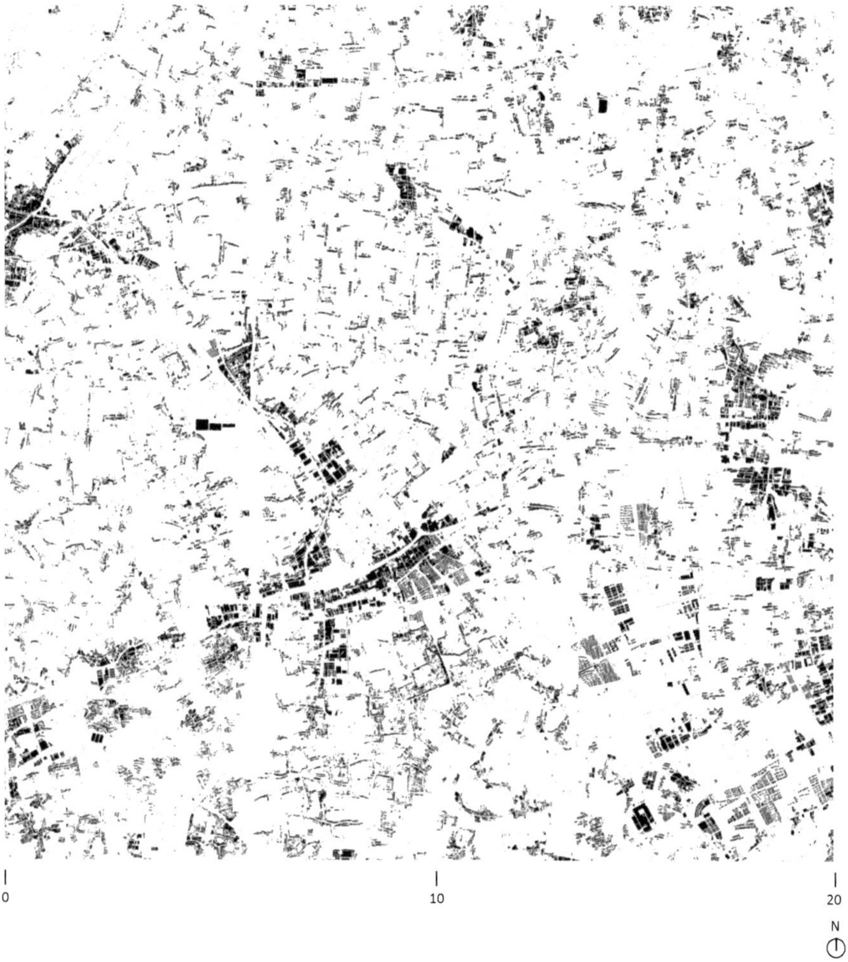

Fig. 34.2 Built fabric of the area between Hangzhou and Huzhou (20 × 20 km). Credits: map elborated by Qinyi Zhang

Fig. 34.3 Satellite image of an area (5 x 5 km) located in the northern part of Hangzhou Metropolitan area, in the Tangqi Area.
Source: Google Earth, 2018.

If Kunshan is exemplary as one of the most unique and successful areas in China's urbanization, Tangqi is representative of an ordinary town within the Yangtze River Delta. Tangqi, with an urban population density exceeding 1000 inhab./km^2, is located in the Hangjianghu Plain, a rural landscape characterised by a rich hydraulic landscape (great number of lakes, rivers, and canals). The polders being a typical and historical construction for agricultural production is still very present. The industrialization has been taking place within each polder and village, generating a diffuse system of small factories and workshops

Fig. 34.4 Evolution of the built fabric of an area (5 x 5 km) located in the northern part of Hangzhou Metropolitan area, in the Tangqi Area.
Credits: Google Earth, map elaborated by C. Küpfer. <2009: grey 2009–2018: red

Fig. 34.5 Aerial sequence (drone) of Tangqi area, in the northern part of Hangzhou Metropolitan area. Credits: Hangzhou Yuanlai Cultural and Creative Media Ltd / Yiming Yan, 2015. Video projected for the exhibition "The Horizontal Metropolis: A Radical Project" in occasion of: XV Mostra Internazionale di Architettura di Venezia, Venice, Italy (2016); La Biennale di Architettura di Lugano, Istituto internazionale di architettura, Villa Saroli (2016); and at Archizoom, EPFL (2015). Tangqi, with an enormous number of lakes, rivers, ponds, and canals, presents many common characteristics of the territory of Yangtze River Delta: heavily populated, densely built, well-infrastructured, and with a strong rural-rban image. Rapid changes are taking place including housing densification, upgrade of industries, and transformation of agriculture

Fig. 34.6 Industrial and residential buildings in the agricultural field in Tangqi area, northern part of Hangzhou Metropolitan area. Credits: Qinyi Zhang.
Today, next to the rapid city-centered urbanization in China, urbanization and industrialization in rural areas have been continuing since the 80s. In the fields, small-to-medium sized industrial platforms, together with small family factories and workshops, are growing rapidly; four and even five-and-a-half floor peasants' houses (complex buildings with activities spaces, living spaces and apartments/studios for migrants) are replacing many of the low-rise houses built in the 80 s and 90s

Chapter 35
A. Branzi on *Agronica*

Andrea Branzi

A. Branzi (✉)
Milan, Italy
e-mail: studioandreabranzi@gmail.com

M. Barcelloni Corte and P. Viganò (eds.), *The Horizontal Metropolis*,
https://doi.org/10.1007/978-3-030-56398-1_35

35.1 *Agronica*: an appraisal

In the climate we were experiencing -at the time- of a progressive fluidization of ideologies, territories and even the project concept, we embarked upon a path that sought to overcome the concept of opposition between two territorial realms, agriculture and architecture, which were nothing other than two different "morphologies" of the same logic, namely that of production, profit and the economy.

In a climate of progressive globalization, we needed to start to think that this type of contradiction was more apparent than real, even considering that agriculture was by now no longer the realm of pre-industrial technologies but on the contrary, it was one of the most advanced areas and systems of production, precisely because it uses the energy of nature, of the seasons.

Agriculture features largescale production albeit of diversified series, hence one million pears—for example—actually constitute one million pears, each different from the other, and this is already a much more advanced form of production than that of industry, it is a form that industry has yet to attain.

This type of reading was accompanied by another idea that was central to us, namely that of overcoming the typical building typologies related to the tradition of the modern movement, which aimed at reaching a mobile system consisting of individual architectural components which were aggregated and/or detached according to different needs.

There was already the idea of absorbing the flexibility of computer technology within the building typologies, a flexibility which allows the computer to go beyond having a single function but rather to be "functionoid", capable of offering as many functions as the operator requires.

There was the idea of passing from the model of the Athens Charter, based on a mechanical contrast between specialized zoning, each with its own autonomy, as in the mechanical organisms that together transmit a single performance through friction, towards a more open and flexible model.

We tried to broach subjects that, in our opinion, corresponded to the more expanded and extensive times of the season we were experiencing and which were the typical fluidization of the economy, politics and contemporary sociology; the term "fluidization", has indeed become widespread through the work of Zygmunt Bauman, even though at that time it was little-known. We were not party to the idea of a liquid modernity, but rather to that of an unstable system that no longer required a territorial design but rather a great transformability of architectural aggregations, all temporary.

The fact of beginning to understand that the society in which we lived was a society that no longer had fixed reference points but only a process of expansion and permanent crisis, was an important philosophical point. It was a society that had to continually produce laws, norms, statutes, to adapt itself to a state of perennial uncertainty, of crisis that the modern movement had not foreseen; on the contrary, the specialization, the typologies, the solidity of the search for permanent, definitive solutions, was creating an extremely rigid and hence fragile system with respect to a society that was transforming itself, which had to manage its own state of discomforture.

The absence of a permanent, unique model is also a typical feature of the contemporary era, experiencing full globalization. The combination of these insights has led to the emergence of this concept where agriculture, the countryside and the landscape finally entered into the design culture, something that had not happened up until then. The modern movement had in fact always ignored the territorial ordering of agriculture and its productive systems, even animals. In fact, one of the results that seemed interesting to us when we started to configure the project was the possibility of inserting animals within a landscape that managed to acquire great modernity. What was being proposed to us was a modernity that only developed through geometries, technologies, very well-defined functions, via stable foundations for perennial solutions in an age that—though—was beginning to head in a totally different direction.

Hence it was from this intuition, from this set of portents, that the project emerged; I call them portents because the set of results we would have obtained was not entirely clear to us. When the result emerged clearly, it was acquired by the Pompidou Center and published by Phillips Corporation, we began to understand the roads this experimental project was opening.

It was not a utopian project because it did not propose a better world; the contemporary world would in fact perfectly allow this type of project to be realized, the continuous movement of architectural components, the presence of agricultural technologies within the city, except for the fact that the weight of the legacy of the modern movement, its Calvinism, its quest for definitive certainties, prevented the same from happening.

This project has been and was at length forgotten, except for some minor publication. Today, together with "No Stop City", starting from the great American universities, it is interpreted as one of the prerepresentations of the project in the era of globalization or, if we wish, of the failure of globalization.

The hypothesis of a project that is always incomplete, imperfect, ever transformable, has therefore constructed a type of design that well represents the uncertainties of the society of contemporary capitalism that has not produced the homologation of brains or the market, but a state of perennial circulation of technologies, commodities and relationships.

Today Agronica *is experiencing a very interesting season of re-reading, also because it was not a project made to be actually realised, as "No Stop City" was, they were both mental projects. When mental projects are represented, explained, they begin to exist; even if they do not enter the material scenario, they have a media circulation, they have their own associated realism.*

Agronica *was certainly a project born working on the longterm and this is something I have always considered highly important. It is in fact inconceivable that the design culture might only be developed via commissions (private or public), albeit legitimate ones and part of the discipline's tradition. One cannot renounce to develop the design culture independently, all the while entrusting it to a real estate or building market which does not assume responsibility for it, which indeed cannot be considered responsible for the same.*

After all, modern architecture, that of the twentieth century, is an architecture that -for the most- has never been realized, starting from Le Corbusier's 1933 Athens charter which exercised an enormous influence but which—in fact—has never been put into effect. Ideas count as much as facts do, indeed ideas are facts and hence produce, at least this is the wish, a change in the way of thinking, which to me is the central point.

Up until when the architects continue, legitimately and according to their professional tradition, to entrust the future of this activity exclusively to the opportunities that present themselves on the market, matters will not change. In this sense, the results rendered by the modern movement lie precisely within the brains of architects and designers.

We must not be afraid of projects that can be forgotten for a long time and then, even after many decades, be reread to again note that they contain ferments that can still be exploited. Because the design culture is a part of human history and as history changes, so too does the design culture. Experimental research projects belong to a level of underground elaboration that can re-emerge in time when certain problems begin to manifest themselves very clearly, as is certainly the case for Agronica.

Chapter 36
T. Sieverts on *Zwischenstadt*

Thomas Sieverts

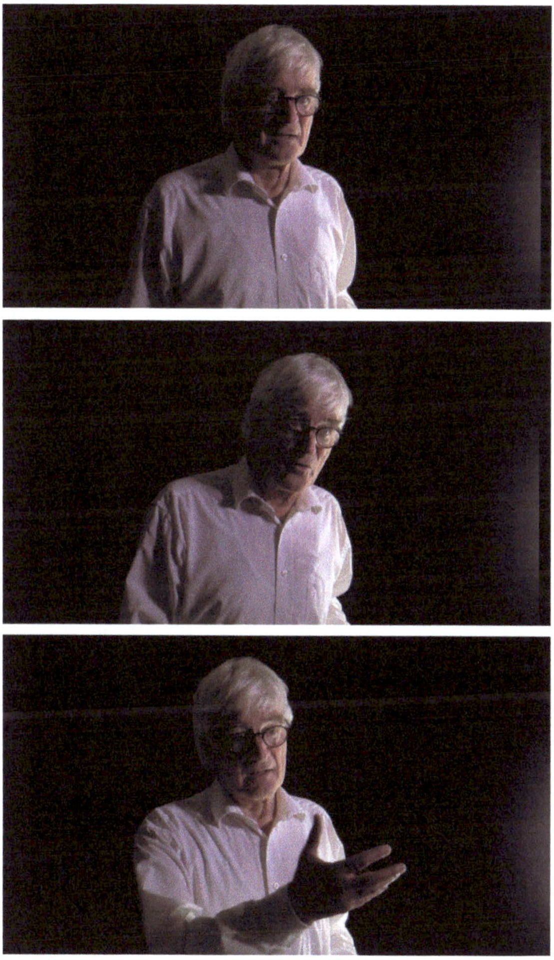

T. Sieverts (✉)
University of Darmstadt, Darmstadt, Germany
e-mail: tom.sieverts@googlemail.com

M. Barcelloni Corte and P. Viganò (eds.), *The Horizontal Metropolis*,
https://doi.org/10.1007/978-3-030-56398-1_36

36.1 *Zwischenstadt*: an appraisal

This all started thanks to an invitation from the "Wissenschaftskolleg", the WIKO, in Berlin. They offered me a fellowship of one academic year from October 1994 until July 1995, to work on a topic of my own choice. At that times I had just finished my directorship at the "International Building Exhibition Emscher Park" in the "Ruhrgebiet", where I worked the five years before. I was—thus—fresh and open for new ideas.

I wanted to do something in relation to my experiences with the "Ruhrgebiet", the largest and one of the oldest industrial regions in Europe, transformed by the Industrial Revolution. It presented a type of seemingly anarchic settlement pattern, quite the opposite of the ideal, well–ordered, old European town.

My curiosity about these kind of anarchic settlement patterns began in Berlin, where I lived from 1957 till 1970, with the analysis, together with my students, of some of the old allotment garden areas there and their spontaneous self—made "people's—architecture", produced without any planning control. Being a child of the 1968 movement I loved these gardens and their self -produced, richly decorated huts, the direct expression of people's life.

When I came back to Berlin in 1994, the spirit of 1968 had disappeared completely. Instead a mood of restauration had become predominant, both in the city's planning administration and in public opinion, also expressed in the postmodernism of new architecture. I disliked this mood and wanted to do something in an opposite spirit.

In the following I shall talk about three topics: Analysing the language we speak about suburbia and the urbanized landscape, the contrasting scales of urban design in the local and regional scale and last but not least about aesthetics and unaesthetics.

I began, intellectually, to dismantle the myths of the "European city" by examining in depth the key terms of these myths, dominating the urban discussion of the 1990s: Urbanity, density, centrality, mixture and ecology.

I tried to disclose the somewhat ideological character of these terms, as they had become rather flat, more or less meaningless conventions:

Urbanity had been degenerated to a kind of superficial "cappuccino–urbanity"; conventional centrality had been submerged into the "network–city"; "density" had been watered down by the still growing specific space we all use and which had tripled in the last decades. "Mixture" had become difficult to achieve with a specific growth of institutions but also with a growing intolerance. Urban ecology was deeply changing: the old contrast between the compact stony city and the open green landscape began to be blurred, instead a new kind of urban landscape emerged.

Once I had deconstructed the old myths of the European city, I had gained a free field to reconceptualise the contemporary metropolis: To start with, I accepted the reality, that the majority of the people of Europe already lived in some kind of urbanized landscape and no longer in concentrated cities and I accepted their reasons for this decision.

I began to avoid the degrading term "Zersiedelung", a term, indicating that low density urbanization destroys a landscape, and spoke instead of a "mutual penetration of town and nature"!

Instead of lamenting the loss of urbanity I spoke of its new meaning in a networked city, concentrating my reflections more on the people and their behaviours, than on the form of the city. Instead of criticising a loss of urban density I spoke of a new virtual density supported by digital means.

In my reflections I contrasted the loss of diversity with combinations opened up by the new digital networks.

In contrast to the thesis of the destruction of nature by low-density suburbanization I emphasised the findings of the urban ecologists, that the urban landscaped-city is much richer in species than the surrounding industrialised agriculture and forestry.

As a result of my reflections, I was questioning fundamentally the old categorical separation of city and landscape. I tried to develop a new understanding of the urbanized landscape/landscaped-city. This new position opened the perspective for a type of city in the form of an extended urban region, with an enlarged catchment area, supported by the "space-jumping" electronic media, (which at the time I wrote the book had just begun to really impact space).

I soon realised, that this enormous extension of the city into the urban region leads to several conflicts: one of these conflicts is related to what the Swedish geographer Gunnar Turnquist called "the conflict between system and agora": The uncontrolled extension of the realm of work and production into the urban region, now organised into an extensive division of labour, will do harm to the "agora" of daily life and spontaneous contacts: There is a danger, that the "system" will take over and the "agora" will suffer.

I tried—thus—to reflect on the type of daily life in an urban region: On the one hand, a neighbourhood, that risks to be isolated in space–time islands, floating in the urban region and, on the other hand, risks to divided by the division of labour, where people are always on the move to take advantage of the region, but do not have any personal localised life any more.

This means: The new regional city must be developed in two directions: On the one hand, into the dimension of the "system" of a network—city, which has no "natural" borders anymore, and, on the one hand, into the direction of the "agora" of the locality, as the realm of daily life, of the family and the extended neighbourhood.

And there is another problem: The extended realm of daily life and work in a networked urban region does not have an equivalent inner representation in the mind of its inhabitants. To develop a sense of belonging, people must have a notion of the area they use in their daily life. Otherwise, the ability of orientation, of finding your way, is completely dependent on printed or electronically presented documents. To realise"instinctively" the use of the extended opportunities of the regional city, people need more than a technical, economic and functional city, they need a "city with a soul".

In this type of city, situated between town and landscape, between an economy serving the locality and international corporations serving world—markets, between intimate relations in the personal neighbourhood and the more abstract, unpersonal relations in the region, there exist tensions, which the old town did not experience.

So, as urban designers, we have to work on both scales: on the one hand we have to work on the regional scale to create a coherent inner regional image, representing a solid network, able to bring into play all the different opportunities and qualities of a region, and on the other hand we have to work on the individual and intimate scale of the locality, on the means of emotional feelings, of being at home, of feeling the soul of a place.

Working as an urban designer on the contrasting scales of the urban landscape/ landscaped—city needs an extended notion of aesthetics, as the conditions of perceiving and realizing the regional scale in the mind versus the local scale are so different. Thinking more about different ways of perceiving the landscaped—city, I came to make a fundamental distinction between aesthetics and unaesthetics: unaesthetics does not mean ugliness, it means, that people "do not perceive at all", or if they see, they perceive with indifference only the functional and economic side of the environment, they do not develop any emotional relations in form of positive, warm memories. I am afraid, the majority of urban environments are of this character.

Therefore, one of the aims in urban designing of the urban landscape should be to help people to develop positive emotions, to feel at home. Our experiences analysing the old allotments in Berlin told us, that as soon as people built their huts and plant their gardens by themselves, they take care of it and love it. We need more of some kind of "hands-on aesthetics" to root people in their locality.

Working as an urban designer on the scale of the urban region should help as orientation. Most important here is to make use of the geographical/morphologic structure of the landscape. Beautiful alleys and architectural dominants with a regional meaning can support this structure of orientation. The urban designer in the scale of the urban region must be a combination of landscaper, urban designer, traffic engineer and architect. At least they must know enough about these professions to serve as a creative curator!

I would like to conclude with saying something about the impact of the book "Zwischenstadt". *I have written this book during my time in the Wissenschaftskolleg in nine months, from the first word to the final touch. Then, when I went to the publisher, I was doubtful it would sell. Once published, a review came out quickly from a colleague, who tried to ridicule the book and dismantle it from the very beginning. This was painful for me, because the author of this review was a close colleague, also a fellow at the WIKO with whom I discussed the book several times before publication, as it is usual in the WIKO.*

The book was—thus—controversially debated: I was called a "traitor of the idea of the 'European City'", supported by a kind of moral judgement: "A matter as uninteresting as ordinary suburbia is not worth writing about!".

Nonetheless, despite this heavy critique, the book began to sell well and to be translated, first into French (in France it became a kind of textbook in schools of architecture) and then into English. A little bit later came the Japanese translation (with a second edition a few years later) and a translation of the first part of the book into American English. In 2015, the Spanish translation was published, mainly for the South American market. The book still sells, more than twenty years after its first publication.

And here I want to pay due tribute to two very dear colleagues, who were pioneers in this field, but sadly are not with us anymore: Bruno Secchi and Francois Ascher. I speak neither Italian nor French, I could not read their publications, when working on my book. That was a great pity. But both colleagues read my book and we became friends, both invited me to talks and discussions. I feel very grateful to them.

Since a few years, younger colleagues took up the topics of the "Zwischenstadt"*: Vera Vicenzotti wrote her doctoral theses "Der* Zwischenstadt *– Diskurs, eine Analyse zwischen Wildnis, Kulturlandschaft und Stadt" (Bielefeld 2011).*

A group of young planners, urban designers and social scientists were funded by the 'Gottlieb Daimler and Carl Benz Stiftung' and worked for three years to apply and test the theses of the Zwischenstadt *at the greater area of Frankfurt am Main under the guidance of some old colleagues and myself. Being already retired, I had a good time discussing my favourite topics with twenty young ambitious researchers and their teachers. In the end, we have produced twenty volumes of research together.*

Meanwhile the term "Zwischenstadt" *is used in German speaking countries as an equivalent to the term "citta diffusa", used in Romanesque countries. And this term has found its way also into the universities: several students have written their master-thesis on that topic. In short: The book might be useful for readers interested in continuing their reflection on the Horizontal Metropolis.*

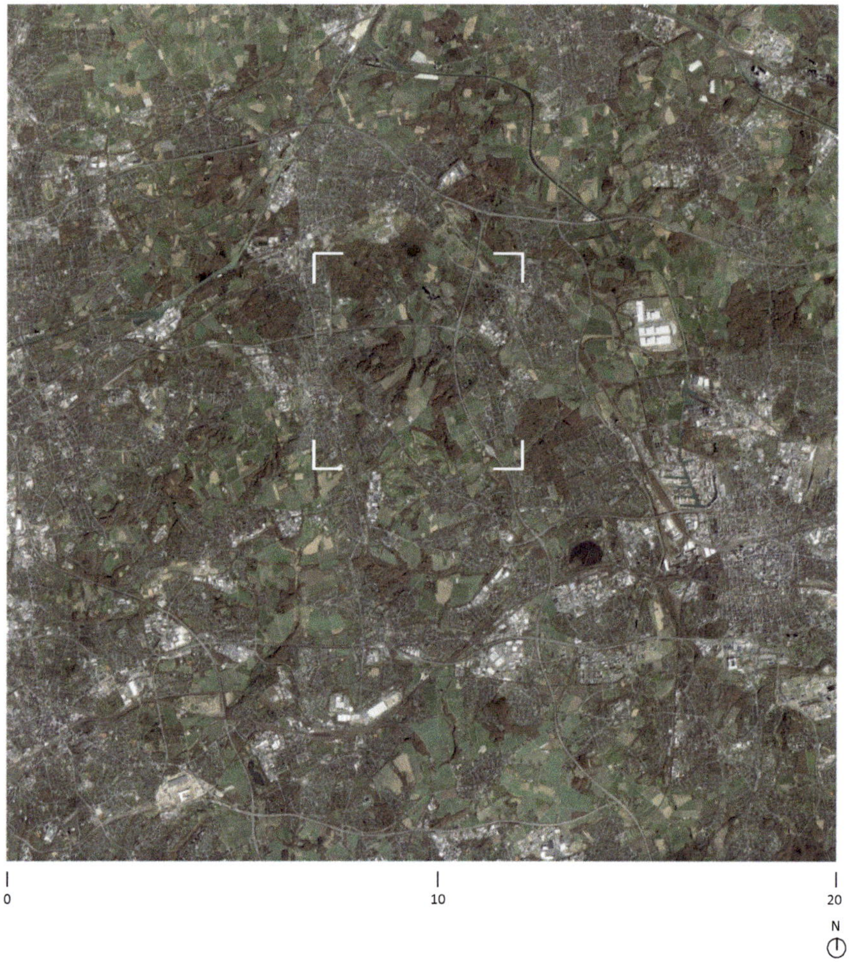

Fig. 36.1 Satellite image (20 × 20 km) of a portion of the Ruhr, on the east of Gelsenkirchen-Bismarck (Sieverts's case study); the area includes Castrop-Rauxel, Bochum Nord, and Dortmund (on the right).
Source: Google Earth, 2018

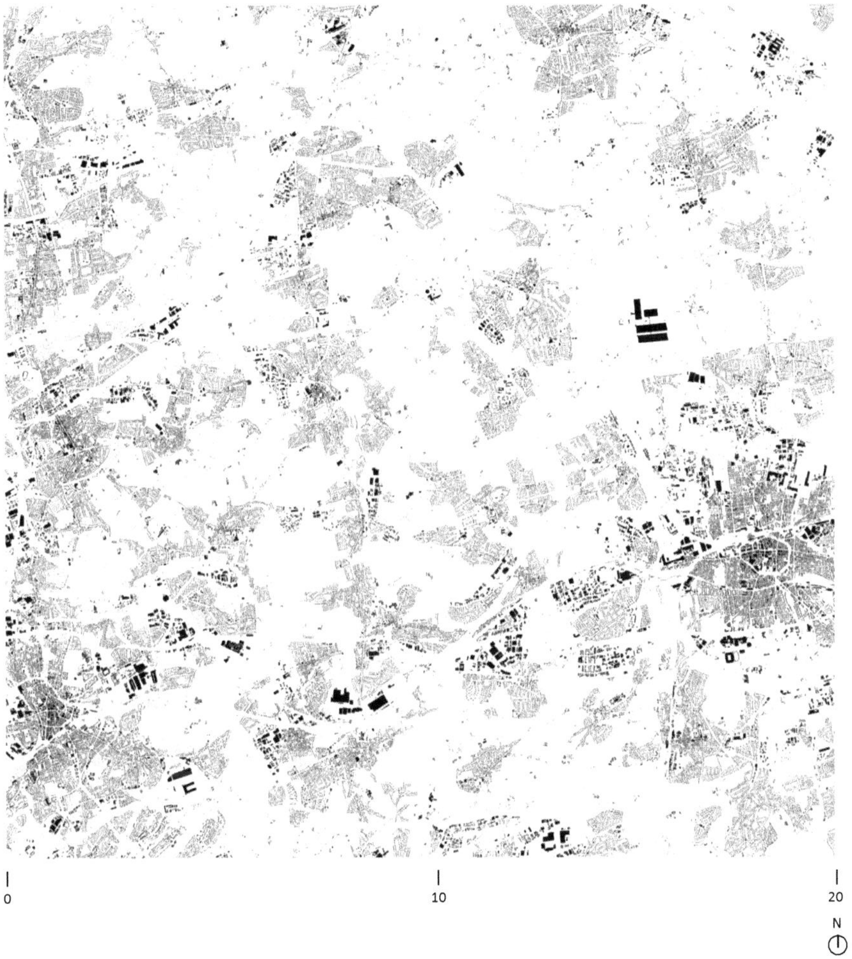

Fig. 36.2 Built fabric of a portion of the Ruhr (20 × 20 km) area in the Ruhr, on the east of Gelsenkirchen-Bismarck.
Credits: GEObasis NRW, dl-de/by-2–0, map elaborated by C. Küpfer

Fig. 36.3 Satellite image of an area (5 × 5 km) located in the Ruhr, including Castrop-Rauxel and Dortmund-Bodelschwingh.
Source: Google Earth, 2018

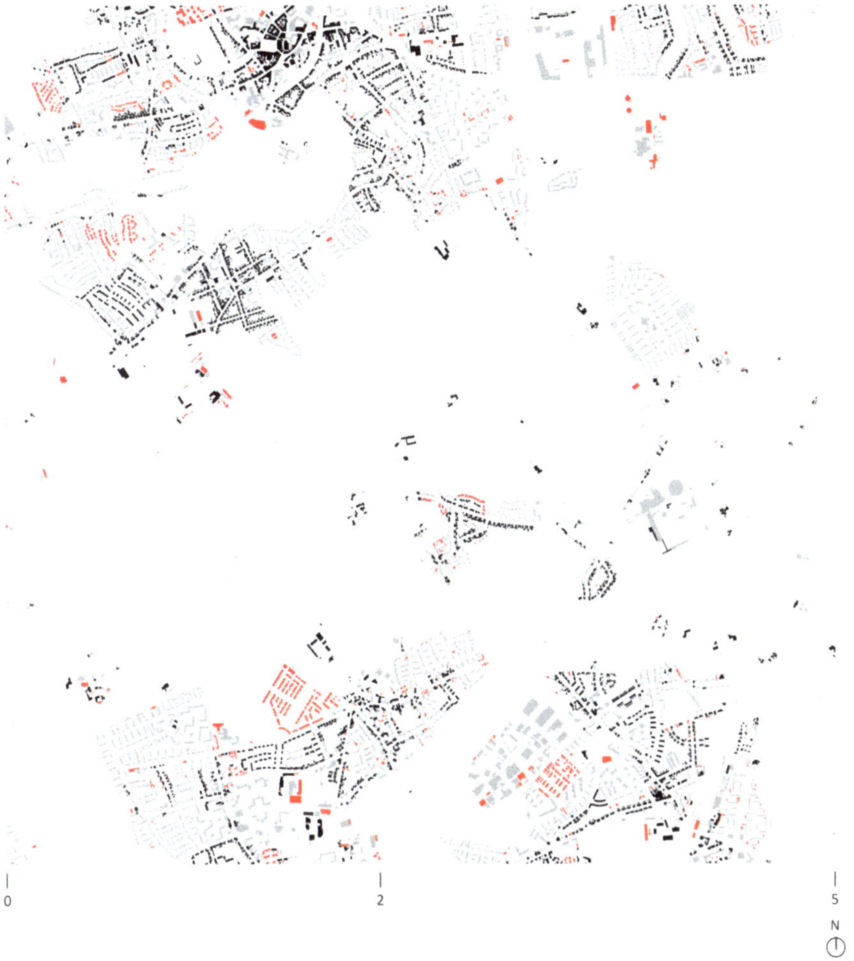

Fig. 36.4 Evolution of the built fabric of an area (5 x 5 km) locatd in the Ruhr, including Castrop-Rauxel and Dortmund-Bodelschwingh.
In black the ancient structure of the built environment, in grey the evolution described by T. Sieverts and in red its subsequent development.
Credits: Hansa Luftbild GmbH, Aerowest GmbH, Regionalverband Ruhr; EUROSENSE GmbH, Hansa Luftbild GmbH, Kampsax, Regionalverband Ruhr; GEObasis NRW, dl-de/by-2-0; map elaborated by C. Küpfer. <1953: black 1953–1998: grey 1999–2019: red

Fig. 36.5 Aerial sequence (drone) of an area between Herne and Castrop-Rauxel, west of Dortmund. Credits: E. Barcelloni Corte, 2020

Fig. 36.6 Diffused and mix-use urbansation on the south west of Castrop Rauxel, west of Dortmund. Credits: E. Barcelloni Corte, 2020

Epilogue

When, at the beginning of 2014, the *Horizontal Metropolis* research was launched, we could never have imagined the acceleration of the debate about the ecological and urban "transition" and the shift, due to the Covid-19 outbreak, in the perception of urban space—public in particular—in its practices and in the associated sense of insecurity. While "modifying our gaze on a number of realities, besides reality itself" (Lévy, 2020), the Pandemic seems—in fact—to have effectively highlighted a number of fragilities within the urban sphere (Latour, 2020), especially within our conception of/relation to open and public space. By radically reducing mobility for all, the Pandemic-related lockdowns have in fact clarified—without nuances—the level of iniquity embedded in large parts of our cities, their spatial injustice and—their deficit in resilience (Honey-Rosés et al., 2020). While providing a mere hint, the tip of the iceberg of deep social, ecological, and economic transformations, the Pandemic seems—nonetheless—to clarify the role that space, and urbanization might play within the ongoing and future "transition".

The effort made with this publication, the last of the three dedicated to the "Horizontal Metropolis", of selecting and sequencing, in a non-linear and open manner, a plurality of heterogeneous yet unexpectedly coherent reflections, reveals an alternative and original line of thought, a set of tracks drafting a broad and robust approach to the City Territory, a source of inspiration (attentive to past cultural contexts' specificities) that could prove to be valuable in tackling and reorienting the challenges that await us ahead.

The necessity of a new scale of observation of the urban phenomena, coupled with the construction of an interdisciplinary gaze; the potential embedded in the spatial hybrid and mix—its heterogeneity—especially if related to the new, structuring role of open space; the idea of redistribution of power and of territorial balance through the re-articulation of strong and weak civic infrastructures; the possibility offered by integration and distribution of functions and land-use in accordance with environmental specificity/variety and against the division of labor; or again the identification of a new space for social emancipation, together with a renewed understanding of security and health, are just some of the themes that have

M. Barcelloni Corte and P. Viganò (eds.), *The Horizontal Metropolis*,
https://doi.org/10.1007/978-3-030-56398-1

animated the minds gathered in this book. This long-term, poorly legible and in some parts forgotten legacy seems—on the contrary—to open up and reveal original and ground-breaking perspectives, able to constitute a framework upon which the next urban and territorial project could be imagined and built.

References

Honey-Rosés, J., Anguelovski, I., Chireh, V. K., Daher, C., Konijnendijk van den Bosch, C., Litt, J. S., Mawani, V., McCall, M. K., Orellana, A., Oscilowicz, E., Sánchez, U., Senbel, M., Tan, X., Villagomez, E., Zapata, O., Nieuwenhuijsen, M. J. (2020). The impact of COVID-19 on public space: An early review of the emerging questions—Design, perceptions and inequities. *Cities Health* 1–17.

Latour B. (2020, March). *What protective measures can you think of so we don't go back to the pre-crisis production model*? AOC.

Lévy, J. (2020, March). *L'humanité habite le Covid-19*. AOC.

MIX
Papier aus verantwortungsvollen Quellen
Paper from responsible sources
FSC® C105338

If you have any concerns about our products,
you can contact us on
ProductSafety@springernature.com

In case Publisher is established outside the EU,
the EU authorized representative is:
Springer Nature Customer Service Center GmbH
Europaplatz 3, 69115 Heidelberg, Germany

Printed by Libri Plureos GmbH
in Hamburg, Germany